# COUIURS
# NTNOIOS*

*\* A curious thing about words - letters can be jumbled inside a word but, as long as the first and last letters are correct, the brain unscrambles them automatically.*

First publication 2021 by Footprint Press, South Africa

email: david@footprintpress.co.za
website: www.footprintpress.co.za

Copyright © Mike Bruton 2021

Cover design and page layout by Anthony Cuerden

Email: ant@flyingant.co.za

Printed by Pinetown Printers (Pty) Ltd

ISBN: 978-1-77632-949-6

All rights reserved. No part of this publication may be reproduced, stored, manipulated in any retrieval system, or transmitted in any mechanical, electronic form or by any other means, without the prior written authority of the publishers, except for short extracts in media reviews. Any person who engages in any unauthorized activity in relation to this publication shall be liable to criminal prosecution and claims for civil and criminal damages

# Curious NOTIONS

## Reflections of an Imagineer

# Mike Bruton

Honorary Research Associate
South African Institute for
Aquatic Biodiversity, Makhanda

## Recent books by Mike Bruton

Great South African Inventions (2010)

Traditional Fishing Methods of Africa (2016)

What a Great Idea!
Awesome South African Inventions (2017)

The Annotated Old Fourlegs.
The Updated Story of the Coelacanth (2017)

The Amazing Coelacanth (2018)

Fishes of the Okavango Delta and Chobe River, Botswana
(with GS Merron and PH Skelton; 2018)

The Fishy Smiths.
A Biography of JLB and Margaret Smith (2018)

Curator and Crusader.
The Life and Work of Marjorie Courtenay-Latimer (2019)

Harambee: The Spirit of Innovation in Africa (2021)

*Dedicated to*

*Professor Brian Allanson*

*who first launched my career*

*in science*

## Cover images

**Front cover:** Clockwise from top left: SKA Reconfigurable Application Board (Skarab; photo: Mike Bruton); Lucky Netshidzati and his 'Talking Glove"; author giving a talk on the dodo in Zanzibar in 2012 (Carolynn Bruton); pangolin; portrait of Nelson Mandela in bottle tops, wood chips and found objects by Ndabuko Ntuli in the Art@Africa gallery in Cape Town (Mike Bruton); two children in front of the Iziko South African Museum (Mike Bruton); street-art entitled 'Digitisation' by Stefan Smit in Woodstock, Cape Town (Mike Bruton).

**Back cover:** Charles Darwin; 'Coelacanth on a Bicycle' by Ellen Marcus (2009); zebra reflections in the Pilanesberg National Park (Mike Bruton); the longest-ology word; 'Here be Dragons', slogan of Mark Shuttleworth's HBD Venture Capital company.

# Table of Contents

| | |
|---|---|
| Foreword by Nicholas Ellenbogen | 8 |
| Introduction | 9 |
| 1. Funny side of science: The ancients | 12 |
| 2. Funny side of science: The moderns | 32 |
| 3. Creativity in the arts and sciences | 48 |
| 4. 'Lunarticks', and Owls | 80 |
| 5. Boneshakers to bloomers: Evolution of the bicycle | 112 |
| 6. Who is South Africa's greatest inventor? | 130 |
| 7. The comical art of naming new species | 164 |
| 8. How well do you know your ologys? | 171 |
| 9. Finding Old Fourlegs: Act 1 | 181 |
| 10. Finding Old Fourlegs: Act 2 | 201 |
| 11. Lessons from the dodo | 223 |
| 12. Which is South Africa's strangest animal? Animals without backbones | 238 |
| 13. Which is South Africa's strangest animal? Animals with backbones, and the winner … | 260 |
| 14. Africa's Nobel laureates | 293 |
| 15. Discussion: What is science? | 321 |
| Acknowledgements | 327 |
| References and further reading | 328 |
| Index | 337 |

# Foreword

**I HAVE A SECTION** in my bookshelf which is made up of books I return to on a regular basis, either for reference or just for the love of reading. It began with books from my childhood like the *A Pageant of History*, Percy FitzPatrick's *Jock of the Bushveld*, Churchill's *The Second World War*, McCall Smith's detective stories and the writings of Gus Mills. A growing section is made up of the wonderful books by Mike Bruton. He has such a quirky style that even his weightier tomes reflect his style of fun in science. Not only is he a recognised giant in the world of natural science, but he also has a love of language and an unusual art of story-telling. I commend him for these attributes.

This collection of essays is a must for those of us trying to make sense of this changing world. It is packed with anecdotes and twists of fate and will certainly be an addition to my favourite 'return-to' books. The breadth of Mike's interest is extraordinary and his ability to hold one's interest has kept me reading when I should have been at my chores. Do I always agree with some of his decisions? Well no, but even science has room for debate.

Mike is a dramatist and enlivens his stories with a theatrical flair. I have written many plays and, although this is a dissimilar art form, Mike has mastered this talent of holding an audience on the page as if they were witnessing a performance on stage - and that is a rare skill.

I had a wonderful few years of playing golf with Mike and that great judge and sports commentator, Pat Tebbutt. It was like Sgt Pepper's Lonely Hearts Club Band as we entered a fantastic world of past, present and future. The golf was truly unimportant; we solved huge problems, often about the natural world. All of it was passionate stuff, never emotional, and enjoyed in the language of the Queen's English. We lost the judge along with the glories of his mind - and a lot of golf balls! But Mike has recalled for me those memorable days in another of his wonderful books. Read this with relish – and don't let the opportunity pass you by.

*Nicholas Ellenbogen*
*Rosebank, Cape Town*
*April 2021*

# Introduction

*'Be yourself; everyone else is already taken.'*
Oscar Wilde

**REJIGGING FROM A RESEARCH SCIENTIST** to a science educator is a challenging task. In the former role one learns more and more about less and less, whereas, in the latter, you are expected to have a broad knowledge about almost everything. This transition does, however, slot nicely into one's second childhood, when in your dotage your childlike curiosity about all things remarkable is brought back to life. Personally, I have enjoyed re-entering the 'kort broek' (short pants) phase of my life!

One of the biggest challenges that confronts a science educator is to offer sensible answers to the perplexing questions that 'fellow children' ask. Kids are not bound by societal norms or a fear of the unknown so they ask searching questions about topics that have often not been the subject of focused research: Why don't fish have necks? Why do animals have faces? Why don't birds give birth to babies? Why do whales breathe air? Why are plants green? Are plants cold blooded? If you plant bird seed, will it grow into a bird?

Adults from fields outside science also ask some challenging questions, unafraid of the 'silos' that are so beloved of scientists. What is the relationship between art and science? Are scientists encouraged to be creative? Why do so few scientists look at the big picture? Why are so many people skeptical about science? Why has science not solved the environmental crisis? Can we learn lessons from past animal extinctions? Is science always serious or do you have fun along the way? Can anyone name a new species? Why has the coelacanth story captured the public imagination? What impact have African scientists had on the world? Who is South Africa's greatest inventor? Which is the greatest invention? What is South Africa's strangest creature?

I have assembled this collection of essays in an attempt to answer some of these questions and share my multiple involvements in science and technology as well as in the arts. The essays cover a wide range of topics with an emphasis on the interesting and unusual, whether they are people, events, discoveries, inventions, words or concepts. To me the history (and herstory) of science and the arts is one of the most exciting episodes in human evolution, peppered with curious individuals and bizarre events that bring these disciplines to life. The two essays on the funny side of science are based on talks that I have given at science festivals, in museums and science centres and at formal dinners. They emphasize that scientists are humans after all and may be more eccentric than the population at large.

The essay on creativity in the arts and sciences arises from my interest in the nature of creativity and the ways in which it is expressed in these creative disciplines. The exploration of the extraordinary Lunar Society of Birmingham was piqued by my interest in the history of science and the origin of industrial revolutions. The members of this society, who were some of the leading luminaries in science and technology at the time, played an important role in launching the industrial revolution in late 18th/early 19th century England. The equally marvellous Owl Club in Cape Town has similar origins but plays a different role in today's post-industrial society.

The essay on the evolution of the bicycle arises from two exhibitions that I conceptualised and developed at the MTN ScienCentre in Cape Town: *Cycology: The Science of the Bicycle* (2002), on the history and design of the bicycle, and *Pedal Power for Africa* (2005), on the diverse uses to which bicycles and pedal-powered machines have been put in Africa. The essay on 'Who is South Africa's greatest inventor?', a controversial topic, arose from discussions that were generated by the exhibitions, talks and interviews in which I have been involved on South African inventions. To me there is a clear winner, but readers might have a different opinion. The two light-hearted essays, on the comic art of naming new species and on my unusual collection of words ending in ology, provide further insights into the lighter side of science.

I have been fortunate during my career to have been involved in one of the most enthralling science stories ever told, the saga of the coelacanth, which never fails to amaze and inspire people. As I have repeatedly been asked to summarise and update this story I do so in the two essays on 'finding old fourlegs'. 'Lessons from the dodo' arises from a series of holidays that my wife, Carolynn, and I spent in Mauritius where I carried out informal studies on the habitats in which the extinct bird once lived, followed up by research on dodo specimens in museums in Zanzibar and England. 'Which is South Africa's strangest animal?' is a playful romp through our country's fauna in search of the creature whose *bauplan* differs most from that of its relatives. It is followed by a more serious discussion on Africa's Nobel Prize winners that begs the question: have the accomplishments of Africa's scientists, writers and peace campaigners been adequately acknowledged by the international community? I end with a brief essay on a perplexing question often asked of me: what is science? and why is it important?

I hope that these essays will further whet your appetite for science and technology, as they have mine. As humankind battles with the conceptual collision between taking advantage of scientific and technological advances while at the same time being better grounded in Nature, we need to recognise that science plays a vital role in both trajectories, and especially in ensuring that the first helps us to achieve the second. Like democracy, science is not perfect but it is our best shot. We need to ensure that it remains in the mainstream of society.

# 1.

# The Funny Side of Science: The Ancients

*'The origin of what we call civilisation is not due to religion but to skepticism. The modern world is the child of doubt and enquiry, as the ancient world was the child of fear and faith.'*

Clarence Darrow, American lawyer and civil rights advocate (1997)

**INTRODUCTION:** Science differs from other fields of creative endeavour in that its subject matter does not derive from the activities of those who practice it. We don't create science like dancers develop dance routines or musicians write music. A 'scientific discovery' may be defined as 'revealing a truth that was already there', although we do derive new methods for finding things out and analyzing the outcomes. Furthermore, in recent decades, new fields of science have developed that determine the impacts of our technologies on the broader environment. So, the mysteries and complexities of the natural (and built) world, and their explanations, are there for us to discover, unravel and understand. In earlier centuries this knowledge was developed so that we could conquer and exploit the natural world but today we have a more nuanced approach that seeks to understand our world and live within its limits.

Science is thus a distinctly human endeavour – no other creatures practice it - and this is where it becomes interesting. The practice of science is ongoing but the parade of humans who carry it out is everchanging. Notwithstanding Newton's telling comment, "If I have seen further it is by standing on the shoulders of giants", the progress of science is determined by the interaction between what we want to know and what we are capable of finding out. The status of science at any given time is determined by the collective knowledge of scientists (and their ability to share it) and by the techniques and materials that are at their disposal. Science does advance in leaps and bounds at times when especially talented individuals (both female and male) emerge from the crowd and exert their influence on lesser beings. These outstanding individuals 'see further' through their unique grasp of the big picture. The theories of relativity, the nature of the atom and the structure of DNA would all have been discovered even if Einstein, Rutherford and Bohr, and Watson, Crick and Rosalind Franklin had never existed. It would just have taken longer.

With our large brains and inquisitive minds, it was inevitable that humans would practice science and that it would be carried out most often in an orderly and rational way. It is therefore typically a plodding slog towards a difficult-to-achieve goal but sometimes, just occasionally, things happen differently. If you dig deeply enough into its history you will find that some peculiar, even funny, things have happened on the way to the lab bench. Furthermore the corridors of science (and technology) have been inhabited by some of the most peculiar humans, many of them with foibles that are more extreme than those of the general population. In these first two chapters some of these eccentric characters are discussed more-or-less in chronological order with a few timely excursions along the way. By mulling over their peculiarities, we learn something about the way in which science works. Furthermore, these examples illustrate how the coincidence of alert minds, nurturing environments and new discoveries in related fields, result in science and technology advancing, not gradually, but in leaps and bounds.

**Early Greeks and Roman natural philosophers**: Democritus (460 - *ca* 370 BC), the 'Laughing Philosopher', took his work seriously but did not take himself seriously, a good lesson for us all. He was so far ahead of his time that few understood and appreciated the importance of his findings. He proposed that space and matter are made up of an infinite number of indivisible and vanishingly small units, which he called 'atomos' (things that cannot be cut or divided), but his particle theory was not supported for over 1 000 years, which was probably why he died laughing.

Democritus, the laughing philosopher.

Aristotle (385-322 BC) had the opposite problem. He was such a respected figure that people believed everything he said, such was Greek elitist culture at the time. He tutored Alexander the Great and in 335 BCE founded his own school, the Lyceum, in Athens. He wrote the first book on meteorology and described the water cycle. He offered sage advice in his book on ethics, "Our happiness is not a state but an activity, and it is determined by our ability to live a life that enables us to use and develop our reason. While bad luck can affect happiness, a truly happy person learns to cultivate habits and behaviours that help him (or her) to keep bad luck in perspective."

But Aristotle made the mistake of insisting that everything could be understood by rational reasoning, although he did believe in empiricism, i.e. that knowledge is derived from personal experience and observation. He did not test his ideas experimentally nor did he encourage others to do so. As a result, many of them were wrong. For example, he proposed that women play no role in conception, that men are better thinkers than women, that thoughts come from the heart and that the role of the brain is to cool the blood. He also proposed that everything is made from fire, earth, air and water, earthquakes are caused by air escaping from the planet, and that the Earth is the centre of the universe. He was right in suggesting, purely on the basis of his 'thought experiments', that the Earth is not a flat disk but a spherical ball. Before we denounce thought experiments too severely we need to remember that Albert Einstein's epic discoveries on special and general relativity were also made using this method as he left others to provide the experimental proof.

Archimedes (288-212 BC) was a remarkable polymath – scientist, inventor and mathematician – who invented the Archimedes Screw, compound pulley and defensive war machines to protect his native Syracuse from invasion. He developed a set of giant 'burning glasses' (mirrors) with which he is reputed to have set the invading Roman fleet of ships alight. In 1747 the Comte de Buffon tested this idea in Paris, setting wooden houses alight with 150 giant concave mirrors, which supports the idea that Archimedes' feat was feasible.

Archimedes was one of the greatest mathematician who, using rigorous, repeatable methods, calculated *inter alia* the area of a circle, the surface area and volume of a sphere, the area of an ellipse and the area under a parabola. Once, as a diversion, he calculated that the number of grains of sand required to fill the universe is $8 \times 10^{63}$. He also derived an accurate approximation for *pi*, explained the theory of levers and built a ship, the largest in classical antiquity, capable of carrying 600 people that also included a garden, gymnasium and temple among its onboard facilities. He devised his screw pump to pump bilge water from ships and it is still in use today pumping water into irrigation canals in the Nile Delta in Egypt or moving coal and grain. The world's first sea-going steamship with a screw propeller, the SS *Archimedes*, launched in 1839, is named in honour of the great scholar.

Archimedes is probably best remembered for devising a method for determining whether King Hieron's crown was made of pure gold. He is reputed to have done this while enjoying a bath when he noticed that the amount of water displaced from the full tub was equal to the volume of his immersed body. He realized that the immersed crown would displace an amount of water equal to its own volume. By dividing the mass of the crown by the volume of water displaced, its density could be calculated. He also surmised that its density would be lower than that of gold if cheaper and less dense metals such as silver had been added. He was so excited by his discovery that he ran down the street naked shouting, "Eureka! I have found [it]". But, as Isaac Asimov pointed out in 1972, the more exciting phrase in science is not, 'Eureka' but, 'That's funny', highlighting the fact that observations that do not fit into the accepted paradigm are the ones to get excited about.

Archimedes died *ca* 212 BC when Roman forces under General Marcellus captured Syracuse after a two-year siege. He was apparently contemplating a mathematical diagram when a Roman soldier commanded him to go and meet Marcellus but he declined saying that he first had to solve the problem. The soldier was enraged and killed Archimedes with his sword.

Pliny the Elder (23-79 AD) carried his dedication to science too far. When he witnessed the explosion of Mount Vesuvius in AD 79 he was so consumed by curiosity that he, against the advice of his colleagues, set sail across the Bay of Naples towards the giant plume of smoke. Even though ash and pumice rained down on him and his crew he persevered, reached the shore and even began walking towards Pompeii. Eventually his crew abandoned him and he eventually collapsed and perished from inhaling sulphur.

**Early Muslim scientists**: During the Golden Age of Islam, from about 850 to 1500 AD, there were many great Muslim thinkers, scientists

al-Jāhiẓ

and inventors, most of them unrecognised today. al-Jāḥiẓ (776-869 AD), the son of a black cameleer, was an important 9th century Islamic scholar who was born in Basra, Iraq. He came from a poor family and started life as a fish seller but educated himself by attending discussions on matters scientific with other youths in the Basra mosque. He eventually became a full-time writer, authoring over 140 books (of which 75 survive) on subjects as varied as the Quran, Arabic grammar, zoology, poetry, lexicography and rhetoric. In Baghdad Caliph al-Ma'mun asked him to teach his children but changed his mind when they were frightened off by his boggle-eyes (جاحظ العينين)!

His epochal *Book of Living* (*Kitāb al-Ḥayawān*), published in seven volumes, comprises anecdotes, poetic descriptions and proverbs on 350 animal species and included some of the first ideas on ecological relationships. For example, he wrote, "All animals, in short, cannot exist without food, neither can the hunting animal escape being hunted in his turn. Every weak animal devours those weaker than itself. Strong animals cannot escape being devoured by other animals stronger than they." He also mentioned biodiversity conservation and offered early interpretations of the struggle for existence and natural selection over 1 000 years before Charles Darwin. Another of his books, *Kitāb al-Bukhalā* (*The Book of Misers*), is a humorous and satirical collection of stories in which he ridicules schoolmasters, beggars, singers and scribes for their greedy behaviour. al-Jāḥiẓ died in his private library when a pile of books fell on his head, killing him at the ripe old age of 92!

In the late 10th century the famous engineer Ibn al-Haytham (965-1040) was invited by the Sultan in Cairo to devise a way of controlling the flooding of the Nile River. He rashly accepted the contract but, after one trip up the great river, concluded that he did not have the means to do so. He also realized that the Sultan was a cruel man who executed anyone who did not obey his commands. Ibn al-Haytham saved his skin by pretending to be insane as he knew that it was illegal to execute mad people in Egypt at the time. He was held under house arrest for about 18 years until the Sultan died and during this time carried out ground-breaking research on optics. He published his revolutionary *Book of Optics* in seven volumes between 1011 and 1021.

Ibn al-Haytham was the first scientist to demonstrate how the human eye works and designed the first camera obscura, the precursor to the camera, as a giant, walk-in model of the eye. He showed that light travels

## 1. Funny science: The ancients

Camera obscura invented by Ibn al-Haytham in Cairo *ca* 1015

in straight lines, explained the rainbow, shadows, twilight and why the moon appears to enlarge shortly before sunset, and split white light into colours about 650 years before Isaac Newton. He was also the first scientist to develop the modern scientific method (described in his *Book of Optics*) in which he emphasized the importance of proposing and testing an hypothesis through careful and repeatable experimentation.

The United Nations proclaimed 2015 as the International Year of Light and Light-Based Industries, one thousand years after Ibn al-Haytham's monumental discoveries, and people throughout the world commemorated his famous work. My contributions to this celebration included reconstructing Ibn al-Haytham's laboratory and repeating some of his experiments in the Bahrain Science Centre in Isa Town, Bahrain, and at the Observatory Museum in Makhanda, South Africa. I also carried out guided tours of the camera obscura at the Cape Town Science Centre dressed in a *dishdāshah* robe, black *agal* cord, *ghutrah* orange-chequered head dress and leather sandals.

**Mediaeval European scientists**: Public brawls between scientists are now rare but were common in earlier times. Today animosities are expressed by way of a discrete disparagement in the committee room when a rival's grant comes up for renewal, but they were settled in more manly ways in the past. Tycho Brahe (1546-1601) was a great 16th century Danish astronomer who once observed a supernova (exploding star) and used his observation to challenge the prevailing view that stars are unchanging and eternal. The Danish King rewarded him for this spectacular find by sponsoring two new astronomical observatories. But Brahe was an argumentative sort and once faced a rival in a sword duel and had the bridge of his nose cut off. Undaunted he repaired to his workshop and fashioned a new nose for himself out of silver. In fact, he apparently made several noses from different metals which he wore on different occasions (observing, walking, attending a banquet, etc.) depending on the formality of the event. When historians questioned the truth of this story and dug up Brahe's grave, they found his silver nose.

Tycho Brahe

In the early 19th century duels were fought with more deadly weapons. A brilliant young French mathematician, Evarioste Galois (1811-1832), had a brief and tragic life as he was killed in a pistol duel at the age of only 21. The night before he died, he feverishly scribbled down new solutions in the theory of quintic and other integral functions which were some of the most intractable mathematical problems of his day. The next morning he faced his rival at 25 paces, was shot in the stomach and died. When his notes were eventually edited and published, the editor noted, "My zeal was well rewarded, and I experienced an intense pleasure at the moment when, having filled in some slight gaps, I saw the complete correctness of the method by which Galois proves his beautiful theorem."

Leonardo da Vinci (1452-1519), the brooding 15th century Italian genius, excelled in painting, sculpture, music and engineering but he was not a particularly successful inventor. Few of his inventions were made in his day, partly because they did not have the materials or knowledge to do so but mainly because they were largely flights of fantasy that would not have worked. His airplanes required the pilots to flap their wings like birds which human musculature cannot do and his 'ornithopter', a helicopter with helical wings, had no tail rotor so it would have spun round and round. If he had been aware of and used some of the innovations made by Islamic engineers (such as Ibn Firnas, the Banu Musa brothers, al-Jazari, al-Rammah and Taqi al-Din) da Vinci might have been more successful. Although the Italian was a man of peace his war machines, including a three-barreled cannon, multi-arrow crossbow and an armoured car propelled by

# 1. Funny science: The ancients

**Sketch of a bird's wing by Leonardo da Vinci**

eight men inside turning cranks, were his most successful inventions.

Da Vinci would have benefited from the knowledge of Roger Bacon (1220-1292), a Franciscan friar who could read Arabic and introduced many Islamic discoveries and inventions, including those of Ibn al-Haytham, to Europe. He was one of the first scientists to wear spectacles (which he co-invented) and did the first tentative experiments with gunpowder in the West. He further developed Ibn al-Haytham's explanation of the rainbow and predicted the invention of hot air balloons, horseless carriages and motorized ships but was eventually jailed for his experiments and supposed disrespect for the church. This was at a time when Muslim military, political and religious leaders embraced their scientists as respected and honorable members of society.

His namesake, the 16th century British philosopher, lawyer and statesman Francis Bacon (1561-1626), is credited with being the father of the scientific method but Ibn al-Haytham is the rightful claimant in this regard. Bacon did publish the first science fiction book, *New Atlantic*, in which he predicted the development of synthetic materials like plastics, and his exciting prognostications helped to stimulate the founding of the Royal Society of London some years later. But Bacon attracted many enemies for, as Robert Hooke later stated, "Being too prying into the then receiv'd philosophy of science". He fell from grace, was banished from London, later returned but sadly came to a sticky end. While travelling with the King's physician he proposed that they try an experiment to test whether ice could preserve the flesh of a chicken. They stopped their horse-drawn carriage, bought a chicken, encased it in ice and found that it remained perfectly preserved. Unfortunately this did not apply to Bacon who caught a chill

Isaac Newton

and died soon afterwards of pneumonia, a true martyr of science.

**17th century**: One of the greatest British scientists, Sir Isaac Newton (1643-1727), was a cranky youth and a cantankerous and ferociously competitive old man. As a kid he released lighter-than-air lanterns containing lit candles which caused pandemonium in his village. He was also absentminded – the family maid once found him in the kitchen staring at a saucepan of boiling water containing his pocket watch while he was holding an egg in his hand! Newton was born on Christmas Day and was only 23 years old when he discovered the Law of Universal Gravitation. He made most of his great discoveries before he was 30 and spent much of the rest of his life in violent arguments with his scientific colleagues, especially Robert Hooke. Newton was eventually forced to publish his work decades after his initial research and would probably not have earned a research grant from any modern granting agency on account of his procrastination. Newton's work defined a whole era of science as he made significant contributions to physics, astronomy and mathematics despite, or perhaps because of, his quirky personality. His approach to science had as much impact as his discoveries as he, like Ibn al-Haytham and Francis Bacon before him, insisted on only proposing hypotheses that could be tested through observation and experiment.

Despite his reputation as an astute experimental scientist Newton dabbled in alchemy, the ultimate pseudo-science, throughout his life. The roots of alchemy can be found in several ancient civilizations – Greek, Chinese and early Islam – when alchemists, drenched in superstition and misled by early interpretations of the nature of matter, tried to find a shortcut to wealth and immortality by transforming base metals into gold and attempting to concoct life-prolonging elixirs. Like all alchemists before him (including Robert Boyle) Newton failed to find the Philosopher's Stone. Late in life he was master of the Royal Mint and was in the habit of lurking in dingy pubs in London trying to catch criminals who cut bits off coins and melted them down to make new coins. He solved this problem by introducing milled edges onto rounded coins.

The Englishman Robert Hooke (1635-1703) was a formidable polymath who made many inventions including the universal joint and iris diaphragm, surveyed and helped to rebuild London after the Great Fire of 1667, and played a leading role in the founding of the Royal Society of London. He was often compared with da Vinci but is hardly remembered today as he was a contemporary of Newton's. He was short, ugly (a victim of smallpox), argumentative and hugely energetic and took part in many of the great scientific debates of the day. Hooke was so paranoid about other scientists stealing his results that he concealed them using codes or deposited his dated observations in an archive. Once, in a paper on elasticity, he stated that the way to compute the velocity of bodies moved by springs is 'ceiiionssttuu' which is an anagram for a Latin code! Another scientist, Robert Boyle (1627-1691), was so neurotic about someone else pre-empting his discovery of phosphorus that he deposited his recipe for its preparation in a sealed envelope with the Royal Society to be opened only after his death.

More recently fierce rivalry surrounded the discovery of the substance with the highest critical temperature as the prospect of great financial reward rested on this revelation. Professor Paul Ching Wu Chu (born 1941) at the University of Houston made the discovery but, in his published paper, he deliberately introduced mistakes into the text so that his competitors could not identify the substance, take the credit and patent it before him. Earlier, in the 16th century, Leonardo da Vinci was also in the habit of introducing deliberate errors into the design of his war machines so that miscreants could not make functional versions of them.

In its early days the prestigious Royal Society of London (founded in November 1660) was quite unethical and politically incorrect in some of its 'research' activities, as pointed out by Bill Bryson in his delightful book, *Seeing Further: The Story of Science, Discovery, and the Genius of the Royal Society*. For example, in 1676 a student agreed to allow two Fellows to transfuse sheep's blood into him to see what would happen. In front of an audience that included the Bishop of Salisbury 14 ounces of blood was pumped from the sheep into the student, fortunately with little effect. Royal Society Fellows, called 'virtuosi' at the time, also carried out bizarre experiments by placing animals in glass vessels and evacuating the air to determine how quickly they would die. They were also not averse to pumping air into dogs to watch them explode, like the fictional experimenters in *Gulliver's Travels*.

Alessandro Volta demonstrates his voltaic pile

**18th century:** Alessandro Volta (1745-1827), a pioneer of research on electricity, demonstrated his revolutionary new battery, the Voltaic Pile, to Emperor Napoleon in 1799 and received a gold medal as his reward. Volta did not, of course, invent electricity, which is a natural phenomenon, but he did devise a way to store it for later use which was a major breakthrough. In fact, crude batteries may have been made previously, perhaps as early as 650 AD. The so-called Baghdad Battery found in modern Khujut Rabu, Iraq, comprises a terracotta pot, a tube of rolled copper sheet and a rod of iron. At the top the iron rod is isolated from the copper by bitumen with both the rod and the copper cylinder fitting snugly inside the opening of the jar. Its purpose is unclear but it might have been used as a battery to generate a small electric current for electroplating but no electroplated objects are known from this period. An alternative explanation is that it functioned as a storage vessel for sacred scrolls.

On 23rd March 2005 the Discovery Channel programme *MythBusters* built replicas of the Baghdad Battery to determine whether it was possible for it to have been used for electroplating. Ten hand-made terracotta jars were fitted out as batteries with lemon juice as the electrolyte. Connected in series, the batteries produced four volts of electricity, enough current to electroplate a small token. This author, while director of the Bahrain Science Centre between 2012 and 2015, also made a replica of the Baghdad Battery that produced a small amount of electricity.

Volta's discovery of ways to generate and store electricity independent of Nature caused great excitement and attracted the attention of many scientists as well as charlatans, and an appetite grew for spectacular demonstrations. In 1720 Stephen Gray "caught an orphan, hung him up with insulation cords, electrified him, and drew sparks from his nose." In 1746 the court electrician to Louis XV in France ordered 148 French Guards at the Palace of Versailles to link hands in a giant circle. When he sent an electric current through them all 148 guards leapt up in synchrony

## 1. Funny science: The ancients

as if in a *corps de ballet*. Everyone was astonished that electricity travelled so fast, but it was another century before James Clerk Maxwell in England demonstrated that it travelled at about the speed of light. In another bizarre experiment an Italian researcher Giovanni Aldini (1762-1834) collected newly chopped off heads at the base of the guillotine during the French Revolution and passed electric currents through them to see how they would react. He reported that the currents provoked grimaces, twitching lips and blinking eyes for several minutes after death.

Napoleon Bonaparte (1769-1821), the controversial French military general, was also a man of science. He was President of the French Academy of Sciences from 1801 to 1814 and organised the Egyptian Campaign which was both a great military adventure and a productive scientific expedition. He took along 154 scientists to investigate Egypt's history and geography and they made some spectacular discoveries, including the Rosetta Stone which would later provide the key to deciphering the hieroglyphics of ancient Egypt.

Napoleon introduced several successful military innovations during the Napoleonic wars including increased maneuverability of his troops, a 'divide and conquer' strategy, concentrated fire power and an ambulance service. In 1791 he introduced to France the metric system of measurement which is now used worldwide except in Myanmar, Sierra Leone and the USA! In 1795 he offered a reward of 12 000 French francs to anyone who could devise a method for preserving food for his troops. Fourteen years later a French confectioner Nicolas Appert unveiled a successful method of preservation using glass jars that were filled with food and then sealed and

**Bizarre electricity experiments at the Palace of Versailles, France, in the 1790s**

heated. Napoleon also awarded a commission to an expatriate American Robert Fulton to build a prototype submarine, the *Nautilus,* and in 1795 one of his military officers, Nicholas Jacques Conté, patented the modern method of kiln-firing powdered graphite with clay to make pencils of any desired hardness. Napoleon even has a mathematical theorem named after him. But, like all creative people, he had his idiosyncrasies. He hated cats, had a horror for open doors and was a poor horseman and a terrible shot, not a great skillset for a military commander!

Interestingly, Napoleon was a boyhood hero of the famous German-born Austrian composer Ludwig van Beethoven who envisaged him leading Europe into a new age of liberty, equality and fraternity. Beethoven originally dedicated his Third Symphony (*Eroica,* the heroic symphony) to Napoleon but angrily withdrew the dedication and tore up the title page when he heard that the military leader had proclaimed himself as Emperor of France and proved to be just another ambitious mortal!

The guillotine (invented by Tobias Schmidt) is named after a French medical doctor Joseph-Ignace Guillotin (1738-1814). He was, by all accounts, a kind and gentle man who advocated its use as it was more humane than the other methods of execution, at that time such as being burned at the stake, pulled over cobbled streets by horses or hanged in a noose. During the French Revolution about 40 000 aristocrats, including royalty, as well as some leading scientists and medical doctors and eventually many revolutionary leaders themselves fell foul of the proletariat and were executed by *Madame la Guillotine*. Amazingly the guillotine continued to be used for public executions in France until 1939 and, behind closed doors, until 1977, when France abolished the death penalty. In Saudi Arabia, where I recently spent nine months, the authorities still chop off the hands of serial criminals. While I was living in Bahrain in 2012 a British tourist had his lower arm chopped off with a sword when he accidentally strayed into a politically charged situation in Manama City.

One of the scientists who lost his head in the French Revolution was the pompous Antoine Lavoisier (1743-1794), a founder of modern chemistry who happily stole the work of others without giving them credit. He formulated the Principle of the Conservation of Matter and put paid to the Phlogiston Theory, the brainchild of a German scientist Georg Stahl (1659-1734, who proposed that phlogiston is a fluid released as a flame when a substance burns. In 1789, shortly after the fall of the Bastille, Lavoisier, in one of his acts of extreme vanity, organised a mock trial of the phlogiston theory at his house. The charge was read out by a character dressed up as 'Oxygen' and the theory was sentenced to death by burning Stahl's books.

Lavoisier, by virtue of his inheritances, had previously extracted massive taxes from the people of Paris before the Revolution and was beheaded as an enemy of the people at the age of 54 years. The French astronomer, mathematician and mayor of Paris from 1789 to 1791, Jean Sylvain Bailly, who mapped the trajectories of the moons of Jupiter, had earlier died at the guillotine during the Reign of Terror in November 1793.

The American scientist/statesman Benjamin Franklin (1706-1790) carried out dangerous experiments with lightning but knew enough about physics to exercise extreme caution and his near-death experience prompted him to invent the lightning rod. He also invented bifocal spectacles, was a signatory of the American Declaration of Independence, American Ambassador to France, a member of the prestigious Lunar Society in England, a leading writer, printer, political philosopher, politician, Freemason, postmaster, humourist, civic activist as well as an avid anti-slavery campaigner, such was the sweep of men of science in those days.

Medical science was sometimes advanced in bizarre ways. An American military surgeon William Beaumont (1785-1853) once treated a young backwoodsman Alexis St Martin (1794-1880) who had been shot at close range by a shotgun. The blast, which blew a hole in his side large enough to put your fist through, miraculously healed but the hole remained, with the wall of the stomach visible from the outside. Beaumont saw this as a research opportunity and, with St Martin's permission, was able to deliver drugs and extract samples of digestive juices and bile from his patient's stomach to determine digestion rates and the effect of temperature on stomach functions. Eventually St Martin grew tired of his role as a walking stomach experiment and retired to the woods, but he returned later with his wife and two children and agreed to act as a medical guinea pig for a few more years. He lived to the age of 86 years secure in the knowledge that over 238 experiments had been carried out through his hole-in-the-stomach by Beaumont, who became known as the 'Father of Gastric Physiology'!

**19th century**: The Russian scientist Dmitri Mendeleev (1834-1907), who invented the Periodic Table of the Chemical Elements in 1869, first formulated his idea in a dream. He was the youngest of 17 children from a poor Russian family who lived in Siberia but overcame his early hardships through sheer hard work. He was addicted to the card game of Patience and wrote the names of the known chemical elements on his playing cards, arranging and re-arranging them endlessly on his desk but unable to find a pattern. Eventually, exhausted, he fell asleep and, in a dream, an organized table came to him in which all the elements fell neatly into place. "When I woke up, I wrote it down and only one fact needed correction." What he found was that, if the elements are arranged by order of atomic weight, their chemical properties recurred at regular intervals. He called it the Periodic Table of the Elements and it made him famous. Furthermore, Mendeleev noticed gaps in the table and predicted that new elements would be discovered to fill those gaps. In one of the most remarkable examplars of the hypothetico-deductive method in science he was even able to predict the chemical properties of the missing elements, one of which, mendelevium, is named after him.

This scenario is not entirely unlikely as several other major scientific insights have been reached while dreaming. August Kekule (1829-1896), one of the founders of structural organic chemistry, is well known for his celebrated dreams. Once on a late-night bus trip he fell asleep and dreamed of atoms gambolling before his eyes. When he woke up he jotted down the structures of several molecules containing carbon. On another occasion, while dozing in front of a fire, he dreamed up the cyclic molecular structure of benzene by imagining a snake biting its tail.

Another famous scientist who had his aha! moment during a dream was the German-born psycho-biologist, Otto Loewi. In the early 20th century a controversy raged as to whether nerve impulses are transmitted by chemical or bio-electrical messengers. In 1921 he carried out a famous experiment by removing beating hearts from two frogs, one with the vagus nerve (which controls the heart rate) attached and the other without the vagus nerve, with both hearts bathed in a saline solution. By electrically stimulating the vagus nerve Loewi made the first heart beat slower then poured some of the liquid bathing this heart onto the second heart, which made it beat slower.

What makes Loewi's experiment remarkable is that the idea for it came in a dream. On Easter Saturday in 1921 he dreamed of a way in which to prove that the transmission of nerve impulses is chemical, not electrical. He woke up, scribbled the experiment onto a scrap of paper and went back to sleep. The next morning, he found to his horror that he could not read his midnight scribbles but that night he had the same dream. This time he immediately went to his lab to perform the experiment. Thirteen years later, Loewi was awarded the Nobel Prize in Physiology or Medicine which he shared with Sir Henry Hallett Dale, a lifelong friend who had helped to inspire the ground-breaking experiment.

Many years later another scientist solved a major problem while daydreaming on a bus trip. Freeman Dyson (1923-2020), one of the best theoretical physicists of the modern era, had recently held discussions with Richard Feynman and Julian Schwinger about their differing views on the rules governing the interactions of subatomic particles. "As we were droning across Nebraska on the third day, something suddenly happened. For two weeks I had not thought about physics and now it came bursting into my consciousness. Like an explosion, Feynman's pictures and Schwinger's equations began sorting themselves out in my head with a clarity they had never had before. For the first time I was able to put them all together. For an hour or two I arranged and re-arranged the pieces. Then I knew they all fitted. I had no pencil or paper, but everything was so clear I had no need to write it down. Feynman and Schwinger were just looking at the same set of ideas from two different sides."

The discoverer of the polymerase chain reaction (PCR) Kary Mullis (1944-2019) made his vital mental breakthrough during a car trip. He was driving to his cabin in California late one night thinking about his sequencing experiments when he had a blinding moment of revelation, an experience granted to few scientists. On the following Monday morning he outlined his ideas to his colleagues. They were unimpressed and it was only when the value of the PCR was proven (and then revolutionized biology, biotechnology, pharmaceutics and agriculture) that they appreciated what a significant insight he had made that night.

The English chemist Sir Humphry Davy (1778-1829), who discovered sodium and potassium and invented the Miner's Safety Lamp, was also the first scientist to show that, when an electric current flows through a wire, its resistance causes the wire to heat up and emit light. Davy was in the habit of conducting public experiments at the Royal Institution in London. One involved feeding laughing gas (nitrous oxide) to visiting dignitaries, which caused extreme flatulence, much to the amusement of the audience. The Gillray cartoon reproduced here shows Davy in 1802 administering the gas to the British diplomat Sir John Coxe Hippisley with dramatic effect. Davy appropriately worked for an organisation known as the Pneumatic Institute.

Cartoon by caricaturist James Gillray lampooning Sir Humphry Davy's laughing gas experiments

One of Davy's greatest contributions to science was to employ a young Michael Faraday (1791-1867) as his assistant. Faraday, who came from a poor family in London, first learned about science by sneaking looks at books while working in a bookshop and became one of the greatest experimenters of all time, but he also had his quirks. When he published his first description of a crude electric motor without acknowledging the contributions of his colleague Hans Ørsted (1777-1851) the resulting fuss caused Faraday to withdraw from research in this field for 10 years as a self-imposed punishment. He nevertheless made significant contributions with his later discovery of electro-magnetic induction and the development of the first dynamo. When he demonstrated this invention to the Chancellor of the Exchequer, he was asked, "What use is it?" and Faraday famously replied, "What use is a baby?" This statement was, however, borrowed from Benjamin Franklin who had used it for the new-fangled French *aerostatique* machines in the 1780s.

Michael Faraday

Some of the most acrimonious scientific debates ever concerned the theory of evolution by natural selection proposed by Charles Darwin (1809-1882) and Alfred Russel Wallace (1823-1913). One of Darwin's strongest supporters was Thomas Huxley (1825-1895), known as *Darwin's bulldog*, who had several memorable confrontations with the Tory Bishop of Oxford, Samuel Wilberforce (1805-1873). Wilberforce was one of the most accomplished public speakers of his day and was known as 'Soapy Sam' because of his slippery tongue. In a dramatic debate in the Natural History Museum in 1860 the Bishop asked Huxley whether he would be proud to trace his descent through an ape, to which Huxley replied, "If the question were put to me, would I rather have a miserable ape for a grandfather, or a man highly endowed by nature ... and yet who employs these faculties ... for the mere purpose of introducing ridicule into a grave scientific discussion, I unhesitatingly affirm my preference for the ape."

This was powerful language to address to a Head of the Church and a member of the audience, Lady Brewster, fainted. There was also much laughter and Huxley clearly won the day. On leaving the Museum the Bishop of Worcester's wife was heard to say, regarding evolution, "Let us hope it is not true. But if it is true, let us hope it does not become generally known." It is notable that the reclusive Darwin played virtually no role in these debates as he preferred to focus on his research and writing from his country home at Down House in Kent.

Darwin is arguably the scientist who did more to revolutionize the worldview of people outside science than anyone else, but he also had his idiosyncrasies. He never worked for a university, museum or government institute (except as a natural philosopher on the *Beagle*) and did all his research at home or in the field. I once played cricket on a field near Down House, now restored and owned by English Heritage, and we had tea in Darwin's lounge. Interestingly, although Darwin's book *On the Origin of Species by means of Natural Selection, or the Preservation of Favoured Races in the Struggle for Life* (1859) launched the modern sciences of evolution and ecology, neither of these words, nor the phrase 'survival of the fittest', appears in his epic tome.

Furthermore, Darwin's *On the Origin of Species*, which changed the course of the biological sciences forever, was originally

**Charles Darwin**

written as an essay with the intention that he would later expand it into a comprehensive book, but this never happened. The essay was prompted by a letter and draft scientific paper that Darwin had received from a colleague, Alfred Russel Wallace, who was then doing field work in Papua New Guinea. Wallace had come to the same conclusion as Darwin regarding the mechanism for evolution and his paper threatened to pre-empt Darwin's ground-breaking theory. The matter was resolved when both publications were presented to a meeting of the Royal Society of London in 1858, with neither of the authors present.

Darwin enjoyed carrying out experiments in his home and garden, often with the help of his children. One involved dousing flying bumble bees with flour using a feather duster and then running after them to determine their home ranges. He used clothes pegs to measure the strength of bean sprout roots and made replicas of the springs found in orchids, that shoot pollen onto a visiting bee, using bits of whalebone (baleen) cut from his wife's corset, as no plastic was available then. He also did experiments to determine whether earthworms can detect vibrations by placing bottles of worms on the piano and observing their behaviour while his wife was playing. He found that they are extremely sensitive to vibrations.

Darwin's paternal grandfather Erasmus Darwin (1731-1802), a medical doctor, botanist, evolutionist and poet, was a Fellow of the Royal Society and a respected if eccentric figure in this time. He was untroubled by patriotic concerns and waxed lyrical about the exciting development of hot air balloons by the French which the President of the Royal Society of London at the time, Joseph Banks, chose to downplay. In fact, Erasmus was one of the pioneers of a balloon powered postal system, but it did not last as hot-air balloons at that time were difficult to control. Erasmus formulated one of the first theories of evolution in his poem *Zoonomia* or *The Laws of Organic Life* (1794-1796) in which he advanced the notion that all animal life arose from 'one living filament' over 60 years before his grandson's famous book.

Another interesting relative of Darwin's was his cousin Sir Francis Galton (1822-1911), the erratic genius who discovered that humans have different fingerprints but also proposed that criminals are born to be wicked and launched the now discredited practice of eugenics whereby humans are 'improved' by selectively mating people with desirable hereditary traits. He was variously described as a statistician, sociologist, psychologist, anthropologist, eugenicist, tropical explorer, geographer, inventor, meteorologist and psychometrician, such were the multiple talents of polymaths at that time.

One of the most famous rivalries in the history of science was between two American palaeontologists, Edward Drinker Cope (1840-1897), who discovered about one thousand new species of extinct vertebrates in the United States, and Othniel Marsh (1831-1899), a professor at Yale College and president of the National Academy of Sciences. They stole, plagiarized, publicly insulted and used every form of skullduggery against one another. Once Cope visited Marsh and briefly watched him examining an exciting new species of fossil lizard. He left the laboratory, telexed an accurate description of the lizard to the *American Naturalist* journal and gleefully stole the credit for the discovery!

Talking about eccentricity, William Buckland (1784-1856), the English theologian, geologist and palaeontologist who wrote the first full description of what would later become known as a dinosaur (*Megalosaurus*), must take the cake, so to speak. He was well known for his anti-evolutionary theories and his efforts to reconcile geological discoveries with the Bible, but he also had a bizarre streak. He and his son Frank were avid carnivores and had an arrangement with the London Zoo that, when an animal died, they would receive a fresh cut of meat to eat. So, his family served up, to themselves and their guests, elephant, dolphin, mole, rat, panther, crocodile and mouse.

Once Buckland was shown by his friend, the Archbishop of Canterbury, a snuff box containing the embalmed heart of Louis XVI which the bishop had bought in Paris during the French Revolution. In an instant, and before anyone could stop him, Buckland saw a unique opportunity and grabbed and ate the embalmed heart! He was also known to give some of his outdoor lectures on geology on horseback and to animate his indoor lectures by imitating the movements of dinosaurs. His passion for things scientific even extended into his home where he had a table inlaid with dinosaur coprolites (fossil droppings).

# 2.
# The Funny Side of Science: The Moderns

*A visitor to Niels Bohr's country cottage, noticing a horseshoe hanging on the wall, teased the eminent scientist about this ancient superstition.*
*'Can it be true that you, of all people, believe it will bring you luck?'*
*'Of course not,' he replied, 'but I understand that it will bring you luck whether you believe it or not.'*

**RADIOACTIVITY AND X-RAYS:** In the late 19th century radioactive materials attracted the interest of many scientists. Marie Curie (1867-1934), who discovered polonium and radium, was the first women to win a Nobel Prize and the first scientist to win two Nobels. Uniquely her husband Pierre and daughter Irène Joliot-Cutie also won the prize. Without knowing Marie carried her dedication too far and regularly handled radium and other radioactive materials during her research and radiological field work in hospitals during World War I. This exposure eventually caused her to die of cancer at the age of 67 years. Even today her laboratory apparatus and notes in the Maria Skłodowska-Curie Museum in Warsaw must be displayed in airtight compartments as they are still radioactive.

Robert Williams Wood (1868-1965), Professor of Physics at Johns Hopkins University, once suspected that his landlady was using chicken bones from the previous day in their morning soup. He sprinkled the chicken with radioactive tracer and then tested the soup the next day. As he predicted, the tracer showed up in the soup but an unfortunate outcome was that his classmates fell ill.

In 1914 the British physicist James Chadwick (1891-1974), who later won the Nobel Prize for discovering the neutron, developed an active laboratory in a German concentration camp to stave off

Marie Curie

boredom. As it was difficult to obtain radioactive materials from normal sources, he used extracts from a popular toothpaste that was advertised as 'radioactive' and really was! It was made by a company that also manufactured gas mantles, which contain radioactive material even today. The toothpaste was later found to contain thorium, which must have done considerable harm to its users. Chadwick's further claim to fame was that, in 1941, he wrote the final draft of the MAUD Report that encouraged the U.S. government to begin serious atomic bomb research efforts.

Some scientists are quite sloppy, yet they still make great discoveries. The Scottish physician and microbiologist Alexander Fleming (1881-1955) spent most of his working life in a laboratory near Paddington Station in London and was known to make bacterial cultures using sputum, tears from his staff and mucus from his nose as well as from blood, semen, ovarian cyst fluid, pus and egg white. He also occasionally used himself as a patient but still developed a reputation as a brilliant researcher. His untidy habit of leaving petri dishes of cultures lying around for weeks led to his two major discoveries – the enzyme lysozyme and the world›s first broadly effective antibiotic, penicillin. In late 1921 he found that one of his agar plates was contaminated with bacteria from the air. When he added nasal mucus to the plate he found that it inhibited bacterial growth, which lead to the discovery of lysozyme.

As Kendall Haven reports in his 1994 book, *Marvels of Science*, Fleming said at the time, "One sometimes finds what one is not looking for. When I woke up just after dawn on September 28, 1928, I certainly didn't plan to revolutionize all medicine by discovering the world's first antibiotic, or bacteria killer. But I suppose that was exactly what I did." His discovery of penicillin has been described as the "single greatest victory ever achieved over disease."

Fleming was knighted in 1944 and jointly received the Nobel Prize in Physiology or Medicine in 1945. In 1999 he was named in *Time* magazine's list of the *100 Most Important People of the 20th century*, and in 2009 he was voted third 'greatest Scot' in an opinion poll conducted by STV, behind only Robert Burns and William Wallace. A statue of Fleming stands outside the main bullring in Madrid where it was erected through donations made by grateful matadors as penicillin had greatly reduced the number of deaths in the bullring. His ashes are buried in St Paul's Cathedral. Was Benjamin Franklin right when he said, "Carelessness does more harm than a want of knowledge?"

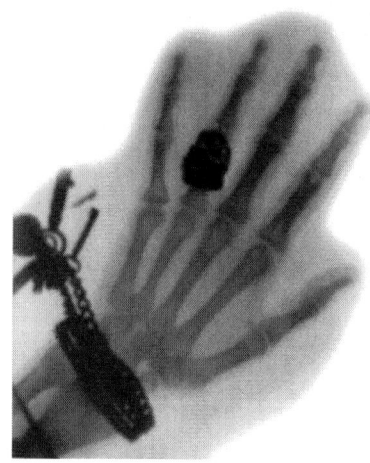

Wilhelm Röntgen's first x-ray, of his wife's hand

When Wilhelm Conrad Röntgen (1845-1923) discovered x-rays by accident while doing humdrum experiments with cathode rays in 1895 he was only mildly excited. He called them 'x-rays' as he did not know their properties and won the first Nobel Prize in Physics in 1901 for his chance discovery (but also for his alert mind). Röntgen was a reluctant hero and maintained a low profile. He built his own research equipment and was a great lover of Nature and a daring mountaineer. He was also a cautious scientist who first x-rayed his wife's hand rather than his own, with the clear image of her finger bones and wedding ring being the first radiograph on record. He died of cancer at the age of 78, probably as a result of his exposure to x-rays.

Soon x-rays were being used worldwide in hospitals and a French scientist even claimed that he had photographed the soul. But, for every great discovery, there are bitter scientists who did not take the gap. Lord Kelvin (1824-1907), one of the leading physicists at the time, was skeptical and called x-rays a hoax. Dr Frederick Smith of Oxford was embittered as he had previously noticed that photographic plates stored near cathode rays became fogged. His reaction, instead of finding out why, was to store the plates further away! Another researcher, Philip Lenard, had made a similar observation to Röntgen's but had not followed it up. He was so distraught that he refused to ever mention Röntgen's name again and became an avid anti-physics campaigner and Nazi.

Many years later South African Allan MacLeod Cormack co-invented the CATscanner, which produces 3D x-rays. His co-inventor, the English bio-medical engineer Godfrey Hounsfield, worked for the company EMI which had a medical research division and a music recording division based at Abbey Road in London. It so happened that, at the time, the company had contracted an insanely profitable band, *The Beatles*, and profits from their records helped to fund the development of the CATscanner and other medical technologies.

## 2. Funny science: The moderns

**Thomas Edison (right) with his new-fangled phonograph**

**Edison, Tesla and Marconi**: Thomas Alva Edison (1847-1931) is arguably the greatest inventor of all time, 'the man who invented the future,' with over 1000 patents to his name. Some of his early inventions were improvements to telegraph machines so it is not surprising that he nicknamed two of his children 'Dot' and 'Dash' and even proposed to his second wife by tapping out the message in Morse Code on her hand! As a youth Edison delivered newspapers on the railroads and was boxed on his ears for being disobedient. This affected his hearing, and he was virtually deaf in later life. To hear the sounds made by his newly invented phonograph he would bite the wooden base of the machine with his teeth and 'listen' through his jaw bones. Edison phonographs with his tooth marks on them can be viewed in the Smithsonian Institution in Washington DC today.

At his incubation hub, the Menlo Park Invention Factory in New Jersey, USA, Edison also invented the movie camera, electricity power station, a method for the mass distribution of electricity and the electric tram. He did not invent the electric light bulb, which is attributed to the British chemists Warren de la Rue and/or Joseph Swan, although he did improve it as he did with Alexander Graham Bell's telephone. The incandescent light bulb, which is widely used as the symbol of innovation, is, in fact, one of the least efficient and most environmentally harmful inventions ever made (on a par with the internal combustion engine and water flush toilet). Only about 5% of an incandescent light bulb's electrical energy is turned into light which is why it is now being replaced by far more efficient halogen, fluorescent and LED lights.

The fabulous world of Nikolas Tesla

The brilliant Croatian-born scientist Nikolas Tesla (1856-1943) rivaled Edison as an inventor, with over 700 patents. Although he was several years ahead of Marconi in devising ways to transmit radio waves, Marconi secured the patents for the radio, but Tesla did build a massive tower on Long Island, New York City, from which he transmitted radio messages, weather and stock market reports, and even images, worldwide. Tesla worked briefly for Edison but they fell out because of Edison's excessively methodical approach, epitomized by his statement that invention is "99% perspiration and 1% inspiration." Tesla argued that, if Edison had used a little more inspiration, it would have saved him 90% of his perspiration. Edison and Tesla also disagreed on how to transmit electricity commercially with Edison favouring direct current (DC) and Tesla alternating current (AC). Although Edison got up to some dirty tricks to discredit Tesla's idea, for instance, by publicly electrocuting animals using AC electricity to show how 'dangerous' it is, Tesla won the argument, but Edison got the credit. When Edison died the lights throughout the USA were turned off for one minute to honour him.

Tesla lacked the business skills of Edison and was always short of funds for his research. He is remembered as a poetic visionary with a very practical touch, but he died a penniless and eccentric recluse with few friends, one of them Samuel Clemens (Mark Twain, who held three patents, one of them for an adjustable strap used to tighten skirts and hold up underpants and corsets!). It is most appropriate that one of the most prominent innovators of our time Elon Musk has named his successful electric car the 'Tesla'.

## 2. Funny science: The moderns

The Italian Guglielmo Marconi (1874-1937) invented the radio in 1902 but he had to fight against considerable odds to prove the usefulness of his new contraption. When news of his success in transmitting a telegraph message from Cornwall to Newfoundland was released, Marconi was expelled from Canada as nobody believed him and a jealous Edison called the report a "newspaper fake," perhaps one of the first accusations of fake news in America. Sir William Preece, then Chief Engineer of the British Post Office, was equally unconvinced

**Guglielmo Marconi**

and decreed, "We have done as much with the wireless telegraph as is likely to be done." Preece had previously distinguished himself as a useless predictor of the value of technology when he stated, with reference to Bell's new telephone, "Americans have need of this invention, but we do not. We have plenty of messenger boys". The Americans were more optimistic as the mayor of Chicago enthusiastically proclaimed, "There will soon be one in every American city."

**Einstein and Kovalevsky**: Albert Einstein (1879-1955), *Time* magazine's Man of the 20th Century, apparently did not speak until he was 3½ years old when his first words were, "The milk is too hot." His mother exclaimed, "But you can talk, why haven't you spoken before," to which he replied, "Because previously everything was in order." His teacher said of him, "This boy will never amount to anything! He is mentally slow, unsociable, and adrift forever in his foolish dreams," and his teacher described him as "a lazy dog who never bothered about mathematics." Although Einstein frequently complained about his poor mathematical skills, it is reliably reported that he devised his own proof of Pythagoras' Theorem at the age of 12. This achievement is eclipsed though by the precocious prowess of the

**Albert Einstein**

Hungarian mathematician Paul Erdős (1913-1996) who could multiply three-digit numbers in his head at age three, manipulate squares and cubes at four and had devised 37 proofs of Pythagoras' Theorem by the age of 14! In keeping with his extraordinary intellect Erdős lead an eccentric lifestyle (*Time* called him 'The Oddball's Oddball'). He published about 1 500 mathematical papers, a figure that remains unsurpassed, and believed that mathematics is a sociable activity and lived an itinerant life with the sole purpose of writing papers with his mathematical colleagues. Possessions meant little to Erdős as most of his belongings fitted into a suitcase and he generally donated awards and other earnings to people in need. He would typically show up at a colleague's doorstep, announce "My brain is open," and stay long enough to collaborate on a few papers before moving on a few days later. Erdős drank copious amounts of coffee which prompted his colleague Alfréd Rényi to say that "a mathematician is a machine for turning coffee into theorems." Erdős had a great interest in promoting mathematics among youngsters. A famous 1985 photograph taken at the University of Adelaide in Australia shows him talking to 10-year-old Terence Tao who would later win the Fields Medal and be inducted as a Fellow of the Royal Society of London at the age of 32 years.

Albert Einstein's late development might have contributed to his great discoveries. In a letter to a friend he wrote, "I sometimes ask myself how it came about that I was the one to develop the theory of relativity. The reason, I think, is that a normal adult never stops to think about problems of space and time. These are things that he has thought of as a child. But my intellectual development was retarded, as a result of which I began to wonder about space and time only when I had already grown up." My friend, the actor David Muller, brilliantly showed, in the science play *Imagining Einstein* that we jointly developed, how Albert's abstract thinking skills were developed while he was a patent clerk in Switzerland. His job required him to conceptualize and often re-describe many of the patent applications that he received. One of the patents that he awarded (in 1900) was for the Toblerone chocolate bar, which apparently mirrors the shape of the Matterhorn.

Einstein never learned how to drive a car, a distinction he shared with Thomas Edison, although he did enjoy riding his bicycle. Henry Ford, inventor of one of the most successful cars ever, the Model T Ford, only obtained his driver's licence in middle age after he had been pulled over by a cop for driving illegally! Other glitterati who never learned to drive include Humphry Greenwood FRS (Head of the Freshwater Fish Section at the Natural History Museum in London when this author did his post-doc there), ex-British Prime Minister Gordon Brown, Ralph Nader, Michael Jackson, David Copperfield, J.K. Rowling, Barbara Streisand, Robbie Williams, Barbara Walters, Spike Lee, Anna Nicole Smith and even the ex-Rolling Stones' drummer Charlie Watts!

Another physicist/mathematician with an interesting story to tell was Sophie Kovalevsky (1850-1891) who made major contributions to the theorem of differential equations as well as to mechanics and the theory of light propagation through crystals. She grew up in a country house in Russia where the walls of her room, when she was eight or nine years old, were covered with makeshift wallpaper made from old lecture notes on differential and integral calculus written by the distinguished Russian mathematician Mikhail Ostrogradsky, left over from her father's student days. Sophie stared at these wallpapers for many months and eventually, after discussions with her uncle, also a mathematician, they began to make sense to her. She rose to great prominence and eventually became the first woman to obtain a doctorate (in the modern sense) in mathematics, the first woman appointed to a full professorship in northern Europe, and one of the first women to work for a scientific journal as an editor.

**Rutherford, Soddy, Pauli and Bohr**: One of the most arrogant but productive physicists was the New Zealander Ernest Rutherford (1871-1937), son of a dairy farmer from Brightwater in New Zealand, who became the first person to split the atom. He did his most important work in Canada and England, in particular at the Cavendish Laboratory at Cambridge University where he became known as the 'Father of Particle Physics' and the greatest experimentalist since Michael Faraday. In 1908, somewhat to his amusement as a physicist, he was awarded the Nobel Prize in Chemistry for the discovery of argon. He is buried next to Isaac Newton in Westminster Abbey.

I once visited his quaint original laboratory Rutherford's Den in the basement of the Canterbury College clock tower in Christchurch, New Zealand. This inauspicious bunker was the crucible of Rutherford's genius until news of his accomplishments saw him move to Cambridge in 1895. The Den, with Rutherford's original apparatus, fortunately survived the devastating 2011 Christchurch earthquake. In fact, it benefitted from the disaster as a well-funded project restored the clock tower building which re-opened in 2016 with excellent interactive displays on his life and work.

I disagree with one of Rutherford's most famous quotes, "Physics is the only science, everything else is stamp collecting," as there are 'stamp collectors' and cutting-edge scientists in every field of scientific endeavour. Rutherford was also wrong when he proclaimed that any thought of commercially exploiting nuclear energy was "moonshine". I support some of his other statements, though, such as, "We didn't have money, so we had to think," and "If a principle of physics cannot be explained to a barmaid, the problem is with the principle, not the barmaid."

Alchemy is the ugly sister of chemistry as its adherents believe, *inter alia*, that a substance must exist that can transmute base metals such as lead into silver or gold. Notwithstanding this mistaken belief alchemists have made some significant discoveries, such as the first isolation of phosphorus. When Rutherford and Frederick Soddy (1877-1956) made the staggering discovery that radioactive thorium can give rise, under certain conditions, to a radioactive gas similar to argon, which they called radon, they were concerned that they would be labelled as alchemists. Soddy said at the time, "I was overwhelmed by something greater than joy, a kind of exaltation, intermingled with a certain feeling of pride that I had been chosen from all chemists of all ages to discover natural transmutation. I shouted, 'Rutherford, this is transmutation: the thorium is disintegrating and transmuting itself into an argon gas,'" to which Rutherford replied, "Don't call it transmutation - they'll have our heads off as alchemists." Rutherford subsequently received the Nobel Prize for the discovery but Soddy never overcame his resentment even though he later won a Nobel for his discovery of isotopes; he ended his life in embittered and paranoid solitude.

**Wolfgang Pauli**

The great Austrian physicist Wolfgang Pauli (1900-1958) was a blunt man who once commented on a colleague's publication, "This paper isn't right, it isn't even wrong." He once said of a young scientist, "So young and already so unknown." One of Pauli's assistants once remarked that you could ask him any question without worrying whether he might think it idiotic as he found all questions to be idiotic. When a

Soviet physicist Lev Landau asked him whether his ideas were nonsense Pauli exclaimed, "Not at all, not at all. Your ideas are so confused I cannot tell whether they are nonsense or not." Other ideas he simply dismissed as *ganz falsch* ("utterly wrong").

Once, when a new assistant arrived to take up his duties, Pauli asked, "Who are you?" "I am Weisskopf, your new assistant." "Oh", he said, "I wanted Bethe, but you will have to do." This reminds me of an incident when I was a first-year zoology student at Rhodes University in Grahamstown (Makhanda). On the occasion of my first meeting with the legendary ichthyologist Professor JLB Smith I was ushered into his laboratory where he was examining a fish under the microscope. He asked who I was and we chatted about ichthyology for 20 minutes, then he told me to leave. Not once during the conversation did he look up from his microscope.

Pauli, who is famous for the Pauli Principle (that no two electrons in an atom can occupy the same quantum state) was equally well known for the Pauli Effect (that his approach spelt destruction for any scientific equipment or mechanical device in his wake). He was aware of his reputation and was delighted whenever the Pauli Effect manifested. These strange occurrences were in line with his controversial investigations into the legitimacy of parapsychology. Furthermore, he was such a bad driver that most of his colleagues refused to travel with him. Pauli was in the habit of carrying a stub of chalk with him wherever he went so that he could jot down any ideas that came to mind. Once, while walking in Paris, an inspirational thought came to him which he scribbled on the back of a nearby hansom cab, which was soon covered in complex equations. Just as his calculations reached a climax he was horrified to see the cab speed off into the distance carrying with it the solution to his problem.

In contrast the habit of the French scientist and inventor Dominique François Arago (1786-1853) to scribble diagrams onto whatever surface was available proved useful to him. Once out on a walk in Paris Arago described to his friend, the French physicist and astronomer Jean-Baptiste Biot (1774-1862), a new device that he had invented, the photometer. Biot was skeptical so Arago pulled out his keys and scratched a drawing of it onto a nearby building column. The following day Biot double-crossed his friend and described, to the French Academy of Science, the photometer as his invention. Incensed, Arago rose to make his case but no-one believed him until he took them to the column where he had scratched his diagram.

Pauli and the Danish physicist and philosopher Niels Bohr (1885-1962) often came into conflict. Once Pauli gave a lecture on how he and Werner Heisenberg had solved all the unsolved problems in elementary particle theory. Asked to comment, Bohr said that, as a radical theory, "It was crazy, but not crazy enough." He argued that any theory that completely overturns current thinking, such as relativity and quantum theory, must completely violate common sense at first sight. Pauli then argued that his theory *was* crazy enough. Bohr had many eccentricities. When he spoke he waved his hands around wildly, placing strong emphasis on important words, but his English was so heavily accented that once, at a conference in the USA, he delivered his talk in English but there was simultaneous translation - into English. Many people referred to his language simply as 'Bohr'.

During World War II Bohr had to beat a hasty retreat when the Germans arrived in Copenhagen. At that time it was illegal to export gold yet he had his Nobel gold medallion received in 1922 with him. He decided to dissolve it in acid and leave it behind in a jar. When he returned after the war the jar was still there. Bohr recovered the gold and the Nobel Foundation agreed to recast his medal!

**Absentmindedness, humility and productivity**: Many top scientists are renowned for their absentmindedness. Sir Nevill Mott (1905-1996), the Cavendish Professor of Physics at Cambridge, who was no clown as he won a Nobel Prize for research he had done after retirement, was notoriously so. He once found himself travelling on a train from London to Bristol when three profound truths came to mind:

1. He no longer lived in Bristol, but in Cambridge,
2. He had traveled to London earlier that day by car,
3. He had been accompanied by his wife.

One of my favourite inventors is the American John Bardeen (1908-1991), a quietly spoken, humble man. One evening he returned home and said to his wife "We discovered something today" but never mentioned it again. Six years later while she was scrambling eggs for breakfast the news came over the radio that he had been co-awarded the Nobel Prize for the discovery of the transistor. Bardeen played golf once a week with the same partners for 20 years and, during their rounds, they chatted about their lives. At his funeral one of his golfing friends was heard to remark that, over those 20 years, Bardeen had not once mentioned that he had co-invented the transistor and won, not one, but two Nobel Prizes.

## 2. Funny science: The moderns

**John Bardeen (left)**

Many years earlier the German scientist Heinrich Hertz (1857-1894) displayed rare humility shortly before his premature death at the age of 36 due to blood poisoning. Hertz, after whom the unit of frequency is named, was the first scientist to prove the existence of electromagnetic waves that had been predicted by James Clerk Maxwell. He wrote to his parents, "If anything should really befall me, you are not to mourn; rather you must be proud a little and consider that I am among the specially elect destined to live for a short while and yet to live long enough". This is reminiscent of the words of the great Italian physicist Enrico Fermi, creator of the world's first nuclear reactor and widely known as the Architect of the Nuclear Age. He died at the age of 53 but said that he did not mind too much because most of what he could do he had already achieved.

In all areas of scholarly endeavour the advancement of knowledge is dependent on the publication of original peer-reviewed articles. Yuri Struchkov (1926-1995), a director in the Academy of Sciences in Moscow and one of the world's best crystallographers, added new meaning to the statement 'publish or perish' as he continued to publish long after his death. He authored or co-authored over 2 000 papers, an average of about one every four days during his working life, with publications bearing his name continuing to appear for 15 years after his death!

The 'publish or perish' frenzy is also evident in the modern world. In October 2020 the *Times Higher Education* supplement reported that Mark Griffiths had published 161 papers in 2020! Griffiths holds a distinguished professorship at Nottingham Trent University where he is Director of the International Gaming Research Unit and conducts research on addictions to gambling, video games, the Internet, exercise and sex. In total he has published over 1 200 papers with 898 different co-authors and garnered over 80 000 citations but acknowledges that he benefits from working with a large stable of collaborators, mainly his ex-PhD students.

**South African innovators**: The laser is an interesting invention that was initially just a laboratory curiosity, 'a solution in search of a problem,' as no-one knew what to use it for. Even its name was an embarrassment as its original acronym was 'loser', for 'light oscillation by the stimulated emission of radiation,' but a bright marketing man changed it to 'laser' by substituting 'amplification' for 'oscillation'. One of the most significant recent advances in laser technology was the digital laser invented in 2013 by South African Sandile Ngcobo (born 1980) under the supervision of Professor Andrew Forbes at the CSIR in Pretoria. He created the first laser whose beam could be controlled and shaped digitally using holograms but Sandile's role in the invention has subsequently been disputed and he has, wrongly in my opinion, apparently been side-lined during the subsequent development of this important innovation.

Other South African inventors who have travelled rocky roads during their careers include Lee Dickman, inventor of the first automatic telephone answering machine, the Colindictor, patented in 1958. While marketing the device in England he decided to try his luck and requested an appointment with the Duke of Edinburgh, a renowned tinkerer, in Buckingham Palace. To his astonishment he received a call from the duke's secretary, an RAF group captain, who told him that he had been called to a meeting the following afternoon. To confuse matters he also received a call from the British Post Office (BPO) extending an invitation for him to meet the postmaster-general immediately. At this meeting he was told that the BPO knew about his appointment at Buckingham Palace but said that they had also been working on a telephone recording device of their own. They felt that it would be fair for them to have the first opportunity to demonstrate their device to the duke. Accordingly, they had arranged to demonstrate their equipment one hour before the South African presentation.

Dickman arrived at Buckingham Palace just in time to see the departing BPO executives and their cutting-edge equipment, transported in three trucks. In contrast he arrived with one second-hand Colindictor in a battered briefcase. His demonstration nevertheless went perfectly and the duke was extremely impressed with the device. Dickman subsequently received a letter from the palace thanking him for the gift of the Colindictor and, on his return to South Africa, a telegram asking him to air freight a second Colindictor to them "as the queen has pinched mine." Sadly this was the only recorder that he sold in England as the BPO stole his idea and the emerging technologies of transistors, printed circuitry and microchips sent the development of the answering machine in a different direction. As a result the Colindictor never reached world markets.

**Ig Nobel Prize:** The Ig Nobel Prize is a satirical prize awarded annually since 1991 to celebrate ten unusual or trivial achievements in scientific research, with its aim being to 'honour achievements that first make people laugh and then make them think.' Notable winners have included Dr Chittaranjan Andrade, a distinguished psycho-pharmacologist from the Institute of National Health and Neurosciences in Bangalore, India, who found that 80% of teenagers in India are nose pickers. In his acceptance speech, which was attended by several Nobel Prize winners, he stated, "Some people poke their noses into other people's business. I decided to poke my business into other people's noses." He is planning further research on trichotillomania (compulsive hair pulling) and onychophagia (nail biting).

**Ig Nobel Prize emblem**

Another winner Donatella Marazziti, professor of biological psychiatry at the University of Pisa in Italy (where else?), has investigated the biochemistry of romantic love. By monitoring serotonin levels in the blood she found that romantic love is biochemically indistinguishable from obsessive-compulsive disorder. She also confirmed biochemically what poets and writers have known for centuries, that the first bloom of love fades quickly and the blood soon acquires the boring characteristics of long-married couples.

Dr Peter Fong of Gettysburg College in the USA contributed to our understanding of happiness in clams by feeding them Prozac, an antidepressant drug used by humans, and then monitoring their reaction. He presumed that their reaction to Prozac at the cellular level would be the same as in humans but was surprised to find that the clams reproduced at ten times their normal rate after treatment. In his acceptance speech he gave an insight into how great science takes place, "It was late one night, and I was sitting alone in my laboratory feeling pretty depressed. Rising from my chair, I clumsily knocked over my prescription of Prozac and watched helplessly as several capsules fell into an aquarium full of clams. To my amazement, within minutes, the clams began producing copious amounts of sperm and eggs into the water. Suddenly I was no longer depressed."

Dr Peter Barss, a young Canadian doctor who worked in Papua New Guinea, initiated a study on the main causes of injury in that tropical country, including death by coconut. He started by defining the Newtonian characteristics of a falling coconut, "When the coconut's mass is accelerated by gravity, falling from a 10-storey height, and then comes to rest by being suddenly decelerated onto someone's head, head injuries result. If a coconut weighing 2 kg falls 25 m, the impact velocity is 80 km/hr."

Robert Matthews of Aston University, England, a Fellow of the Royal Astronomical Society and of the Royal Statistical Society, studied Murphy's Law in relation to falling toast and found that toast does fall more often on the buttered side. He first demonstrated, using complex statistical analyses, that his finding was not due to the toast becoming asymmetrical as the weight of the butter (4 g) is small compared with the weight of the toast (35 g), and its contribution to the total inertia of the toast and thus its effect on the toast's rotational dynamics, was negligible. He furthermore found that the maximum height of humans, which determines the distance of the dropped toast, is a function of three fundamental constants of the universe, and therefore concluded, *inter alia*, that toast lands butter-side down at a rate of 62% "because the universe is made that way." Many years earlier Louis Pasteur had studied the same profound problem and reached a conclusion that was consistent with the laws of physics: the toast fell equally on both sides as his children buttered their toast on both sides!

Another winner of the Ig Nobel Prize is Hyuk-ho Kwon of Kolon Company in Seoul, South Korea who invented a self-perfuming suit that allowed businessmen to return home smelling sweet even after a hard night's drinking and smoking. The fabric of his suits is soaked in micro-encapsulated scent which is available in pine, lavender or peppermint - every time the fabric is rubbed more scent is released.

The ultimate example of cheeky innovation was perpetrated, of course, by an Australian. John Keogh, a freelance patent attorney in Hawthorn, Victoria, discovered that the wheel had never been patented. To test the

## 2. Funny science: The moderns

Australian Patent Office he registered an 'innovation patent' for a 'circular transportation facilitation device', and to his amazement it was granted. This happened notwithstanding the fact that the Australian Commissioner of Patents makes it clear that, 'To obtain a patent the applicant must make a declaration that they are the first inventor' and the website of the Australian Patent Office advises applicants to carefully examine the record of patents already on file so as '... not to re-invent the wheel.'

The Australian Commissioner of Patents at the time Vivienne Thom stated, "To obtain the patent for a wheel would require a false claim, which is a very serious matter and would certainly invalidate the patent as well as amount to a misrepresentation on the part of the applicant and unprofessional conduct by any professional advisor." Indeed. The episode nevertheless caused the Australian Patent Office considerable embarrassment as it was subsequently awarded an Ig Nobel Prize!

So, when you next look around at your relatives, friends and colleagues, forgive those who are dreamers, eccentric, absentminded or slightly weird, who are competitive risk-takers, obsessive, maybe even opinionated, paranoid, pompous, childish and cheeky, for they are probably the creative ones. The ones you need to be worried about are those who are behaving normally.

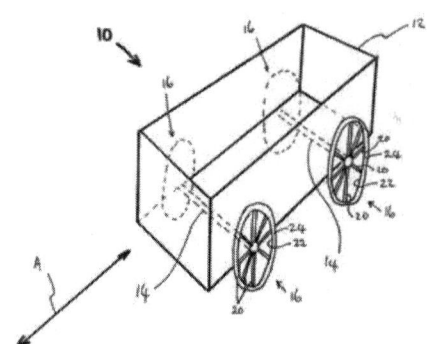

**John Keogh's drawing for his patent of the wheel**

# 3.

# Creativity in the arts and sciences

*'Among all highly civilized peoples the golden age of art has always been closely coincident with the golden age of the pure sciences, particularly with mathematics.'*

Emil Lampe, German mathematician

INTRODUCTION: I have always been interested in the ways in which creativity is expressed in the arts and sciences. As director of a research institute in Makhanda (Grahamstown) I aspired to create the optimal environment in which our scientists could reach their full potential. I realised that being a faithful formalist who cautiously toes the line and does not make waves would not lead to novel ideas and discoveries, so I encouraged my researchers to be risk takers who occasionally made mistakes but always explored and tested big ideas. I noticed that art and science have not only the procedure of trial and error in common but also the processes of experimenting and creating a perspective that is new and innovative. I discovered that art and science are much more than bodies of knowledge, they are ways of thinking and finding things out. I decided to compare creativity in the arts and sciences and learn more about the synergies between them.

The most stimulating expression of creativity in one of the arts (music) that I have heard was a talk by Maestro Bernhard Gueller, principal guest conductor of the Cape Town Symphony Orchestra, to the Owl Club,

> "It is not my intention to start a kind of a competition between science and music – it's not possible. I would never dare to ask what is greater - Don Giovanni or the Uncertainty Principle by Heisenberg. I want to show the difference between the two – the difference for me is that the breath-taking achievements of science are of this world – earthly, you can see them, you can measure them, you can weigh them. …
>
> Great art, especially great music … achieves the creation of different new worlds. And that is, I think, a very unique ability of our species.… Art

– and here I mean the entire creative spectrum, literature, the visual arts, all performing arts and music – makes us human beings, that very unique species. … But no other art form comes from so deep inside the human psyche, and can go that deep into a human psyche, than music. No other art form can create such ecstasy, happiness, sadness, grief, desperation, and joy. Music can be a medium which articulates emotions you aren't able to express with words, and an emotion not expressed remains, to me, unredeemed. …"

When Wagner's *Tristan and Isolde* appeared on stage for the first time, there were people in the audience who couldn't bear the emotional stress that this music expressed in a way they had never heard before, and many of them simply freaked out. There are even reports of suicides.

No other composer before him, no matter how big the name, had even come close to the ecstasy that Beethoven achieved in the second movement of the *Eroica*. Haydn was in the audience and said, "From now on nothing will be the same anymore." I would give a year or two of my life to go back in a time machine to the premiere of the *Eroica*. Not to see how the audience reacted, but to see how I would have reacted. It must have been an overwhelming experience when you are told in an understandable language something completely new, unknown, with emotions you never thought possible, and all of that at the highest artistic level.

Beethoven's Ninth Symphony, especially the last movement, makes one jump up in the air with joy. You are convinced that there is nothing more perfect, more glorious than mankind. A bit unrealistic but, at that very moment in the concert, that is how we feel …. This conflict, between showing us a fraction of the idea of a promised land and knowing that it is unreachable, stirs up our emotions. The fact that we have felt, on rare occasions, that we were closer to this ideal, thrills us, but it also makes us sadder, happier or more depressed. It sharpens our awareness."

Maestro Bernhard Gueller, talk to the Owl Club on 17th September 2019

Artists, including painters, sculptors, musicians, writers and dancers, use their imagination to create new works that entertain and inspire us whereas, in the sciences, we do not need to create the objects of our attention as they are already there. Scientific facts are therefore revealed, not discovered, and scientists use their imagination to devise novel ways to make discoveries and place their findings in a broader context. Like music, science exists in the domain of the mind and is driven by restless and curious individuals who are ill at ease with the status quo. As the American sociobiologist E.O.

Wilson once said, "In the early stages of creation of both art and science, everything in the mind is a story." I was once accused by a scientific colleague of having "a rather unscientific willingness to dream," but I took it as a compliment. Although modern scientists have moved far beyond the pencil-and-paper, typewriter and homemade equipment era in which I first worked, and have far more complex technology at their fingertips, they still need that creative spark to visualize their storyline, appreciate the significance of their findings and place them in the proper context. To be a good scientist you also need to be an artist.

I would argue that fine art, like music, also transcends cultures, languages and disciplines and can convey messages and articulate emotions that are not easily expressed in words. Fine art and music remind us that there is something more to life than just surviving. I believe that the arts, without diminishing their intrinsic value, represent a powerful means by which we can convey scientific messages to millions of people who choose not to listen to scientists or science educators.

**Art and science**: Art and science are two of humankind's greatest achievements, yet they are peculiarly remote from one another in our everyday lives. Some art aficionados even brag about their lack of interest in science as if that were a positive attribute, yet science is more important for our survival, and that of the other nine million plus species on Earth, than ever before. Fortunately, in this modern, highly connected world there is an increasing interface between art and science. At the 2018 South African National Arts Festival in Makhanda over half the arts and culture works on display or performed addressed a scientific topic.

While science can help us to soften our environmental impact and force us to change our ways, the problem is that it is mainly scientifically orientated people who listen to the voice of science. This is where art comes in. Art can convey complex and emotional messages to people whether they are interested in science or not. It can help people to change their mindsets, their behaviour and their lifestyles, and to tread more softly on the planet. Most importantly it can encourage them to become ambassadors for sustainable living who help us to spread the message that we need to make wholesale changes to our lifestyles *now*.

The environmental group Extinction Rebellion has predicted that humans face mass starvation within the next 50 years if we continue to disrupt Nature's balance at the current rate. It is already obvious that we have upset the planet's climatic system, which has resulted in the 10 hottest years in history occurring this century, and the amplitude of extreme events, such as floods, droughts, storms and hurricanes, has increased dramatically.

Over 40 years ago, when I visited the Smithsonian National Air & Space Museum in Washington DC, I saw space rockets for the first time and touched a bit of real moon rock. I came away filled with emotion and ambition but what I remember most from that visit was an art exhibition on the meaning of space exploration and the uniqueness of planet Earth. One painting portrayed an ancient alligator in the Everglades watching a rocket take off from Cape Canaveral; it made me wonder what its tiny reptilian brain was thinking about. We need new ways of communicating the importance of the scientific message to the public and, to me, art can be this messenger and agent of change.

This idea is not new. For centuries artists, especially in Africa, have been using art as a creative tool for change, for resistance, even rebellion. Recently the Covid-19 pandemic has encouraged artists to step out of their comfort zones and address this particularly challenging issue. In South Africa, at the start of the pandemic in March 2020, everyone was challenged to submit videos of themselves performing the national anthem. The result was a remarkable 1 027-piece virtual choir that included people from every walk of life performing together in a show of true unity. The once-in-a-lifetime choral tapestry was led by well-known South Africans such as Busiswa, ProVerb, Leah, Yvonne Chaka Chaka, Andre Schwartz, Donald, Katlego Maboe, Zolani Mahola, Tumi Morake, Khaya Dladla, Danny K, Relebogile Mabotja, Caroline Grace, Moonchild Sanelly, KB and others. In the video, the virtual choir appears together on screen and then slowly morphs into a striking image of the national flag in full furl.

To further my research on creativity in the arts, I decided to explore some of the ways in which it has been used in Africa to raise awareness of the most pressing medical, scientific, heritage conservation, human rights and political challenges that we face on the continent.

**Art preserving and promoting our African heritage:** Although the performance of the multinational supergroup United Support of Artists in Africa (USA in Africa) in 1985, and the recording of their hit single, *We are the World*, to support famine relief in Ethiopia, will probably never be eclipsed many African writers and artists have performed or produced work that uplifts African communities and makes them proud of their heritage.

The late Algerian singing star Idir (Hamid Cheriet; 1949-2020), who started life as a geology student with little musical training, was propelled to worldwide acclaim as one of the originators of a new musical genre New Kabyle Song (*la nouvelle chanson Kabyle*). The Kabyles are one of North Africa's indigenous Berber people who are distinguished by their unique language and vibrant cultural traditions. Idir was inspired by the songs that he grew up hearing his grandmother sing while she was weaving, churning butter or rocking a child to sleep. He cloaked his songs in Western harmonies, arranged them for guitar and other instruments, collaborated with an Algerian poet Ben Mohamed on the lyrics and took them to audiences across North Africa and Europe. His hit song *A Vava Inouva* (Oh My Father), inspired by a traditional story about a young girl trying to save her father from danger, launched his international career in 1973, selling over 200 000 copies by 1978 (Goodman, 2020).

Idir and Ben Mohamed were inspired by the First Pan-African Cultural Festival held in Algiers in 1969 where they witnessed dynamic cultural performances from troupes across Africa and heard talks by leading postcolonial luminaries such as Joseph Ki-Zerbo and René Depestre. By setting traditional Kabyle songs in contemporary musical language he gave them new life and propelled songs that had only been sung by women in remote villages into a far wider circulation. In Algeria Idir's songs gave Kabyles a stronger sense of identity and made them realize that Amazigh customs and traditions are not backward but are part of the modern Algerian culture (Goodman, 2020).

Algerian singing star, Idir

Africa has many traditional dances, from the signature Masaai *Adamu* (the jumping dance performed by young men at coming-of-age ceremonies and to impress potential brides) to the *riel* dance. Known as *Ikhapara* by the Nama, the *riel* is one of the continent's oldest indigenous dance styles that was originally performed by Khoisan hunter/gatherers around a fire after a successful hunt or forage. Today the *riel* is a celebration of ancient traditions that finds new expression in contemporary forms with distinct Irish and Scottish folk music influences as well as Dutch folk songs and American minstrel lyrics, all performed to the beat of *boeremusiek* (farmer's music), a wonderful blend of different cultures. *Riel* includes movements, such as the *bokspring* and *kapperjol*, that imitate the behaviour of butterflies, snakes, birds, antelopes and horses and employs rapid foot movements that kick up dust, hence its nickname, *stofskoppers* ('dust kickers') (Javan, 2017).

***Riel dance*** **by Terence Visagie**

The most famous *riel* performers are probably the Nuwe Graskoue Trappers from Wupperthal in the Cederberg, South Africa, who won awards in 2015 at the World Championship of Performing Arts in Los Angeles. The Brandvlei Stofskoppers, a talented school troupe from the Northern Cape trained by Thomas Arendse, is unusual in that it has performed at prestigious events, such as the National Arts Festival in South Africa, as the cultural representative of the Square Kilometre Array radio telescope project that is headquartered in Carnarvon in the Karoo. Another accomplished troupe of *stofskoppers* is the Kingdom Connected Campers (KCC) from Richmond in the central Karoo.

> "We formed the KCC in 2017 to address crime among our youth in our little town. As in so many farming communities, poverty leading to crime is a major issue affecting poorer families and youth who are trying to survive. Dance, and in our case, riel, has given youth a reason to believe that there's more to life than the incredible hardships they experience, and it provides a temporary reprieve from their harsh realities, giving them a sense of pride and purpose."
>
> Thys Bouwers, founder of the Kingdom Connected Campers ( Bouwers, 2020)

Guinea's most famous musician, Mory Kanté, was deeply immersed throughout his life in the culture of *griot* (hereditary musicians) and *jeli* (traditional storytellers) of West Africa. When he died in May 2020 he was celebrated as a legendary singer and master of the *kora*, the West African harp. The commercial success of his 1987 dance hit *Yéké Yéké*, the first recording from Africa to sell over one million copies, encouraged record companies to back African artists. With its hard-edged, driving rhythms and relentless disco back beat *Yéké Yéké* captured the spirit of global 1980s pop and raised the profile of African popular music worldwide (Durán, 2020).

> "You know, I went to three schools: the white man's school, the Koranic school and the griot school … but [the griot school] was the most intensive training I had. The griots are anti-tribalist … they're the opposite of racist. The griot is there for the people."
>
> Mory Kanté, Guinea's most famous musician (Durán, 2020)

Kanté's father was a blacksmith-turned-musician who mainly played the *balafon*, a wooden xylophone with 18 to 22 keys. From childhood Kanté performed in family musical ensembles at a time of intense artistic creativity in Guinea whose people were developing a growing spirit of independence. In the 1960s Kanté's ensemble performed a popular style of music called Apollo, after the space mission. "It was futuristic music. No one believed the spaceships would ever reach the moon. Only the griots knew … because the griots understand man's potential." Unlike his contemporaries, who played traditional music on Western instruments, Kanté played Cuban and American music on traditional instruments, such as the *kora* and *balafon*. As an artist he was restless and ambitious, always looking for new ways of playing old music, dressed in his trademark all-white outfit.

Nolufefe 'Fefe' Mtshabe was a life force who changed the careers of those around her for the better and brought African culture to the attention of the world. Born in the village of Tsomo in the Eastern Cape, South Africa, she became one of the most influential singers and composers of her generation and strongly influenced the careers of such superstars as Pretty Yende, Pauline Malefane, Pumeza Matshikiza and Monde Masimini. Fefe, who died at the age of 55 in February 2020, was the creative energy that gave urgency and wit to an African *Carmen* and *Magic Flute*, having translated Mozart and Bizet's works from German and French into isiXhosa. Both performed works express a wisdom and flare that allow them to surpass the confines of Europeanness or Africanness and endow them with a rare universality (Sassen, 2020).

> "Singing is my life. It has allowed me to transcend all barriers. Through my teaching, I have, in turn, been able to change people's lives for the better."
>
> Nolufefe Mtshabe

Fefe was also a gifted soprano who performed in Gabriel Faure's *Requiem* in St George's Cathedral in Cape Town in 1997 and in the African opera *Princess Magogo* composed by Mzilikazi Khumalo. In 2010 she took her Masiyile Choir to the USA for a year at the invitation of US conductor Kamal Khan, where it honed its skills and collaborated with Ray Charles, Stevie Wonder and Simon Estes. She also formed the award-winning Heavenly Voices choir with Estes.

The father of Ethiopian jazz Mulatu Astatke started his musical education in London where he heard performances by Caribbean and West African musicians that awoke his memories of the big bands that he had enjoyed at home in Ethiopia. His education continued in Boston and New York City where he participated in American Jazz, interacted with Latin musicians, performed in concerts and made records. By the time he returned to Ethiopia in the late 1960s he had developed the concept of ethio-jazz (Ethiopian jazz) and was actively experimenting with this new, hybrid musical style. Ethio-Jazz draws on multiple trends in the American jazz scene, including bebop and modal jazz, and combines them with melodies and harmonies based on the Ethiopian modal system. Astatke's innovative style was also informed by his childhood memories of traditional Ethiopian secular and church music and by the highly original mixture of sounds that he had experienced on his life journey (Shelemay, 2016).

His signature piece *Yekermo Sew* (A Man of Experience and Wisdom) takes its title from a traditional Ethiopian Christian New Year's blessing in Amharic and brilliantly captures the essence of his musical style. The piece was used as the soundtrack for the 2005 film *Broken Flowers* starring Bill Murray and became

**Mulatu Astatke**

known worldwide. Although he travels regularly Astatke has maintained a home in Ethiopia and established an African Jazz Village from which regular radio broadcasts are made of ethio-Jazz and traditional Ethiopian folk music. He has become a legend among his own people as well as with the Ethiopian diaspora (Shelemay, 2016).

Angélique Kidjo, a Beninese singer-songwriter, actress and activist of Nigerian descent, is famous for her diverse musical influences and creative music videos. She started singing in her school band *Les Sphinx* and found success as a teenager with her adaptation of Miriam Makeba's *Les Trois Z*. Her astonishing musical range includes Afro-pop, Caribbean zouk, Congolese rumba, jazz and Latin and she sings in five languages (Fon, French, Yorùbá, Gen, Swahili and English). Since 2002 she has been a Unicef Goodwill Ambassador and, in 2007, *Time* called her 'Africa's premier diva'.

Political conflicts in Benin prevented Kidjo from being an independent artist in her home country so she relocated to Paris in 1983 and, by the end of the decade, she was one of the most popular live performers in the French capital. Since then she has sung at many prestigious events around the world and in famous venues such as Carnegie Hall, Royal Albert Hall, Sydney Opera House and the United Nations General Assembly. In November 2018 she performed with Yo-Yo Ma and Renaud Capuçon to 70 Heads of State and a television audience of millions under the Arc De Triomphe in Paris, and in September 2016 she shared the stage at the opening ceremony of the National Museum of African American Culture and History in Washington DC with Angela Basset, Oprah Winfrey and Barack and Michelle Obama.

When the founder of Ladysmith Black Mambazo, the world famous South African *isiCathamiya* group, Joseph Mxoveni Mshengu Bigboy Shabalala died aged 78 years on 11th February 2020 tributes poured in from around the world. Shabalala's decades-long quest from the 1960s to improve on *isiCathamiya*, an acapella sound that had metamorphosed from Zulu wedding songs, eventually took his group onto the world stage. The group's first album *Amabutho*, released in 1973 by Gallo, sold over 25 000 copies and they eventually won three Grammy Awards, produced over 50 recordings and sold millions of records. Their big breakthrough came in 1981 when they were invited to perform at a music festival in Cologne, Germany, and footage of their performance was seen by Paul Simon. This planted the seed that lead to their collaboration with Simon on the famous

*Graceland* album, released in 1986 (Sosibo, 2020).

Musicologist Sazi Dlamini believes that Shabalala's main legacy is the way in which he expanded his music's visibility through his openness to collaboration. "They were persistent and prolific in the 1970s when Radio Zulu was prominent, and *isiCathamiya* grew as a genre at the time. This is partly because it was non-threatening and benign music. Censorship was at its prime at that time and *isiCathamiya* was docile and associated with religion through the missionary infrastructure. The migrant labour system also gave it impetus as a method of reinforcing homeboy networks", stated Dlamini. Today Ladysmith Black Mambazo, which wears the colourful presidential shirts made famous by Nelson Mandela, continues to perform and tour, headed by some of Shabalala's sons. Their music has changed but they have stayed true to the maxim of their founder, "We are not singing this kind of music to make ourselves famous - we are singing to remind people of who they are" (Sosibo, 2020a).

The legendary Nigerian trumpeter and composer Victor Olaiyo, 'Africa's Satchmo', was born in 1930 in Calabar, the 20th child in a family of 24. As a child he took to music secretly for fear of being castigated by his rich family as music was not a respected profession at the time and musicians were regarded as beggars. In the same way that the late Afro-beat legend Fela Anikulapo-Kuti rejected his parents' advice to study medicine, Olaiyo shunned engineering for music. He avoided the bohemian lifestyle and worked hard to become a musician of repute. His big break came when, at the age of 29, he performed at Nigeria's Independence Day celebrations on 1st October 1960 where he led other notable musicians including Zeal Onyia, Bobby Benson, Charles Nwegbue and Eddy Okonta at the Independence Day Ball (Balogun; Omajola 2020).

Advert for Victor Olaiya performance

Olaiyo sang in the major Nigerian languages of Yoruba, Igbo, Efik and Ibiobio and used his music to promote the rights of ordinary people and the downtrodden. He was a maestro in highlife, a genre of West African music that originated in Ghana in the late 19th century and spread to other West African countries in the 1950s and 60s. According to Sola Balogun, an expert on

West African music at the Federal University in Oye Ekiti, Nigeria, his work combined the melody and rhythm of Akan music, but he improved its texture and aesthetic quality by introducing the trumpet. Olaiyo also contributed to the development of Afro-beat, a hybrid of highlife that had been pioneered by Kuti.

When Louis 'Satchmo' Armstrong visited Nigeria in 1960 Olaiyo played alongside the master in some memorable performances and also formed a remarkable duo with the late Ghanaian highlife singer E T Mensah. Olaiyo died in February 2020 at the age of 90 but his legacy lives on through his evergreen songs such as *Ilu le, Omo Pupa, Africa, Kendi Mama, Eko Ile, Se fun me, Oruku tindi tindi* and *Tina mata*.

Mo Laudi exemplifies the borderless modern African musician and DJ whose goal is to take African music to the world. His multi-faceted lifestyle has included starting night clubs, punk bands and record labels and creating iconic sound installations in South Africa, England and France. In an installation for the Investec Cape Town Art Fair in February 2020 he told the story of how sound moves from one country to another and the ways in which, for example, Manu Dibango inspired disco which influenced techno and house music and then spread around the world. His sound installations present what he calls "the journey of African music" through percussive, ever-changing sets that are punctuated by histories of oppression, spirituality and attempts at imagining borderless worlds. His sound layering using vocal clips from political speeches serves to trigger different points of view and ask probing questions about reality (Sosibo, 2020b).

Starting off as a rapper in Polokwane in Limpopo province Mo and his group hosted club nights before leaving for Johannesburg where he studied advertising. By 2000 he was in London trying to subvert the idea of 'exile' by creating club nights that brought music to African expats. "People didn't know much about the South African sound, which at that time was kwaito and house before it became known as Afri-house". His project took off and, at a site near King's Cross Station in central London, the crowds grew to include Nigerians, Zimbabweans, Ghanaians and Londoners. In 2010 he relocated to Paris where he collaborated with Julien Creuzet on an allegorical sculpture of a fantasised Caribbean landscape that highlighted the problem of insecticide pollution of rivers, soil and groundwater in Martinique and Guadeloupe. He also contributed to Ernest Mancoba's modern art retrospective at the Centre Pompidou in an installation that included the chanting of mineworkers at Marikana together with excerpts from Linda's *Mbube*, Mancoba's speeches and clips from Winnie Madikizela-Mandela's funeral (Sosibo, 2020b).

Increasingly musicians are using music to peddle their political agendas in a benign way. Egyptian musician Dina el Wedidi's journey began when she sang Egyptian folk songs as a member of a Cairo theatre group. When she formed her own band in 2011, she melded heritage sounds with influences from jazz, underground rock and electronic music. A modern operetta in which she participated in 2011 *Khalina Nehlam* (Let us Dream) captured the mood of the 2011 Arab Spring protests by encouraging young people to aim high and never lose hope. In 2014 her *Turning Back* album was a trip through history, mining Egyptian folk music and Arab poetry, and her collaborative Nile Project crossed borders and connected artists along the great river's course in the fight to preserve its life-saving water. In *Slumber*, released in 2018, she uses the sounds and sensations of Egyptian trains to create harmonies, melodies and rhythms that invite listeners to enjoy a nostalgic trip through the ancient kingdom (Hincks, 2019a).

**Dina el-Wedidi**

The Zimbabwean singer, composer and architect of popular African liberation music Dorothy Masuka died in 2019 but her spirit lives on. Because of her radical, pan-Africanist spirit she was exiled from Zimbabwe and South Africa and spent 16 years in Zambia where she wrote her most famous song *Ghana*, an ode to the first ladies of Africa. This upbeat, infectious and optimistic melody was described by John Samson as "a joyous celebration of the winds of change blowing through Africa" that crossed borders with respect to its sentiments and its music, including a jaunty guitar rif played in the West African highlife style. Masuka composed and recorded 30 singles, several of them major hits, and produced over 100 compositions in several African languages during her lifetime (Ansell, 2019).

Ndivhudzannyi Ralivhona, known as *Makhadzi*, took Limpopo house music to the world with her smash hit *Matorokisis* which was hugely popular in South Africa, Botswana, Mozambique, Nigeria, Dubai, Singapore and China. The song uses a train and its carriages to talk about how a man cannot love a woman without loving her children and YouTube is filled with images of strings of people dancing in trains to the song. "Everyone holds on to each other when they hear my song," Makhadzi says. "You can even hold your enemy, whether you are ANC, EFF, Kaizer Chiefs or Orlando Pirates!" she laughs. Limpopo house music is a melting pot of the bubblegum pop of the 1980s and the cultural 'riddims' and melodies of the Pedi, Venda, Tsonga and Shangaan. It is dominated by elastic synchs and heavy snares and is squarely aimed at the dance floor. "I sing in Venda," says Makhadzi, "But these people all over the world love my song even though they don't understand my language, which means they love the beat and the melody" (Gedye, 2020).

**Art raising awareness of HIV/Aids:** Oliver Mtukudzi, known as *'Tuku'* or Zimbabwe's 'man with the talking guitar', was a cultural icon, activist, philanthropist and a goodwill ambassador for Unicef. He was deeply loved for his unique *tuku* music, a blend of traditional and modern African musical instruments and traditions including *mbira*, *mbaqanga* and *jit* and the drumming styles of the Korekore people of Zimbabwe. He had his first successes shortly after Zimbabwe's independence in 1980 and his debut solo album *Africa* was an immediate hit. The prolific Kutu released his 67th album *Hanja 'Ga* (Concern) in 2018 (Chinyamurindi, 2019).

Many of Mtukudzi's songs confront the challenges facing African people. *Todi* (What shall we do?) reflects on the devastation caused by the scourge of HIV/Aids and *Mabasa* (Works) paints a dire picture of how young people are the first to die, leaving the elderly to fend for themselves. In *Ndagarwa nhkaka* (Inheriting) he highlights the importance of Shona cultural traditions and in *Dzoka uyamwe* (Come and suckle) he bemoans the experiences of a person who suffers from prejudice based on how they look. His *Wasakira* (You are old) was seen by some as a reference to then-President Robert Mugabe's advancing age and in *Magumo* (The end) he highlights poignant life lessons on the importance of ideals such as humility. Although Kutu died from diabetes on 23rd January 2019 (exactly a year after his close friend, musician Hugh Masekela) his message lives on through his songs and music.

**Art raising awareness of the coronavirus pandemic:** Ugandan pop star Bobi Wine has spent his career singing about social injustice but in 2017 he decided to take things further by running for, and winning, a seat in the Uganda parliament. The 38-year-old singer, whose real name is Robert Kyagulanyi Ssentamu, also ran for president in 2021, taking on incumbent President Yoweri Museveni who had ruled for 34 years, but he lost a disputed race. Although Wine has been jailed, beaten and charged with treason he ran on a platform of reform, making optimal use of the pop appeal of his flashy red 'revolutionary' beret, like Julius Malema in South Africa. In 2020 Wine led a star-studded galaxy of African musicians and politicians who composed and performed songs about Covid-19, including George Weah the soccer star who is now president of Liberia and singers and songwriters from Senegal, Gabon, Nigeria, the DRC, South Africa and Ghana. Wine's track was made with longtime collaborator Nubian Li and features a fusionist dancehall style that mixes African musical traditions with modern Jamaican riddims.

**Bobi Wine**

"Sensitise the masses to sanitise, keep a social distance and quarantine," sings Wine, whose hit soon passed a million clicks on YouTube and earned international kudos.

Wine also launched a collective call to action *#DontGoViral* and invited content creators from every musical genre 'and creatives from all over the world' to share their work on *#ShareInformation*. He argues that in this time of crisis humanity needs artists and cultural entrepreneurs to bring people together, to activate their collective intelligence and shared humanity and to translate public health information into language that everyone can understand. He also emphasizes that, while it is important to prevent Covid-19 from spreading, it is equally important to prevent misinformation and discrimination from going viral.

Musicologist Dominic Makwa at Makerere University in Kampala conducts research on the impact of pop music on society. He found that music created an increased awareness of the HIV/Aids epidemic and also provides psycho-social support for stigmatised victims; he sees no reason why this should not also be the case for the Covid-19 pandemic. Songs can not only communicate information about a disease but also shape popular opinion and sensitise people about how to avoid contracting the disease. They can also be a mechanism for counselling due to the power of metaphor and their ability to turn despair into hope. Makwa has therefore recommended that the Ugandan government should include Bobi Wine as well as Nubian Li, Bebe Cool and other musicians in their official campaign against Covid-19 (Makwa, 2020).

Song and dance can convey a variety of messages but none as strange as that of the Dancing Pall Bearers of Ghana and their founder Benjamin Aidoo. The comedic grim reapers in black suits, sunglasses and leather shoes, who groove to a techno beat while carrying a coffin, have become the accidental face of the lockdown movement with their stark message, "Stay at home or dance with us," forcing mourners to grin through their grief but also reflect on their own mortality. A video on the Dancing Pall Bearers attracted millions of clicks worldwide and Aidoo is now planning to expand his strange venture internationally (Sullivan, 2020).

In Cape Town an electronic group The Kiffness, founded by David Scott in 2011, has embraced the challenges of the Covid-19 pandemic and parodied well-known songs to raise awareness of the disease. They include *Quaranqueen* (*Dancing Queen*; I've lost track of time and my life), *Yesterday* (Yesterday, Covid-19 seemed so far away, now it looks as though it's here to stay), *Lockdown Rhapsody* (*Bohemian Rhapsody*; This is the real life, this isn't fantasy, caught in a lockdown, no escape from the quarantine) as well as *The Sound of Sirens* (*Sound of Silence*) and *Sweet Corona* (*Sweet Caroline*).

In Kenya there was serious concern that the arts would suffer during the Covid-19 lockdown so Kenyan President Uhuru Kenyatta announced in April 2020 that he was awarding a US$900 000 grant to the arts to help them survive the crisis. In South Africa an online event *Lockdown Legends* was launched to raise funds to support the South African music industry. The ongoing event is hosted by Covid-Zero and features musical performances by top South African pop and rock stars and DJs from the 1960s to the 90s. The campaign aims to support the unsung heroes, artists and musicians whose songs and music kept the nation (and the continent) entertained prior to the pandemic.

**Art highlighting racial issues:** In 2010 the Wole Soyinka Prize for Literature in Africa was jointly won by two medical doctor/authors, Nigerian Wale Okediran and South African Kopano Matlawa. Matlawa was recognised for her debut novel *Coconut* published in 2007 which addresses issues of race, class and colonization in modern Johannesburg. The book also won the European Union Debut Fiction Award and has sold over 30 000 copies. Other books by Matlwa include *Spilt Milk* (2010), on South Africa's Born Free generation who became adults in the post-apartheid era, and *Period Pain* (2017), about the heartache and confusion that so many South Africans feel who have been subject to xenophobia, rape, corruption and crime.

Nigerian author and feminist Chimamanda Ngozi Adiche is arguably the most famous African woman of her generation. As a modern, post-colonial author and columnist she has been translated into 30 languages, had two of

**Chimamanda Ngozi Adiche**

her novels adapted for screen and is a celebrated international speaker, sometimes sharing the stage with Michelle Obama. She writes and talks about feminism, racism, colonialism, African leadership and dictatorships and has become a fierce critic of racism in both North America and Africa. Her books include *Purple Hibiscus* (2003), *Half of a Yellow Sun* (2006), *The Thing around your Neck* (2009), *Americanah* (2013) and *We should all be Feminists* (2014), with the title of the latter being used, with her permission, in a hit song by Beyoncé.

**Art highlighting frustrations and hardships:** Although their bizarre name resembles that of the Dancing Pall Bearers, the Soweto Skeleton Movers have a different origin and message. The four dancers from Soweto - Jabulani, Junior, Molefi and Topollo - made their start busking on the streets and trains of Johannesburg (*Jozi*). In addition to entertaining their audiences they highlighted the frustrations and hardships of black commuters in Gauteng. Then, in 2016, they were hailed as the best act at the world's largest hip-hop dance festival, Breakin' Convention, presented annually by Sadler's Wells Theatre in London under the artistic direction of dancer Jonzi D. At the festival they delighted their audience with their joyous *pantsula*, a South African street-dance style that originated in the 1950s and was inspired by commuters jumping on and off and 'surfing' (leaning out the open doors) of moving trains (*isparapara*). As its name implies *isparapara* also has affinities with the *para para* synchronised dancing style from Japan.

Kwaito

**Art celebrating freedom of expression:** *Kwaito* is a music genre that emerged from house and hip hop in *Jozi* during the 1990s. According to music writer Setumo-Thebe Mohlomi (2019), "House music was an instrumental chromosome in the creation of the *kwaito* which captured the spirit of a recently emancipated Black youth in the 1990s." Today *kwaito* has evolved away from the simple, repetitive lyrics of old into a more complex and nostalgic expression of the South African *zeitgeist, with* the early 2000s welcoming a surge of new artists who have been influenced by it. Stiff Pap play a blend of *kwaito* called *gqom*, hip-hop and industrial sounds and artists ranging from Kwesta, Stilo Magolide and Kid X to DJ Sliqe, Cassper Nyiovest, Riky Rick and AKA have all produced *kwaito*-inspired tracks that manifest the newfound freedom of expression of South African youth (Krige, 2020).

**Art as an agent for peace and reconciliation:** The Mozambican artist Malangatana Ngwenya was active in the nationalistic Frelimo guerrilla movement in his youth but became famous for his epic paintings and murals that implored people to find a peaceful solution to the Mozambican civil war. When Frelimo came to power he was instrumental in founding the Mozambican Peace Movement and later became a Unesco Artist for Peace.

Zenzile Miriam Makeba's global hit song *Pata Pata* (Touch, touch) has been described as the 'world's most joyful song' and is widely regarded as the opening musical riff in a revolution that eventually lead to the end of apartheid in South Africa. Makeba, known as Mama Africa, was an outspoken voice in opposition to apartheid who was forced into exile in the USA in 1960. There she spared no effort in highlighting the crime of separate development and salted her concerts with harrowing accounts of growing up under white minority rule. She returned to South Africa in 1990, picking up where she had left off 30 years earlier and using her music as a balm for the country's wounded soul (Baker, 2020).

The genocide engineered by the late Perence Shiri, then Minister of Lands, Agriculture & Rural Resettlement in Zimbabwe, to quell dissent resulted in over 20 000 civilians, mainly Ndebele, being killed by his North Korean-trained Fifth Brigade between 1983 and 1987. The massacre was named *gukurahundi*, a Shona term referring to the early summer rains that remove chaff and dirt from the fields. In the aftermath of *gukurahundi* former president Robert Mugabe enforced 'collective forgetting' of this period in Zimbabwe's history, writing it off as a 'moment of madness' and suggesting that discussing the event would undermine attempts to nurture national unity, but the opposite has happened. Writers in Zimbabwe have decided that the best way to banish silence and avoid suppressing history is to write about gagged events.

Several books on the massacre have appeared in indigenous languages including *Uyangisinda Lumhlaba* (This world is unbearable) in Ndebele by Ezekiel Hleza and *Mhandu Dzorusununguko* (Enemies of independence) in Shona by Edward Masundire. Books in English about the horrors faced by villagers include Yvonne Vera's novel *The Stone Virgins* (2002), *Running with Mother*, a 2012 novel by Christopher Mlalazi, and Peter Godwin's largely autobiographical *Mukiwa: A White Boy in Africa* (1996), which paints a picture of *gukurahundi* from the perspective of a young white journalist. Novuyo Rosa Tshuma's descriptions of the genocide in her 2018 novel *House of Stone* are particularly graphic and ghastly.

Godwin Ncube has pointed out that visual artworks have also engaged with *gukurahundi* such as the 2010 exhibition *Sibathontisele* by Owen Maseko. *Sibathontisele* is a Ndebele word meaning 'we drip it on them' and refers to an infamous torture method used by the Fifth Brigade whereby they dripped hot, melted plastic onto victims. Unlike literary texts, which have remained unbanned and uncensored, Maseko's exhibition was banned by Zimbabwe state security the day after its opening at the National Arts Gallery in Bulawayo and the artist was arrested but it has since been hosted outside Zimbabwe. Visual art, it seems, is deemed to be more subversive than written texts! (Ncube, 2020).

The legendary Nigerian poet Harry Olúdáre Garuba, who died in February 2020 at the age of 62 years, is regarded as the best Nigerian poet writing in English since the national bard Christopher Okigbo, who was killed in the 1967-1970 Nigeria-Biafra civil war. While Okigbo was the leading light of the first generation of Nigerian poets, together with Gabriel Okara and Wole Soyinka, Garuba was the brightest star in the second generation, which also includes Pol Ndu, Odia Ofeimum and Femi Oyebode, that emerged post-Biafra to tackle the legacies of civil war and the failures of independence. In *Estrangement: Kano '78*, Garuba wrote, "Surely the poet is/ estranged who cannot share/ his people's fount of being". He wrote poetry that was shot through with sensitivity and tenderness, wit and wisdom, loss and longing, irony and pathos. In his lyrics he addressed the tragedies of history, including slavery, colonialism and civil war, in poetry of uncommon grace and originality (Himmelman, Sarmiento & Thipe, 2020).

**Art promoting women's and gay rights**: Nawal El Saadawi, the Egyptian psychiatrist, feminist and novelist, was a fearless campaigner for women's rights in Egypt in the 1970s and 80s. Her fundamental work *Women and Sex* (1972) cemented her reputation but led to her being jailed in 1981 for 'crimes against the state' for her outspoken views, especially her criticism of female circumcision (*khifaḍ* in Arabic). For El Saadawi the prison sentence was a clear demonstration of the link between political power and patriarchy. While in prison she wrote, with eyebrow pencil on a toilet roll, her famous *Memoirs from the Women's Prison* which was smuggled out and published in 1983 and shaped the discourse on women's liberation in the Arab world for years. She died on 21st March 2021 aged 89 years (Jamodien, 2021).

**Nawal El Saadawi**

"To be creative means to connect. It's to abolish the gap between the body, the mind and the soul, between science and art, between fiction and nonfiction."

Nawal El Saadawi

Tsitsi Dangarembga became famous as a writer, film maker and activist in Zimbabwe and gained international acclaim with her debut novel *Nervous Conditions* (1988). It was the first English novel by a black woman from Zimbabwe to be published and the BBC listed it as one of the top 100 books that have shaped the world. According to Rosemary Chikafa-Chipitro, a lecturer at the University of Zimbabwe, what distinguishes Dangarembga from other Zimbabwean writers, such as Yvonne Vera, NoViolet Bulawayo, Novuyo Rosa Tshuma and Petina Gappah, is her focus on the freedom of women in Zimbabwe's patriarchal milieu. *Nervous Conditions* was followed by *The Book of Not* (2006) and *This Mournable Body* (2018) to form a memorable trilogy by Dangarembga (Chikafa-Chipiro, 2020).

*Nervous Conditions*, with its girl child protagonist Tambudzai, is a representation of how people, mainly women, coped with oppression during British colonial rule. In *The Book of Not* Tambudzai transforms from a disciplined, rural girl who fought for her education to a 'non-person' who undergoes psychic self-annihilation as she struggles to cope with the racial exclusions at her white boarding school. In *This Mournable Body* Tambudzai is like a wounded animal and the text resonates with the tragic stories of individual Zimbabweans who are spectators to rape and trauma and develop an indifference to violence and abuse. In early 2020 Dangarembga was arrested for demonstrating against the government and, shortly afterwards, *This Mournable Body* was one of six books shortlisted for the 2020 Man Booker Prize. Previously, in 2013, NoViolet Bulawayo had been the first black African woman to be shortlisted for the Booker Prize for her novel *We need new Names*.

Zainab Fasiki, a comic illustrator and author of graphic novels in Morocco, uses her special talents to fight against street harassment and gender-based violence. When she first left her conservative hometown of Fez for college in Casablanca, she thought that life would be easier but soon found that catcalling, bag snatching and bullying were commonplace. In her latest graphic novel *Hshouma* (Shame), published in French and Arabic in 2019, she pairs her playful illustrations with hard-hitting discussions on sexuality, gender-based violence and censorship. "I want every Moroccan to read it as we have nothing on these topics at school or at home," she says (Hincks, 2019b).

Fasiki's conservative critics say that her images, some of which feature society's obsession with how women present themselves. "I'm trying to say you can see a naked woman and you can find it normal. It has nothing to do with sex. It's just a body." In early 2018 she launched a mentorship campaign for female artists to help them navigate exploitation in Morocco's art industry. Inspired by Egypt's feminist Nawal El Saadawi and French Iranian graphic novelist Marjane Satrapi, Fasiki is also countering Western narratives that portray the Middle East as either hypersexual or repressed.

Kenyan filmmaker Wanuri Kahiu's 2018 film *Rafiki* was initially banned in her country as it depicted a same-sex relationship between two young black women in Nairobi but it was the first film from Kenya to premiere at the Cannes Film Festival. She is currently involved in bringing Octavia Butler's *Wild Seed* to screen and fans of Butler can be assured that the beloved Anyanwu, the centuries-old, shape-shifting West African woman who travels part of the Middle Passage as a dolphin, will come to life in Kahui's magic hands.

**Art promoting education**: At the age of 19 Nigerian poet and playwright Harry Garuba produced a one-act play *Pantomine for Saint Apartheid's Day* which would set the anti-establishment tone of his writing for the rest of his life. After completing his PhD and teaching for 16 years at the University of Ibadan he moved to the University of Zululand and then the University of Cape Town, where he spent the rest of his career. The titles of his most famous essays and poems reveal that he felt deeply about the plight of underprivileged people - 'Negotiating the (post)colonial impasse', 'How not to think Africa from the Cape', 'Chinua Achebe and the struggle for discursive authority in the postcolonial world', 'Race in Africa' and 'The ghetto in the ivory tower'. Above all Garuba understood the profound significance of education and stated in his 2016 lecture, 'No easy walk to education', "Because historically for marginalised people, for oppressed people, for minorities, racial and ethnic all over the world, education has always been deeply connected to the question of freedom" (Himmelman *et al.*, 2020; Omoyele, 2020).

**Art against abuses of power:** Fela Anikulapo-Kuti, who died in 1997, was known in Nigeria as the 'King of Afro-beat', a fusion of jazz, funk, psychedelic rock and traditional West African chant and rhythms. He developed this genre with his equally famous drummer Tony Allen through their band Africa'70. Their art has fuelled a new generation of creatives, including Wizkid and Wyclef Jean who are keeping the spirit of his music alive. But Fela's biggest impact has been as a revolutionary who used his music to spread hope in Nigeria during the dark days of military rule in the 1970s. On stage his grinding, bass-heavy rhythms anchored his strong political rhetoric and he became enough of a threat to the authorities that they tried to shut him up. During a raid on his house in Lagos his mother was injured and subsequently died. His brave response was to deliver her coffin to an army military barracks and write the song, *Coffin for Head of State*. When he died of HIV/Aids in 1997, with a military man still in the President's palace, his funeral, which was attended by nearly one million people, became a nationwide form of protest (Robinson, 2006).

***Things Fall Apart*** **by Chinua Achebe**

The Nigerian novelist, poet and critic Chinua Achebe, who died in 2013, is best remembered for his first novel *Things Fall Apart* (1958), probably the most widely read book in modern African literature. It is the story of the incursion of Western missionaries into an Igbo community in 19th century Nigeria that has long gripped readers around the world. It is a work of extraordinary power and insight that challenges the skewed view of the continent in the Western literary tradition. According to Wole Soyinka (2006) his confident narratives of the lives that were destroyed under the colonialist mandate in this and other works continue to serve as models of historical restoration and stylistic mastery.

The Nobel Prize-winning South African author Nadine Gordimer (1923-2014) campaigned strongly for freedom and democracy and against abuses of power through her 14 novels and 11 story collections. Her proudest moment was not winning the Nobel Prize in 1991 but testifying at the Rivonia treason trial in 1964 on behalf of 22 South African anti-apartheid activists. At the trial she was no literary bystander as she had joined the ANC well before it was legal to do so and once hid its leaders in her home. When Nelson Mandela was released from prison in 1990 she was one of the first people he asked to see. Her last book *Get a Life* traced the life of a cancer-stricken ecologist who, while battling a planned nuclear power plant, must undergo radiation therapy, making him a health hazard to his family. The book is a deft portrait of the dilemmas facing South Africa after apartheid had been banished and more mundane problems had intruded. She also persuaded 20 major authors to write short stories for a 2004 book for the Treatment Action Campaign which lobbies for HIV/Aids funding. "If musicians can get up and sing," she said, "We can get up and write" (Morrison, 2006).

The murder of the renowned Ethiopian singer and activist Hachula Hundessa in Addis Ababa in July 2020 lead to riots and the death of several hundred protestors. His murder illustrated the total enmeshing of cultural, political and economic challenges in a country that is experiencing seismic changes. Hundessa, who was only 34 years old, was an extremely popular figure who used his music to advocate for the rights of the Oromo, the largest ethnic group in the country that has been systematically suppressed by the Ethiopian government for much of the country's modern history. The riots lead to an internet blackout and the closure of the Oromia Media Network by the government at a time when people most needed access to information (Zelalem, 2020).

Hundessa's 2015 hit single *Maalan Jira* (What existence is mine) is an uplifting lament about the historical injustices suffered by Oromo farmers who have been displaced from their land. It has an autotuned vocal track with the faint sound of a *massenqo* (one-stringed spike fiddle) in the background. One reason Ethiopians love Hundessa's music is his use of the literary device of 'wax and gold' in his lyrics. This term, which derives from Amharic, the official language of Ethiopia, refers to a poetic and literary sensibility that presents deep truths within superficial trappings. It is unquestionably at play in his popular melodies which convey deep messages about sovereignty and independence. In contrast, in his 2017 song *Jirra* he embodied a collective optimism that Oromo culture was no longer in jeopardy. Prime Minister Abiy Ahmed, himself an Oromo but affiliated to the Amhara ruling party, commented in a tweet after Hundessa's murder, "We have lost a precious life today" (Zelalem, 2020).

Lina Attalah in Cairo is one of a new breed of fearless online journalists who wages campaigns against injustice and prejudice through her website Mada Masr, one of the few truly independent newsrooms in a country that is known for its suppression of the media. Under Attalah's leadership the website has earned a reputation for being fearless, running blockbuster corruption investigations and revelations of regime purges and covering the war against Isis in the Sinai Peninsula which the authorities have attempted to shield from public view. The Egyptian government blocked Mada Masr in May 2017 but could not shut it down. The newsroom still publishes every day although readers in Egypt must read the articles on Facebook or by accessing the site with a VPN. The organization works under constant threat from the authorities but Attalah is determined to keep it going (Malsin, 2017).

South African *enfant terrible* film maker Jahmil XT Qubeka comes from Mdantsane near East London where, he says, he was "raised by gangsters". His second feature film *Of Good Report* sent the censors into a spin with its poignant but potent look at an exploitative relationship between a teacher and a student. The film was banned so Qubeka appeared at the 34th Durban International Film Festival, where it was due to be screened, with his mouth taped shut in protest. His latest movie *Knuckle City*, a crime/gangster/boxing drama filmed on location in Mdantsane, was spawned by his fascination with the fact that 17 boxing world champions have been produced by this hard-scrabble township since 1994 yet most of them fell from grace after their glory days were over. In the film Qubeka highlights the uniqueness of township life. "Because everything spills out onto the street, one of the things that it brings is a lack of privacy. We all live in such close proximity to one another; you're always hearing everyone else's business so you know exactly what the neighbour is doing and what the issues and challenges are," he says.

*Knuckle City* is audacious in its determined probing of the rough edges of life on the streets and in the underbelly of a world that has largely been ignored by filmmakers. It unflinchingly exposes the darkest corners of society, dredging up all the misogyny and toxic masculinity, violence, abuse and criminality that pervades the world that he grew up in. The film impels us to reflect on the plight of township dwellers in South Africa and elsewhere in Africa and to improve their lot (Hampton, 2019).

*Knuckle City*

**Graphic novels as agents of change:** Although graphic novels originated over 40 years ago (and comics over 90 years ago) they have recently become a major force among millennials and Generation Xers who are accustomed to images being combined with text. Almost all graphic novels written by African authors carry positive messages of hope, redemption and resilience and include strong take-home messages on topical issues such as child trafficking, disability rights, crime, corruption and poverty. As they mainly reach younger readers they are an important vehicle for conveying information on these adult issues.

Roye Okupe's passion for animation lead him to found YouNeek Studios in Lagos in 2012 to allow him to contribute to the superhero genre that is at the height of its popularity. With the aim of creating a connected universe of superheroes in African locations his debut graphic novel *E.X.O. The Legend of Wale Williams Part One* (2015) is set in Nigeria in 2025 with themes of redemption, hope and overcoming evil forces. In *WindMaker: The History of Atala - Art Inspired by African History* (2017), illustrated by Godwin Akpan, Okupe creates a breath-taking mythical kingdom *Atala* and a new universe YouNeek YouNiverse with heroes, dragons, mythical relics, magical swords and a feuding royal family and turns the fantasy genre on its head by placing familiar concepts in awe-inspiring African settings. His graphic novels have received wide critical acclaim and he has also written, produced and directed 2D/3D animated short films and music videos. In 2016 he was listed by *Ventures Africa* as one of the 40 African Innovators to Watch and as one of *New African* magazine's 100 most influential Africans.

*Aya* (2007), written by Marguerite Abouet and illustrated by Clément Oubrerie, is set in Yop City, Côte d'Ivoire, in 1978 and tells the story of its nineteen-year-old heroine, the studious and clear-sighted Aya, her easy-going friends Adjoua and Bintou and their meddling relatives and neighbours. It is a breezy and wryly funny account of the desire for joy and freedom and of the simple pleasures and private troubles of everyday life in Yop City. It is an unpretentious and gently humorous story of an Africa we rarely see - spirited, hopeful, resilient and free of war and famine. *Aya* won the 2006 award for Best First Album at the Angoulême International Comics Festival.

*Karmzah* by Farida Bedwei tells the story of Morowa Adjei, an archaeologist with cerebral palsy who gains amazing superpowers through her crutches and uses them via her *alter ego* Karmzah. The author also has cerebral palsy and is a disability rights advocate in Ghana. *Lake of Tears* by Etor Fiadzigbey and Kwabena Ofei is about child trafficking and child labour and chronicles the fates of Kyei, Anima and Aya who work together to rescue other trafficking victims. This astounding graphic novel won the 2018 Nommo Prize for the best graphic novel. In *EL30SBA*, by John Maher, Maged Refaat and Ahmed Rafaat, an Egyptian god Horus is reincarnated and mounts a campaign against crime and corruption in modern-day Egypt with the help of Mariam (a microbiologist with healing powers), Kaf (a spell-casting intelligence officer), Microbusgy (a shape-shifting bus driver) and Al Walhan (a reformed terrorist).

**Art campaigning against environmental degradation**: Sokari Douglas Camp, the Nigerian-born metal sculptor now based in London, produces evocative works that are inspired by her Kalabari heritage. She was born in Buguma in the Niger Delta and one of her most famous works *Green Leaf Barrel* (2014) was inspired by the struggle of her home town against rampant pollution and a lack of jobs. The figure represents a woman god who is creating growth from an oil barrel split in two. In 2012 her sculpture memorial to commemorate the abolition of slavery *All the World is Now Richer* was exhibited in the British House of Commons and she was awarded the CBE in 2005.

Camp has been awarded many commissions for public memorial sculptures, most notably *Battle Bus: The Living Memorial to Ken Saro-Wiwa* (2006). Saro-Wiwa was a member of the Ogoni people, an ethnic minority in Nigeria whose homeland Ogoniland in the Niger Delta has been targeted for crude oil extraction since the 1950s and which has suffered extreme environmental damage from decades of petroleum waste dumping. Saro-Wiwa lead a campaign against environmental degradation in Ogoniland and was an outspoken critic of the Nigerian government which he viewed as reluctant to enforce environmental controls on foreign petroleum companies. At the peak of his non-violent campaign he was tried by a military tribunal and hanged in 1995.

Sculptor Kara Schoeman is passionate about biodiversity conservation. In her father's foundry in Bloemfontein, South Africa, she melts down the aluminium components of car engines as her revenge against these pollution-belching machines and casts the liquid aluminium into beautiful artworks and bowls with evocative names such as *Leave-Let*, *Birdnest* and *Coral Bleach*, all addressing a conservation theme. A Cape Town craftsman

Davis Ndungu collects discarded flip-flop sandals that are washed up on the beach, glues them together into multi-coloured blocks and then carves highly innovative sculptures of animals, mythical creatures, boats and other objects from the blocks (Haynes, 2020).

Another South African artist Ndabuko Ntuli collects discarded plastic bottle tops and fashions them into elaborate, three-dimensional artworks including human portraits, animals, and landscapes. His 'bottle top' portrait of Nelson Mandela, on display at the Art@Africa gallery in Cape Town, has brought him international acclaim and he has received lucrative commissions for further artworks using recycled materials. Kenelioe-Mpho Mazibuko of Johannesburg uses thrown-away and recycled objects such as tyres, dustbin lids, buckets, plastic bottle tops, pans, baths and suitcases to create artworks that celebrate 'the beauty of the discarded'. Storytelling and poetry, and the agony and ecstasy of everyday life in South Africa's townships, drive Mpho's creativity (Haynes, 2020).

Pilato Bulala is famous for his scrapture creations that are made from scrap metal. He works as a motor mechanic during the day but after hours creates remarkable sculptures using old metal car and bicycle parts, welding them together into what he calls "stories of democracy", epic tales of the triumph of the human spirit over adversity. His recent works include South Africa's coat-of-arms and a giraffe made from spanners, cogs and bicycle chains. "It is all about imagination. I go to my brother's house to plan, where he has electricity to weld. I also sell my work on Facebook and Instagram. I started in 2011 and evolved from earrings made from old cans to bigger sculptures", he says. Chenserai Mutato, a scrapture artist from Zimbabwe now working in Hout Bay, Cape Town, also creates amazing sculptures of real and mythical animals from scrap metal.

**Art celebrating technology:** Although street art was kickstarted in Los Angeles in the 1950s it has been part of the African scene for many decades. In February 2020 twenty-five street artists produced 100 colourful murals as part of the International Public Arts Festival in Salt River, Cape Town, where public art is deeply rooted in community life. The theme of the 2020 Festival, now in its fourth year, was 'Digitisation' and the artists were challenged to address the theme but also express community values, enhance the environment and transform the landscape. Stefan Smit produced a large orange-and-blue face and commented, "It's amazing but I just wanted to give a different perspective … Blue represents the digital era we're going into and the orange being the warm familiar old-school vibes. So I wanted to have both of those colours shining in the face of my figure so that it shows we're influenced by the old and the new and we're not just one-dimensional" (Mgujulwa, 2020; Mtuta, 2020).

Co-organiser of the festival, Alexandre Tilmans of Baz-Art, states that his goal is to provide work for street artists, make communities more cohesive and prouder, and create conversation pieces that stimulate debate on topical issues. Cape Town mayoral committee member for community services, Dr Zahid Badroodien, said, "Art is not for a select few, and it matters to all of us because our communities gain cultural, economic and social value through public art". Jakes Mbele, who is recognised as one of the best street and graffiti artists in South Africa, has launched a company Jakes Mbele Communications that produces stunning, large-scale murals and graffiti for corporate clients, municipalities and street festivals.

**Arts promoting a positive world view of Africa**: The Triggerfish Foundation was established in Cape Town by Triggerfish Animation Studios, Africa's most successful animation studio, to upskill animators, foster the development of animation networks, bring African animation to global audiences and, above all, contribute to a positive world image for Africa. Their programmes build on the success of their fruitful partnership with Disney and currently support animators from 30 African countries through digital learning platforms, drawing workshops in schools and townships and pan-African webinars and talent competitions on animation.

The vision of Triggerfish, which is based in Cape Town, is to increase

Triggerfish Animation Studios' animated film *Khumba*

diversity and gender equality in the animation industry and generate innovative content for over 400 million African children. The company, in partnership with the Goethe-Institut and the Federal Ministry of Economic Cooperation & Development in Germany, has also created the Triggerfish Academy, a free digital learning platform for anyone wanting to learn more about career opportunities in animation. Their website features 25 free video tutorials, quizzes and animation exercises that introduce animation and explain the principles of storytelling and storyboarding. The platform was created by Tim Argall, animation director on Triggerfish's third feature film *Seal Team*, Malcolm Wope, character designer on Netflix's first animated original film from Africa, Mama K's *Team 4*, Daniel Snaddon, co-director of the multi-award-winning BBC adaptations, *Stick Man* and *Zog*, and Mike Buckland, head of production at Triggerfish.

**Placing African art in an international context**: Hamady Bocoum, director of Senegal's Museum of Black Civilisations, has big ambitions for his new repository of African art, culture and history. For decades, the most prominent sanctuaries for African art have been museums in Europe but he hopes to reclaim some of the continent's lost wealth by demanding the restitution of artworks 'stolen' during colonial times. The US$34 million project is designed to be a creative laboratory that will help to shape the continent's future sense of identity and many of its galleries are deliberately empty, ready to accept newly reclaimed artworks. Some European countries, including France, were prompted to lend pieces for the opening (Rockwood, 2018; Baker, 2019).

William Kentridge's audio-visual installation at the MOCAA (2020)

## 3. Creativity in the arts and sciences

The Zeitz Museum of Contemporary Art Africa (Zeitz MOCAA), which opened in September 2017 on the Victoria & Alfred Waterfront in Cape Town, is a public not-for-profit art museum which collects, preserves, researches and exhibits art from Africa and its diaspora. Hailed as 'Africa's answer to the Tate Modern', the MOCAA has over 6 500 m² of gallery space plus a rooftop sculpture garden and is housed in a stunning building that was, until recently, a complex of abandoned grain silos. Forty-two 30-metre-tall concrete silos have been artfully sliced in different ways to create unique spaces, including shafts for the lifts, transforming it into a cathedral-like space bathed in light from over 100 faceted glass windows. The opening collections include a stunning visual art display by William Kentridge, cowhide sculptures of the female form by South African artist Nandipha Mntambo and multi-media installations by Zimbabwean artist Kudzanai Chiurai. At the opening ceremony Archbishop Desmond Tutu pretended to take a phone call from heaven, chatting with former President Nelson Mandela. "Yes", Madiba said, "This is what we were fighting for!" (Rockwood, 2018).

Dirk Durnez has always been brimful of creative ideas. After founding a company Monex Themed Environments (now MTE Studios) that themed the 128 000 m² Canal Walk shopping mall and its associated theme park, Ratanga Junction, in Cape Town, he developed a variety of other creative initiatives. Then, in 2016, after 31 years' involvement in developing interactive museums and edutainment experiences, Dirk and Henk van Aswegen established the cutting-edge Art@Africa gallery

Dirk Durnez in the Art@Africa gallery

Mike Bruton interpreting Kosie Thiart's charcoal drawing *Totem of Resurrection* in the Art@Africa gallery in Cape Town

on the Waterfront in Cape Town to promote the work of African artists. Their stable of artists currently includes Ndabuko Ntuli, Kenelioe-Mpho Mazibuko, Andries Visser, Barney Bernado, Caelyn Robertson, David Griessel, Eben, Kara Schoeman, Kosie Thiart, Lauren Redman, Maureen Quinn and Talita Steyn and they are constantly on the lookout for more artists from Africa (Haynes, 2020).

Art@Africa seeks to create a significant positive experience and address the issues that face modern African society, such as gender equality, racial discrimination, homophobia, African identity and biodiversity loss, through its art. It recognizes that artists are often the first to oppose binary prejudices like racism and rather see it as an opportunity to portray the kaleidoscopic continuum between extremes. They also believe that there is an increasingly strong interface between art and science, especially environmental science, and that art can convey messages and articulate emotions that we are not able to express easily in words (Haynes, 2020). The theme of Art@Africa's mid-2020 exhibition *The Blue Dot*, reflected on humankind's perspectives on Earth and referred to the iconic photograph of the planet, the Pale Blue Dot, taken by NASA's *Voyager 1* space probe on 14th February 1990 from a distance of six billion kilometres. The gallery also prides itself in the development of digital experiences including augmented reality that allow its clients to discover and interrogate art in other realms.

Situated in Lomé, Togo, the Palais de Lomé is a unique art and culture centre that opened to the public in late 2019. The building, formerly the Governor's Palace, has been restored and transformed into a landmark addition to the cultural scene in West Africa. According to the director Sonia Lawson, this ambitious renovation, cultural and environmental project, which is funded by the Togolese State, is dedicated to showcasing the diversity of Togolese and African culture in visual arts, design, new media, science and technology, culinary arts and the performing arts and music.

*Protea humanoid* by David Griessel on display at the Art@Africa gallery in Cape Town

**Discussion**: It is clear from this brief overview that Africa has a very dynamic arts and culture scene and that art, in all its forms, can be used to convey messages to ordinary citizens and to those in power that the *status quo* is not good enough. In contrast, science is a relatively austere field of endeavour that is typically inhabited by bright but reclusive characters who could do with the injection of some pizazz into their efforts to bring their scientific findings to the attention of a broader public. Science makes an enormous contribution to the development of Africa and to the well-being of its people but, to many, it is the continent's best kept secret. Art and science are both avatars of human creativity but their different attributes need to be used in complementary ways to address the challenges that face the continent.

> *"Comparing the capacity of computers to the capacity of the human brain, I've often wondered, where does our success come from? The answer is synthesis, the ability to combine creativity and calculation, art and science, into a whole that is much greater than the sum of its parts."*
>
> Garry Kasparov, chess grand master

# 4.

# 'Lunarticks' and Owls

*'We, here at the Owl Club do what the world needs so badly – we communicate ideas. ... Brother Owls, in the Owl Club it is always like old times. But the secret is this, that the ideas are usually new.'*

Owl 'Bags' Baigrie in a talk to the Owl Club in 1969

**INTRODUCTION:** One of the benefits of living in Cape Town is that one can meet fascinating people and indulge in all manner of activities through the variety of clubs and societies in the metropolis. I have chosen to be active in a few clubs and societies, including the Friends of the Cape Town Science Centre, Friends of Iziko South African Museum, Friends of the East London Museum, Whale Coast Conservation in Hermanus, Royal Society of South Africa, Society of Model and Experimental Engineers in Rondebosch (the oldest such society in the world), and, most interestingly, the Owl Club. I was nominated for membership of this club by the then-director of the South African Museum Michael Cluver, seconded by astronomer Brian Warner, within a year of arriving in Cape Town and it has been one of my most invigorating experiences in the Mother City. In this chapter I discuss the Owl Club as well as a remarkable society that preceded it, the Lunar Society of Birmingham. Although the two clubs had similar origins and characteristics, there is no evidence that the Lunar Society, which terminated in 1813, led to the establishment of the Owl Club, founded 81 years later.

Men have met to chat over an evening meal for millions of years. Cave men gathered in an inner circle around the fire discussing the day's hunt and the prospects for tomorrow while gazing into the fire for long periods without talking. The women and children in the outer circle chatted about the day's harvest of berries, fruits and roots and how they would prepare them for dinner. Early Greek scholars gathered in the shade of trees and scratched patterns in the sand as they discussed geometry, military campaigns, politics, and philosophy. When the Lyceum and Acropolis were built Aristotle, Plato, Socrates and others gave talks and held discussions. They studied and discussed the natural world in pursuit of wisdom and understanding and to be inspired, with relative indifference to the utility of their knowledge. They debated, for instance, the criteria for defining beauty and truth.

The Romans gathered in forums to plot the expansion of their empire and early Muslims assembled in Houses of Wisdom where they translated the works of the ancients, shared their knowledge with other cultures and made great advances in science, mathematics and engineering. The Middle Ages in western Europe were characterized by a slavish adherence to sterile, untested ideas until the Enlightenment and Renaissance during which discussions evolved towards both scholarly and utilitarian matters, such as: How can we gain control over Nature? How can we strengthen our arm through the invention of machines? The age of discovery and the new scientific revolution ignited by Muslim and European intellectuals in the late 15th century pivoted the debates even further towards science and technology. At that time universities were centres for teaching whereas academic clubs and societies, such as the Royal Society of London were the focus of innovation and research.

In these thinktanks, as we would call them now, intellectuals and entrepreneurs, some of them quite radical, met to debate a common question: How could the quality of life of society be improved through alternative political systems or the intervention of science and technology? The American philosopher, statesman and inventor Benjamin Franklin, whose life work was focused on revolution and on applying scientific knowledge to practical social change, set the pattern in North America when his private study group in Philadelphia evolved into the famed American Philosophical Society. In Europe there were precipitous, sometimes calamitous, changes that resulted in the French Revolution (1789-1799), the French Revolutionary Wars (1792-1802) against Great Britain, the Holy Roman Empire, Prussia, Russia and other monarchies, the Industrial Revolution and the sudden development of science- and technology-based industries in England.

One of the most important of these early 'brain trusts' was the Lunar Society of Birmingham whose members cheerfully referred to themselves as 'lunar men' or 'lunarticks' (although the word 'lunatic' has an earlier origin as a form of insanity reputedly brought about by the phases of the moon).

**Lunar Society of Birmingham**: I first read about the Lunar Society when I was researching Charles Darwin for the bicentenary of his birth in 2009 and discovered that his paternal grandfather, Erasmus, was a member of the society. The precise origins of the Lunar Society are difficult to pinpoint but appear to lie in a friendship that developed from 1757 between two very different characters - the cultured, dreamy poet/scientist Erasmus Darwin and the flamboyant, hard-headed industrialist, Matthew Boulton. Boulton left school as a 14-year-old and established a manufacturing business in silver, plate and ormolu, a gilding process. When he inherited a button and buckle factory from his father, he transformed it into the renowned Soho Works where he built James Watt's revolutionary steam engines, Through their partnership they installed hundreds of Boulton & Watt steam engines throughout Britain that mechanized factories and mills and kickstarted the Industrial Revolution. Boulton also applied new techniques to the minting of coins, striking bronze farthings, half-pennies and pennies and silver shillings for Britain and coinage for other countries. He made money, and his ambition knew no bounds.

Erasmus Darwin

Erasmus Darwin was quite the opposite. He was a man of extraordinary intellectual insight, a philosopher, inventor, poet, anti-slavery campaigner and medical doctor who was sufficiently well respected to have been invited to be physician to King George III of England, an offer he declined. He was also a prolific author who wrote books on topics as varied as the classification of vegetables, plant reproduction and physiology, evolution, and female education. He was a larger-than-life figure (in more ways than one; he became so fat that he had to cut a semi-circle into his dining table to accommodate his stomach; Uglow, 2002) who dabbled in many of the fashionable topics of the day and was ahead of his time in many. In 1794 he published a book of poetry, *Zoonomia; or, The Laws of Inorganic Life*, that includes early ideas on evolution. He suggested, for instance, that all species have a common ancestry, that competition

Matthew Boulton

# 4. 'Lunarticks' and Owls

**Boulton and Watt's steam engine factory at the Soho Manufactory**

between species due to overpopulation may have prompted them to evolve and predicted that apes and humans may have a close evolutionary relationship. He died nine years before Charles was born but his writings did influence his grandson's thinking.

Some members of the Lunar Society, such as Erasmus Darwin, stemmed from the landed gentry with ten becoming Fellows of the Royal Society of London whereas others, including Boulton, came from humble origins and worked their way to the top. Despite their different backgrounds they shared a common interest in experiment and innovation with Darwin's theoretical understanding complementing Boulton's practical, hands-on experience. Soon they were conducting joint investigations into topics as varied as electricity, meteorology and geology and their application in industry which Darwin dubbed as 'a little philosophical laughing'. Many Lunar Men were 'non-conformists' or 'free-thinkers' who were outside the Establishment, which proved to be a strength as they were not hampered by stuffy institutions or outdated traditions of deference (Uglow, 2002). Furthermore, in those far-off days, there was little separation between the arts and the sciences as the arts comprised the 'fine arts' as well as the 'mechanical arts' and men could be poets and engineers or artists and inventors at the same time. When the British Museum first opened its doors in 1759 it displayed cultural and natural history objects together.

The Derby-based clockmaker John Whitehurst soon joined the group through his involvement in supplying clock movements to Boulton's factory. In 1758 Whitehurst excitedly shared with Darwin and Boulton one of his latest inventions, the pyrometer, a remote-sensing thermometer used to measure the temperature of distant objects. Another early member was the remarkable John Michell, an English natural philosopher (scientist) and clergyman who provided pioneering insights into a wide range of scientific fields including astronomy, geology, optics and gravitation. Michell is widely considered to be one of the greatest unsung scientists of all time (although that term was only coined in the 1830s) and was the first to propose the existence of black holes and to suggest that earthquakes travel in waves. He invented an apparatus to measure the mass of the Earth, designed the torsion balance used by Henry Cavendish and described how to make an artificial magnet. He has been called the father of both seismology and magnetometry and was so far ahead of his time that most of his ideas languished in obscurity until they were re-discovered more than a century later. He remains virtually unknown even today, partly because he did little to promote his own pioneering ideas.

Michell introduced Boulton, Darwin and Whitehurst to the American statesman and inventor Benjamin Franklin when he visited Birmingham in 1758 "to improve and increase acquaintance among persons of influence". Franklin was a leading author, political theorist, politician, freemason, scientist, inventor, civic activist and diplomat who invented the lightning rod, bifocal spectacles and the Franklin stove. He became a corresponding member of the society and returned to Birmingham in 1760 to conduct experiments with Boulton on electricity and sound. On another visit in 1765 he met, through the Lunar Society, the radical polymath and preacher Joseph Priestley, the visionary leader of the *Rational Dissent*, with whom he compared notes on electricity. Franklin was impressed by Priestley's research and encouraged him to publish it, which he did in 1767 in a landmark book, *The History and Present State of Electricity, with Original Experiments*.

Benjamin Franklin

Priestley went on to make many other notable discoveries, demonstrating in 1771 that plants convert carbon dioxide into oxygen and isolating oxygen (1774), such was the impact of members of the Lunar Society.

The Lunar Society was formally constituted in Birmingham in about 1765 and reached its zenith during the heyday of the Industrial

## 4. 'Lunarticks' and Owls

**A meeting of the Lunar Society of Birmingham**

Revolution when the most prominent contributors to the Midlands Enlightenment were active members and huge advances were made in science and engineering. In just a few decades technology advanced from hand-operated spinning wheels to steam engines and steam-driven heavy machinery. Consequently, hordes of people left the countryside to work in factories in the cities which created significant socio-economic opportunities as well as problems.

The membership of the Lunar Society was never large, a maximum of about twenty at a time, but it was exclusive and ever-changing. The exact nature of the Society is difficult to discern as it had no office bearers although it did have leaders, and no formal records, minutes, constitution or reports on its activities have survived. What we do know about it has been gleaned from the prolific correspondence of its members for that was the golden age of letter writing. Its name, which was only formally adopted in 1776, was derived from their habit of meeting in the early evening on the Sunday nearest the full moon as the extra moonlight made the journey home by horse-drawn cart easier and safer in the absence of street lighting.

They initially met monthly at Matthew Boulton's home Soho House in Handsworth, Birmingham, or at Erasmus Darwin's home in Lichfield. The society reached its peak in the tumultuous decade of the 1780s during which the Treaty of Versailles was signed, William Pitt the Younger became the youngest Prime Minister of Great Britain, the Montgolfier brothers launched their hot-air balloons, Mozart produced *The Marriage of Figaro*,

Soho House, where the Lunar Society first met

King George III went mad, the Bastille was stormed in Paris, a bill to abolish the slave trade was defeated in the British parliament (but approved in 1807) and George Washington was sworn in as the first president of the new United States of America.

During their meetings and in correspondence with one another members of the Lunar Society had lively discussions and debates on philosophy, politics, art, culture, science, technology, commerce and manufacturing, and on the challenging technological and social problems of a rapidly industrialising society. They shared a childlike fascination for discovery and invention and designed and built steam engines, iron bridges and boats, canals, aqueducts, lighthouses and factories, flew hot-air balloons, named new minerals, gases, plants and animals, discovered new medicines, advanced science across many fronts, managed world-class businesses that changed the face of Birmingham, the Midlands and England and introduced industrial innovation to the modern world. They even dreamed that their technological fix would light the fuse of democratic change.

Invitations to meetings were informal and witty. For example, on 25th February 1776 Boulton wrote to Watt,

> "Pray remember that ye celebration of ye 3rd full moon will be on Sunday March 3rd. Darwin and Keir will both be at Soho. I then propose to make any Motions to Members respecting new Laws, and regulations, such as will tend to prevent the decline of a society which I hope will be lasting. Pray bring Mr Wilkinson, I think he will make a good member."

The nature of the group changed significantly with the move to Birmingham in 1765 of the Scottish physician William Small who had been Professor of

Natural Philosophy at The College of William & Mary in Williamsburg, Virginia. There he had been a major influence on Thomas Jefferson and had formed the focus of a local group of intellectuals. His arrival, with a letter of introduction from Franklin, had a galvanizing effect on the Society as he was an organizer *par excellence* and a lively debater. The next new member to join was the famous potter Josiah Wedgwood, a close friend and business colleague of Erasmus Darwin who would later become Charles Darwin's maternal grandfather. Erasmus' son, Robert,

Josiah Wedgwood

married one of Wedgwood's daughters Susannah and the inheritance that he received after Josiah's death allowed him to partly fund Charles Darwin's chosen vocation in natural history. Wedgwood developed advanced pottery techniques through systematic experimentation, perfecting his world-famous Jasper porcelain in 1778, and was a leader in the industrialisation of European pottery as well as a respected social reformer.

Another new member was Richard Lovell Edgeworth, the prominent Anglo-Irish politician, writer and inventor who devised novel educational methods, designed carriages and surveying equipment, invented the optical telegraph and caterpillar tracks and managed to father 22 children with four wives along the way! He, in turn, introduced his friend and fellow Rousseau follower and abolitionist Thomas Day, an author and lawyer, to the society. In 1767 a prominent Renaissance man, the Scotsman James Keir, a chemist, industrialist, politician, author, translator, geologist, metallurgist and military captain, visited Darwin and subsequently moved to Birmingham and joined the group.

By 1768 a core group of nine individuals had assembled with William Small at their heart. In the same year the brilliantly perceptive steam engineer James Watt visited Birmingham and became a corresponding member. He invented his revolutionary steam engine in 1788, at the peak of the society's influence, as well as a copying press. He was

James Watt

also the first to propose the concept of horsepower and promoted the importance of patents. Unfortunately, Small achieved the unusual feat of dying of malaria in England in 1775. As he had been a key link between members his demise caused the group to re-organize itself. Boulton took the lead but reliance on him proved to be a weakness as the 1880s coincided with the apogee of his work building up his steam engine empire and he was frequently absent.

Another prominent member at the time was John 'Iron Mad' Wilkinson, the English industrialist who pioneered the manufacture and use of cast iron. He also developed one of the first machine tools, a precision instrument for boring into cast iron cylinders, and, together with architect Thomas Pritchard, built the famous Iron Bridge across the River Severn at Coalbrookdale, completed in 1779. It was the first major bridge in the world to be made of cast iron and its success inspired the widespread use of this metal as a structural material; today the bridge is celebrated as a symbol of the Industrial Revolution. In the late 1780s the nature of the society changed again with the move to Birmingham of Joseph Priestley. He had been associated with the group's activities for over a decade and was a strong advocate of the benefits of learned societies. Shortly after his arrival the meetings moved from Sundays to Mondays to accommodate Priestley's duties as a clergyman while the society's dependence on Boulton was lessened by holding meetings at other members' houses.

Other prominent members of the society during this productive era included William Withering, botanist, geologist, chemist and physician who brought *digitalis* into mainstream medicine, another botanist and physician Jonathan Stokes, Samuel Galton, Jr., a Quaker and gun maker, and William Herschel, a musician and astronomer who built monumental telescopes and discovered Uranus in 1781. They also included John Smeaton, the English physicist and engineer who designed bridges, canals, harbours and the Eddystone lighthouse, who is widely regarded as the 'Father of Civil Engineering'. Another was James Hutton, the not-so-humble Scottish experimental agriculturalist, physician, geologist and chemicals manufacturer. He developed the theory of uniformitarianism, now a fundamental principle of geology, which postulates that the Earth's crust is constantly being reformed by natural processes such as erosion and sedimentation over geological time. His work, which established geology as a 'proper' science, influenced Charles Darwin and, much later, James Lovelock, who developed the Gaia Hypothesis which proposes that the

## 4. 'Lunarticks' and Owls

Earth is a self-regulating system.

Other luminaries among the membership of the Lunar Society, or frequent attendees at their meetings, included:

- Sir Joseph Banks, English explorer, naturalist, botanist and a leading international patron of the natural sciences who took part in Captain James Cook's first great voyage (1768–1771) and advised King George III on the development of the Royal Botanic Gardens at Kew. *Inter alia*, he was president of the Royal Society of London for 41 years and a trustee of the British Museum for 42 years. About 80 species of plants bear his name and he is credited with introducing *Eucalyptus*, *Acacia* and *Banksia* species, all highly invasive, to the Western world, a dubious honour.
- Richard Kirwan, an Irish geologist and chemist who worked with Antoine Lavoisier, Joseph Black, Joseph Priestley and Henry Cavendish but was an anti-evolutionist and one of the last supporters of the discredited phlogiston theory.
- Petrus Camper, the Dutch anatomist, anthropologist, palaeontologist and naturalist who was one of the first scientists to dissect a rhinoceros and orangutan, proposed the term 'extinct' and had interests in architecture, mathematics and sculpture. Famously, he investigated how head shape might affect mental development and formulated 'Camper's angle' using measurements from the ear, nose, chin and forehead.
- Rudolf Erich Raspe, German librarian, writer and scientist who is best known for his collection of tall tales, *The Surprising Adventures of Baron Munchausen*, originally a satirical work with political intent.
- John Baskerville, English printer and creator of a typeface that bears his name whose works include some of the finest examples of the art of printing.
- John Wyatt, the English inventor who developed the first mechanical spinning machine.
- Joseph Black, the Scottish physician and chemist, who discovered magnesium, latent heat, specific heat and carbon dioxide.
- Jean-André Deluc, a Swiss meteorologist, physicist and geologist who invented a portable barometer and hygrometer.
- John Ash, a prominent physician who established the Birmingham General Hospital and is reputed to have treated his own mental illness

using a cure he found through the study of mathematics and botany.
- Ralph Griffiths, an English editor and publisher of Welsh extraction who founded London's first successful literary magazine, the *Monthly Review*.
- John Roebuck, an English inventor and industrialist best known for developing the industrial-scale manufacture of sulphuric acid.
- Thomas Percival, an English physician and health reformer who wrote an early code of medical ethics.
- Daniel Carlsson Solander, a Swedish naturalist who was the first university-educated scientist to set foot in Australia when he accompanied Joseph Banks on James Cook's first voyage to the Pacific Ocean aboard the *Endeavour*. He was the inspiration behind the name Botany Bay in 1770 and made and described an important collection of Australian plants while the *Endeavour* was beached at the site of present-day Cooktown for nearly seven weeks after being damaged on the Great Barrier Reef. Solander was an apostle of the renowned Swedish taxonomist Carolus Linnaeus and supported his novel binomial system for naming plants and animals.
- George Fordyce, a prominent Scottish physician and physiologist who was educated at Aberdeen and Edinburgh and became an authority on fevers at St Thomas's Hospital, London. Sadly his eccentric lifestyle and lack of social graces reduced his standing among his medical colleagues.
- Louis Joseph d'Albert d'Ailly, a French aristocrat, chemist and natural historian who conducted pioneering experiments on the formation of carbon dioxide.
- Barthélemy Faujas de Saint-Fond, a French advocate, chemist, volcanologist, traveller, author and early supporter of hot-air ballooning.
- Johann Friedrich August Göttling, a prominent German chemist who designed and sold chemical assay kits and developed processes for extracting sugar from beets (to supplement his meagre university salary). He also studied the chemistry of sulphur, arsenic, phosphorus and mercury and was one of the first scientists in Germany to take a stand against the phlogiston theory and support the new chemistry of Antoine Lavoisier.

They were undoubtedly a 'who's who' of Enlightenment scientists, engineers and entrepreneurs.

While it may come as a surprise that these dedicated capitalists were part of a revolutionary cabal, we should realize that, in 1785, capitalism itself was still a relatively controversial concept. The idea that industrialists and businessmen should exchange plans and plot the future together with philosophers, engineers and scientists was even more far-fetched but that is exactly what these new-age revolutionaries did. When they consciously joined forces it was because they wanted to shape a decent life for everyone. Can you picture being in a room with these imagineers, fired up by their vivid imaginations but challenged by the technological and social challenges of their time? What ran through their thinking was the simple understanding that the good life is more than material sufficiency.

The Lunar Society lost several stalwarts during the 1780s although the increasing number of corresponding members ensured that their debates became increasingly multi-national. Edgeworth ceased regular involvement in 1782, John Whitehurst died in London in 1788 and Thomas Day passed away in 1789. Most significantly, Erasmus Darwin moved to Derby in 1781 but complained of being "cut off from the milk of science" so he continued to attend meetings until 1788.

The outbreak of the French Revolution in 1789 and the post-Revolution wars in Europe caused political strains between members but it was the Priestley Riots of 1791 in Birmingham itself that saw a decisive falling off of the Society's spirit and activities. The rioters' main targets were religious dissenters, most notably the controversial Joseph Priestley who was driven from the town, leaving England for the USA in 1794. In addition the house of William Withering was invaded by rioters and Boulton and Watt had to arm their employees to protect the Soho Manufactory.

The letters and papers exchanged between Matthew Boulton and other members of the Lunar Society cover an extraordinary range of topics from Watt's steam engines, the importance of canals (which Wedgwood, Darwin, Boulton, Small, Galton and Watt all encouraged and helped to finance), to improvements in road transport, the importance of capital investment, technological innovations and new manufacturing methods, scientific experiments, health and working conditions, politics, medicine, sanitation and improvements in education.

Robert Schofield in his book, *The Lunar Society of Birmingham: A Social History of Provincial Science and Industry in eighteenth-century England*, highlights the significance of Boulton: "The most important of these men was Matthew Boulton and the entire history of the Lunar Society is marked by characteristics impressed upon it by Boulton's personality". He characterized Boulton as "a born promoter" who jumped from one project to another, often before the first was completed or before its success was guaranteed. Although he possessed a quick mind, he worked best at the elaboration of other peoples' ideas.

Two of Boulton's friends left revealing statements about this trait. James Keir wrote, in a memoir on Boulton,

> "Mr Boulton is a proof how much sound knowledge may be acquired without much regular study, by means of a quick and just apprehension, such practical application, and nice mechanical feelings . It cannot be doubted that he was indebted for much of his knowledge to the best preceptor, the conversation of eminent men.'

To which James Watt responded,

> 'Mr Boulton possessed in a high degree the facility of rendering any new invention of his own or others useful to the public by organising and arranging the processes by which it could be carried on. His conception of the nature of any invention was quick and he was not less quick in perceiving the uses to which it might be applied and the profits which might accrue from it."

In *ca* 1766 Erasmus Darwin wrote to Boulton,

> "I have got with me a mechanical Friend, Mr Edgeworth from Oxfordshire - the greatest conjuror I ever saw ... He has ye principles of Nature in his Palm, and moulds them as He pleases. Can take away Polarity to give it to the Needle by rubbing it twice on ye Palm of his hand And can see through two solid Oak Boards without Glasses, wonderful! astonishing! diabolical!!! Pray tell Dr Small He must come to see these Miracles."

On 2nd October 1772 James Keir wrote to Boulton,

> 'Yours I had the pleasure of receiving yesterday. Your orders shall be executed as speedily as possibly especially I hope for your own Experiments as I well know the Impatience of my fellow-schemers, and I should also be sorry to check by delay your present hobby-horsicality for chemistry .'

In July 1777 Keir wrote again to Boulton,

> "I had a letter yesterday from Dr Darwin in which he says he longs for a little philosophical laughing. - therefore when you are at leisure, some full moons Sunday, I hope you will indulge the Dr and let us both know in time, that we

may lay aside our patients and Glass making to attend you."

In the 1790s society meetings were continued by the younger generations of the families of the founders, including Gregory Watt, Matthew Boulton Jr, Thomas Wedgwood and James Watt Jr. From 1800 onwards the frequency of meetings declined but as late as 1809 Leonard Horner described the remnant of the Lunar Society as being very interesting. While individual members continued to produce work of importance the collaborative activity that had been a hallmark of the society rapidly decreased. The society had collapsed by 1813 as, in August of that year, Samuel Galton, Jr. won a ballot for possession of the scientific books from its library.

The many tributes to the society and its members include a series of memorial stones, the 'Moonstones', two statues of James Watt, a statue of Boulton, Watt and Murdoch and especially the Soho House Museum that celebrates the life of Matthew Boulton, his partnership with James Watt, his membership of the Lunar Society and his contribution to the Midlands Enlightenment and the Industrial Revolution. This museum, and the Birmingham Central Library, also houses an irreplaceable collection of the letters exchanged between Lunar Society members which defined the vigour behind the industrialisation of Britain in the late 1700s. The memorial to James Watt in Westminster Abbey hails him as among the most illustrious followers of science and the real benefactors of the world (Uglow, 2002).

The sense of adventure and enthusiasm with which those eminent men at the vanguard of the Industrial Revolution threw themselves into their scientific experiments and business transactions, and the skill with which they organized significant social, industrial and economic change, has probably never been equalled. The caucus of eminent scientists and technologists who were active about a century later, including South African James Greathead, Karl Benz, Thomas Edison, Nikolas Tesla, Albert Einstein, Niels Bohr, Max Planck, Ernest Rutherford, Alexander Graham Bell, the Wright brothers, Robert Goddard and others also brought about a revolution but they were able to build on a much stronger technological foundation.

Of all the philosophical societies that formed in the USA, France, Germany and England the Lunar Society of Birmingham was probably the most important because it was not just provincial. All the world came to Soho to meet Boulton, Michell, Darwin, Watt or Small who were, in turn, acquainted with the leading men of science throughout Europe and America. Many historians believe that the meetings of a select few fertile minds in the Lunar Society who focused on the *application* of science and technology to industry changed an age and provided the 'engine' for human ingenuity during the Industrial Revolution.

The Lunar Society is not entirely dead as it was revived in a modern form in Birmingham in 1990 by the author and consumer activist Dame Rachel Waterhouse, who received the reconstituted society's first Lunar Medal in 2006. Like its illustrious predecessor the new apolitical society aims to bring together those best able to contribute innovative and practical ideas to shape the economy, culture, environment and development of the West Midlands region around Birmingham. It draws its membership from all walks of life - city, industry, business, commerce and the universities - to form a forum that promotes an interdisciplinary approach to the new challenges of a rapidly changing world.

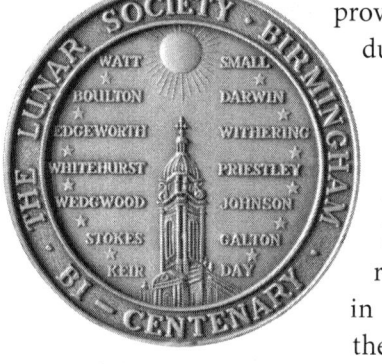

Medal commemorating the bicentennial of the Lunar Society in 1966

In a passage in the first newsletter of the new Lunar Society, Sir Adrian Cadbury wrote:

"I see the eighteenth-century Lunar Society as providing us with a model and as being the inspiration for our activities. In that sense, we are not attempting to recreate what was literally a unique gathering of great minds, but rather to honour their example. They mustered an extraordinary range of talents and experience because of the breadth of their individual interests. To follow that pattern we (now) need a much wider membership, but with the same aim in view - that of crossing the boundaries which divide us. This is made all the more necessary by the increase in specialisation since the time of the Lunar Society, when Erasmus Darwin could combine the callings of doctor, botanist, scientific classifier, inventor, geologist, educationalist and poet."

At the University of Birmingham another Lunar Society meets every Thursday to debate and discuss all manner of topics from current events to historical precedents and, in Italy, the Lunar Society Italia (aka

Astronomers without Borders) with an astronomical bent was formed in 2009 to communicate science and astronomy to the broader public.

The triumph of the Lunar Society was based on the multidisciplinary nature and conviviality of its members. That is also one of the key success factors of the Owl Club.

**The Owl Club**: The Owl Club of Cape Town was formed in 1894 to provide a social meeting place for those with an interest in the liberal arts and sciences. The members are entertained, informed and encouraged 'to let their hair down ever so slightly' by a tradition of speakers, musical performances, lively conversations and occasional themed exhibitions. Unlike the Lunar Society the origins and history of the Owl Club have been comprehensively documented by James Malcolm (1959; from 1894 to 1900), WE Ranby (1952; 1894-1950), Eric Rosenthal (1982; 1951-1981), James Henry (1996; 1982-1994) and Tony Murray (2019; 1994-2018) and meticulous minutes have been kept of every meeting.

The diligent way in which the Owl Club has documented its proceedings means that its deliberations could be available to anyone who wants to familiarize her/himself with the changing mores of society over a period of more than 125 years in Cape Town. Its records represent an important slice through history from colonial to post-colonial times and into the new millennium. Such 'time capsules' are rare and valuable especially when they represent the views of hundreds of erudite people from diverse walks of life rather than those of a few historians who are pushing a particular ideology or world view. Also, unlike the Lunar Society, which dissolved after 48 years, the Owl Club is still thriving 127 years after its founding.

The late University of Cape Town (UCT) paediatrician, singer and songwriter Maurice Kibel, summed up the spirit of the Owl Club in his witty ditty, *'Get Me to the Club on Time'*, with acknowledgements to Stanley Holloway,

**Emblem of the Owl Club, Cape Town**

I'm off to Owldom in the evening,
T'wit t'woo will be the loud refrain,
Call me a taxi
Mini or maxi
Or get me to the club by train.

I gotta be there in the evening,
Black tie, and looking in my prime,
Spruced up and frisky,
Can't miss my whisky,
So get me to the club on time.
If music's ghastly,
There's the next bar,
If speaker's boring
Whewt him out the door!

I gotta be there in the evening,
Black tie and looking in my prime,
Kick up a rumpus,
But don't lose the compass
And get me to the club
Get me to the club,
For Pete's sake, get me to the club on time.

On 5th December 1921, the *Cape Times* described the Owl Club as 'a refuge where men of solemn walks of life among us may forget for an evening now and then their necessary dignity and be just their natural selves.' In 1994 Owl Brian Warner paraphrased the Reverend Sydney Smith when he surmised that

'... the Owl Club has had dignity without dullness, erudition without ennui, scholarship without scatology, anecdotes without animadversions - and a lot of other things without a lot of other things.'

In an entertaining talk to the Club entitled, 'An Owl in a Sack Troubles no Man' (Henry, 1996), Owl Warner further noted that the membership of the club includes,

'well-versed poets and dramatic playwrights; case-hardened judges and creditable bankers; heavenly astronomers and criminal lawyers; first-rate admirals and particular Generals; patient doctors and fully-booked librarians' as well as 'reams of writers and quires of clergy; an exaggeration of fishermen; a cure of doctors; a corps of anatomists; a flutter of cardiologists; and even a smear of gynaecologists.'

The wife of the first Secretary of the Club, CG Lowinger, is credited with coining the name. Soon after the idea for the club was hatched four men were sitting in the twilight in the Lowinger's garden discussing what it should be called, when, "Into the gathered darkness came Mrs Lowinger with the natural inquiry, 'Why are you all sitting here in the dark like a lot of owls?' The very name for our club, they cried, and the Owl Club it was called and remains" (Malcolm, 1959). The first formal meeting took place on 1st October 1894 in the Victoria Hotel in Grove Street (now Parliament Street), Cape Town.

The club subsequently met in Poole's Hotel in New Street (now Queen Victoria Street), had a brief bohemian existence in the Theatre Building on Grand Parade, a few years at 42 Burg Street overlooking the Town House garden and then rooms at the British Medical Association in Wale Street. A stable nest was found for 45 years at the historic Lodge de Goede Hoop from 1930 to 1975 and, from 1976 to 1998, meetings were held at the prim-and-proper City and Civil Service Club (now the Cape Town Club), a Herbert Baker-designed building erected in 1878 in Queen Victoria Street. Since then, the Owls have contentedly met and dined at the Kelvin Grove Club in Newlands, originally the home of gold mining magnate J.C. Rimer.

'Black tie' is the customary plumage. although this ruling was relaxed slightly at the March 2021 AGM to include 'elegant attire'. At the beginning of each meeting the parliament of Owls is invited to 'take their perches' and, at the end, they are wished 'a safe flight home'. The club has a Moulting Fund, which enables it to assist Owls who may find the dinner costs unaffordable, and Owls who have died are referred to as members who have 'flown to a higher perch'. The nicknames of some of the office bearers continue the ornithological theme: Secretary Bird (secretary), Fiscal Shrike (treasurer), Chorister Robin (music organizer), Familiar Chat (talks organizer) and Sociable Weaver (newsletter editor).

The first president of the club, Sir David Gill, Her Majesty's Astronomer at the Cape and a distinguished Fellow of the Royal Society of London and President of the Royal Astronomical Society, presided over the Owls until 1905 and set the august tone. Four successors to Gill as director at the Royal (later Cape) Observatory, Dr (later Sir) Harold Spencer-Jones,

John Juritz, president of the Owl Club in 1977

John Jackson, Richard Stoy and Bob Stobie, also became presidents of the Owl Club. The president at the 50th anniversary meeting in 1944 was the ex-British High Commissioner Sir Edward Harding, and the distinguished astronomer and historian Professor Brian Warner presided over centenary meeting in 1994. Other well-known astronomers and cosmologists in the club's membership roll include Tony Fairall, George Ellis and Ian Glass.

Presidents have also included barrister William Philip Schreiner, Prime Minister of the Cape Colony during the Second Boer War, the author and ultimate 'wise man' Eric Rosenthal and Supreme Court Justice Pat Tebbutt SC, an eminent Owl whom I had the good fortune to know well. His autobiography, *A Life Spiced with Variety. My Memoirs* (2015), relates his remarkable life story as a journalist, sports broadcaster, barrister, businessman and judge but his humility and general knowledge are the traits that I will remember. Towards the end of this life we played golf every Sunday afternoon at Rondebosch Golf Club during which we often discussed the Owl Club but he rarely spoke of his own achievements, which I learned about when I read his book and attended his memorial service.

Murray Wilson, president of the Owl Club in 2000

The Owl Club has also been served by some distinguished and dedicated Secretary Birds including Alfred Holtzer (from 1914 to 1958), Arnold Matthews (1959-1983), James Henry (1983-1993), Murray Wilson (1994-1998), Michael Fisher (1998-2005), Oliver Trevor (2008-2017) and currently Hugh Amoore, ex-Registrar at UCT and president of the UCT Legacy Society (2018-). Holtzer originally trained as a poultry farmer but made his name in Cape Town as a patron of the performing arts and had the production costs of one of his plays, *A Broken Chain*, sponsored by the club. Arnold Matthews OBE served in the British

Colonial Services of over 30 years, including acting as Governor of the Falklands.

Membership is by invitation. Although women have frequently attended club meetings as musicians, singers and speakers they were first admitted as members of the club at the March 2021 AGM, with Sheila Camerer as the first woman member. Originally members of the club were defined as 'gentlemen' whom one Owl defined as "someone who can play the bagpipes, but doesn't". A prospective member must demonstrate some distinction in his/her profession or hobbies but the overriding criterion is his/her ability 'to dwell in harmony' with fellow Owls. So, on occasion, members have been elected more for their 'clubability' than for their distinction. Fledgling members are encouraged to contribute to the club's activities by 'bearing their part in our assemblies'.

Initially medics were excluded from membership but, on 9th January 1895, Owl Frank Bradley proposed the admission to the club of 'the medical and other artistic professions'. The then-Secretary Bird objected on the grounds that the club had been formed to bring together votaries of the aesthetic arts as distinguished from the technical arts and suggested that medics should only have honorary status, but his objection was overruled. Today members are classified into one or more of five faculties which best describe their main interests: art, drama, literature, music and/or science. The category might mirror a member's profession but the catholicity of interests of the Owls means that many members have an involvement in one or more of the membership categories without engaging in it professionally.

In the early days of the club the hosting of distinguished guests from abroad was an important activity, although 'distinguished' was broadly interpreted as the guests at a meeting in 1898 included the Edwardes Gaiety Comedy Club! Over its long history the club has hosted many prestigious writers, artists, admirals, bishops, scientists, diplomats, musicians, actors, explorers and politicians but it remains characteristically discreet about its status. These guests have included Samuel Clemens (Mark Twain), F.W. Reitz (ex-President of the Orange Free State Republic, who recited his Afrikaans translation of Robert Burns' poem, *Tam O'Shanter*), Sir Alfred Milner, Sir Arthur Conan Doyle, Lord Baden-Powell, Joseph Chamberlain, the renowned architect Sir Edwin Lutyens, the pianist Mark Hambourg, Dame Sybil Thorndike, Field Marshal Jan Smuts, Field Marshal Lord Roberts, Lord Kitchener, the long-serving Oxford don Reverend William

Spooner (after whom spoonerisms are named), 'Jimmy' Logan (laird of Matjiesfontein), the poet Roy Campbell, David Vaisey, librarian at the world famous Bodleian Library, and that doyen of the English language, Professor Guy Butler.

Rudyard Kipling regularly visited Cape Town on holiday between 1898 and 1908. Although he was invited to visit the Owl Club on several occasions he never did so although he did sign the visitor's book on 13th April 1898 (it was taken to him at the Vineyard Hotel by the Secretary Bird!). The tradition of hosting distinguished guests has continued but the end of the colonial era, combined with the advent of scheduled air flights to South Africa from the 1930s and the discontinuation of the weekly Union Castle mail ships in 1977, shifted the 'Gateway to South Africa' from Cape Town to Johannesburg and foreign dignitaries became less available.

Samuel Clemens' social visit on 13th July 1896 was the first grand occasion hosted by the club. He was welcomed 'with great acclamation' by 150 members and guests at De Goede Trouw Refectory, presented with an *Album of Cape Views*, and responded with a humorous speech. "After a capital programme the meeting broke up at about 12.30 a.m., all present joining in singing *Auld Lang Syne* and *God Save the Queen*, in which the guest of the evening most heartily joined" (Rosenthal, 1982). At an Owl Club meeting in 1898 the English writer and war correspondent Edgar Wallace recited one of his poems, *In the Presence of the Queen*, which described the feelings that overpowered 'Tommy Atkins' (slang for a common British soldier) on having the Victoria Cross pinned to his breast by Queen Victoria. In 1900 Lord Roberts, Commander of the British Forces in the Anglo-Boer War, dined at the club ten days after handing over his command to Lord Kitchener, who was a guest in June 1902.

Discourses on pioneering explorers and epic journeys are greatly appreciated by the Owls, some of whom are intrepid adventurers themselves. Who could forget Owl Joe Tyrrell's riveting account of Joshua Slocum's daring round-the-world yacht trip from 1895 to 1898, Owl Ian Woods' philosophical musings on mountaineering, or the brave escape by German prisoner-of-war Gunther Plüschow recounted by Owl John van der Linde? In 1910 Captain Robert Falcon Scott together with four members of his Terra Nova Expedition Edward Wilson, Teddy Evans, Lawrence Oates and Henry 'Birdie' Bower dined at the Owl Club *en route* to Antarctica on their ill-fated attempt to be the first to reach the South Pole. Many years later guest speaker Robert Swan gave a mesmerizing talk, 'In the Footsteps of Scott', on his pilgrimage that retraced Scott's epic journey. Another memorable talk enjoyed by Owls at the time was given in 1960 by Professor Reginald James, a physicist at UCT, who took part in Ernest Shackleton's heroic 1916 expedition to the Weddell Sea on the *Endurance*. Although James delivered his address 44 years after the event, "… he managed, without any histrionics, to bring the icy Antarctic wastes and their incredible dangers right into the midst of the gathering that listened to him, completely enthralled" (Henry, 1996).

In another fascinating talk, 'Reminiscences of an Old Coot', the British astronomer Patrick Moore discussed his interests in astronomy as well as his engagements with oddballs interested in intergalactic travel, colonizing the moon, UFO sightings, flat earth theories and fairies. He also shared his encounters with Orville Wright, Albert Einstein (for whose violin playing he occasionally provided piano accompaniment), Werner von Braun, Yuri Gagarin and Neil Armstrong. At other meetings Uys Krige discussed Afrikaans literature, TV Bulpin reminisced about the history of the Cape, Paul Mills reviewed 'Book Collectors and their Habits', actor Ralph Lawson pontificated about 'Early Cape Theatre', Geoffrey Wittenberg decoded 'The Translator's Art' and John Juritz enthused about 'Music: State of the Art'. In 1967 the British historian Peter Quinnell divulged the details of the legal proceedings concerning the banning of D H Lawrence's *Lady Chatterley's Lover*, in which he had given evidence.

Musical performances have been at the heart of the Owl Club from the outset. At the first meeting in 1894 eight of the twelve members present contributed to the evening's musical entertainment! At a later get-together, the Owls were entertained by the delightful singing of a Professor of Paediatrics accompanied by a Professor of Physics playing a harpsicord made by a Professor of Astronomy! In 1955 a sublime Cape Malay choir sang to the Club and, in May 1956, Major-General Sir Clarence Bird entertained the Owls with an excellent violin solo. Many Owls remember with delight the spell-binding performances on the mouth organ by the former Attorney-General Owl 'Broeksie' Broeksma QC, who had apparently developed his *fluitjie* skills while confined to his hotel room during tedious circuit court travels. During one of his performances Broeksie's exertions on the mouth organ led to his false teeth coming loose and having to be hastily pocketed with the minimum of interruption. Owl Keith Jewell, who was accompanying him on the piano, noticed his predicament and acting in the true spirit of Owldom pocketed his own set in sympathy (Rosenthal, 1982)!

Musicians entertain the Parliament of Owls

The Owls have been entertained on a variety of musical instruments, not only their own treasured Blüthner piano (bought in 1961 and still in use today after several 'face lifts') but also on the pianoforte, organ (at St George's Cathedral), violin, viola, cello, double bass, clarinet, saxophone, trumpet, bassoon, trombone, oboe, flute, guitar and drums. Other instruments have included the piano accordion, euphonium, spinet, harpsichord, harmonica, banjo, mouth organ, recorder, bagpipes, ukulele, lute and voice and even a carpenter's saw artfully bowed by Owl Mike Hill, a cellist.

On the Owl Club's 70th birthday on 24th October 1964 the *Cape Times* reported,

> "The Owls' purpose sounds a little solemn: 'to provide opportunities for social intercourse among members of the various professions and those qualified in the liberal arts and sciences.' But solemnity is certainly not the keynote of their informal monthly meetings – evenings given over to music, recitation and the arts of the raconteur, with refreshment for body as well as spirit – and the diverse membership is one of equals. They are equals in the sense of preparedness to say, play, sing, or recite one's piece expected of all; no Owl in fact may be a drone. And it is surely this easy sense of shared participation which, even more than the distinction of many of its guests, keeps the Owl Club vigorously alive."

Today the Owl Club meets on the evening of the third Tuesday of each month (although that was not always the case) except in January. Between 65 and 75 members and their guests, with an average age of over 60 years, habitually perch at the dinners. In the past the membership included a distinguished mix of governors-general, earls, diplomats, servicemen and military leaders, including the Marshal of the Royal Air Force, Sir Arthur 'Bomber' Harris. Another prominent early member (from 1927) was the physicist Sir Basil Schonland, famous for his research on lightning, his involvement in the development of radar during World War II and for being the first president of the CSIR and first chancellor of Rhodes University; he was also posthumously voted as South Africa's 'Scientist of the Century' in 1999. Two other early members who particularly interest me are the erudite entomologist Dr S H 'Stacey' Skaife, a prolific author of scientific and popular books, broadcaster and conservationist who played a leading role in the establishment of the Cape Point National Park (member from 1927 to 1956) and Maciek Miszewski, a remarkable man who had represented Poland at yachting, been awarded the Polish equivalent of the Victoria Cross in World War II and designed the Artscape Theatre, Constantia Wine Museum and many other important buildings (1970-2010).

Peter Spargo, president of the Owl Club in 1989

The current membership comprises an eclectic mix of active and retired academics, judges and other legal eagles, priests, musicians, writers (in English, Afrikaans, French and Spanish) as well as poets, librarians, actors, diplomats, medics, engineers, imagineers, scientists, accountants, yachtsmen, entrepreneurs and other engrossing people. The club has both town and country members, with some of the latter, resident in New Zealand, the USA and Europe, occasionally becoming 'swallows' who fly to Cape Town for the austral summer and attend meetings. The Owl Club enjoys reciprocal relationships with the Savage Club (since 1925) and Eccentric Club (2010), both in London, and with the Savage Club in Melbourne (2001).

While enjoying pre-prandial drinks Owls sometimes view special anniversary exhibitions which have included celebrations of the centenary of the Relief of Mafeking, 400th anniversary of the founding of the Dutch East India Company, centenary of the end of World War I, bicentenaries of the births of Charles Darwin and Robert Gray (the first Bishop of Cape Town), the bicentenary of the Astronomical Observatory in Cape Town and

Bust of Ludwig van Beethoven at the Owl Club meeting in December 2020

the 250th anniversary of the birth of Ludwig van Beethoven. Exhibitions have also been mounted on the travels or pastimes of Owls such as 'Owl Migrations' (they travelled to 337 destinations on all continents between 2017 and 2019), 'Books by Owls' (over 100 books published since 2010) and 'What Owls Collect' (from editions of *Don Quixote* to fountain pens, dictionaries, stamps and ancient maps). Interestingly, the first display mounted by the club, at a meeting on 31$^{st}$ May 1899, was an exhibition of paintings by British artists called a *Conversazione* which took place two years before dinners became regular fare. In 1904 a live demonstration of fencing with foils was part of the entertainment for the evening.

Initially only the committee and their guests attended dinners but meals were later opened to all members. Today dinner is a three-course meal punctuated by a toast to our guests during which the lights are switched off, a statuary owl blinks and the nocturnal Owls hoot, more-or-less in tune, "Tu whit tu whit tu whoo, Tu whit tu whit tu whoo, Tu whit tu whit tu whoo, our guests", borrowing the chorus from the poem *Christabel* by Samuel Taylor Coleridge. The blinking owl nicknamed 'Chalkis' (a mythical owl referred to in the *Iliad*) was presented to the club on 22$^{nd}$ February 1922, an appropriate date when one considers its onomatopoeic echo of

Dinner at an Owl Club meeting in 2019

## 4. 'Lunarticks' and Owls

the toast to our guests!

The guest speaker is then introduced, with some flourish, and invited to give her/his 20-minute talk, which is followed by an interval during which the Owls roost around the bar. This is followed by an informal talk by an Owl which, because of its 'purposeful inconsequence' (a delightful phrase coined by Secretary Bird Hugh Amoore) has been nicknamed 'The Wastepaper Basket' (WPB) which successfully maintains the tradition of 'learning lightly borne and skilfully imparted'. As James Henry stated in *The Unjealous Years* (1996), 'There has been little or none of the didactic prolixity which has often been declared the occupational hazard of professors', although, in 1974, Owl Anton Paap, then Professor of Classics at UCT, did treat the Owls to a lengthy discourse in fluent Latin on the analogues of the Owl Club in ancient Greece and Rome! Meetings end with a second musical interlude and a toast to the Owl Club.

By convention guest speakers choose their own topic but are requested to avoid reference to party politics as the podium is not a husting. The ideal talk entertains and informs, is thought-provoking and is not too technical. The press is not invited but the proceedings are recorded and a monthly newsletter, the *Notice*, which includes write-ups of the talks and musical performances, is circulated online to the 200+ members. Several Owls are competent cartoonists and the sketches that they produce of Owls and the speakers are a delightful inclusion in the *Notice*. During the Covid-19 lockdown in 2020 and 2021 the Club hosted virtual meetings which gave members the opportunity to listen to online talks by speakers from elsewhere in South Africa as well as from New Zealand, Madagascar, England and Canada. In addition, country Owls were able to present WPBs to members.

In his address to the club, 'China – Handle with Care', diplomat John Selfe waxed lyrical about the Chinese and agreed that 'they are quite simply the nicest people in the world'. In other talks Owl James Henry rhapsodized about 'The Lighter Side of Money', Owl Hugh Amoore contextualized 'Postal Innovations of the Nineteenth Century', and City Counsellor Owl Owen Kinahan rued the parlous state of museum finances in Cape Town in 'This Little Pig Went to Market'. Palaeontologist and then-director of the South African Museum Owl Michael Cluver discussed dinosaurs and the causes of dinomania, Stephen Banks shared his fascination with Stonehenge, Owl Owen Cardinal McCann, the first cardinal in Africa, discussed the popes he had met and Owl Ken Gunn shared his thoughts on heraldry in 'Hark the Heralds'. Guest Bernard Heuvelmans (referred to as 'Bernard Heugelmann' by Rosenthal, 1982) discussed the fascinating 'science' of cryptozoology which examines evidence for the existence of extremely rare or even fictitious animals (cryptids) such as the Lochness Monster, Bigfoot, *Mokele-mbembe* (the legendary aquatic dinosaur from the Congo) and the yeti. This author's involvement with the International Society of Cryptozoology, which Heuvelmans co-founded, has been through his research on the coelacanth, a so-called 'Lazarus species' that was rediscovered after it was thought to have gone extinct, in common with the takahe, black kokanee, Zanzibar leopard and Grand Comoro scops-owl.

Professor Nick van der Merwe, in one of many fascinating addresses to the club, spoke about 'Magic and Science in Malawi', Leonard Anstey described 'A Day in the Life of a Pathologist' and Professor Roger Short brought real life into sharp focus with a talk on 'The Menace of AIDS'. Brian Warner gave an amusing account of 'University of Cape Town Roots' in which he mentioned a particular 'thorn in the side' of the administration, the professor of zoology, Lancelot Hogben who once stated during a debate on the role of sport in university life that "UCT is an athletic institution where intellectual advancement is not altogether discouraged" and, when the university's new campus was being discussed, "Among the trees now being planted in the academic paradise at Rondebosch we should be careful not to omit the tree of knowledge" (Henry, 1996).

In a WPB in 1992 geneticist Owl Irven DeVore came to the alarming conclusion that males were an evolutionary mistake, which reminds me of Mark Twain's statement that, "God created man because he was disappointed in the monkey". The inimitable Owl Maurice Kibel entranced the Owls with a fantastical journey through the world of physics, Owl Barry Malson enthused about his hobby of phillumeny (collecting matchboxes), a talk described by the committee as 'illuminating;' and 'matchless', and Owl John Gardner addressed the important topic of 'How to have a good Idea'. Champion winemaker Arnold Schickenling assessed the status of the South African wine industry and Anthony St John disclosed the inner workings of the House of Lords from his personal experience. The WPB entitled 'Hiding the Decline and other Mischief', in which maverick Owl 'Taffy' Lloyd, a member of the Intergovernmental Panel on Climate Change that won the 2007 Nobel Peace Prize (with Al Gore), claimed that climate change proponents were distorting the facts was, however, regarded as highly controversial and several 'ecological' Owls did not agree with his conclusions.

Owl Anthony Lister, in an amusing talk on 'calumny, insult, imprecation, incivility and invective', entitled 'Eyeball to Eyeball', commented,

> 'Two of the most famous swordsmen in the business were Bernard Shaw and Winston Churchill; they clashed on one occasion when Shaw, perhaps unwisely, sent Churchill tickets to the opening night of his new play, with the following note: 'Bring a friend … if you have one'. Churchill replied that he was engaged for that night, but requested 'tickets for the second performance … if there is one.'

In a talk deceptively titled 'A Love Affair with Josephine' Owl Michael Ryan enthused about his involvement with Josephine's Mill, part of a working farm with a distillery, brewery and watermill in the 1940s (not to be confused with Mostert's Mill, a windmill built in Mowbray in 1796 and sadly damaged in the UCT fire in April 2021), and engineer Richard Holms discussed his hobby of campanology (bells and bell-ringing), mentioning that the most famous bell in the world, Big Ben, has had a crack in it ever since it was cast in 1850. In a WPB in 2003 chemistry lecturer Owl Jack Elsworth proved conclusively that, if you put all the colours of our rainbow nation together, you end up with a colourless liquid of great transparency! Acclaimed ornithologist Owl Peter Steyn is a prolific Wastepaper Basketeer who has spoken frequently on birds as well as on Ethiopia and Antarctica. In a fascinating talk cultural and natural historian Francis Thackeray weighed the evidence for William Shakespeare's indulgence in cannabis and concluded that it may have been responsible for some of the bard's creativity.

The topics of other talks given to the Owl Club ranged from the sublime to the ridiculous. They included the human population explosion, yachting, hippies in California, Richard III ('Who Killed the Princes in the Tower?'), rude words, origin of horses, Shaka, history of motoring, x-rays, plastic pollution, African presidents, the Piltdown skull hoax, Union Castle liners, magic, tax law, mathematics, importance of humour, crime, futurology, human evolution, ecological intelligence, art, Lady Ann Barnard, Siege of Mafeking, women in the legal profession (Mothers in Law), architecture, music, Arab-Israeli conflict, greed, diamonds, globalization, tourism, Nepad, the internet and space research. They have also included the slave trade, sharks, Mossgas, dams, NSRI, American Civil War, bubbles, evacuation of Tristan da Cunha, Dr James Barry, shipwrecks, Galileo, GM crops, Steven Hawking, the wit of Winston Churchill, baboon management, Cecil John Rhodes, contraceptive pills, biology of race, marijuana, Christine Keeler, DryBath, antique bottles, Eskom, black holes, genetic profiling, Neanderthals, DNA, and even cannibalism (by cardiologist Joe Tyrrell)!

Recently the club has been addressed by, *inter alia*, Daniel Ncayiyana, ex-deputy vice-chancellor of UCT, Helen Zille, then-Premier of the Western Cape, Susan, Baroness Greenfield, Monica Newton, executive director of the National Arts and Science Festivals, Dr Sizwe Mabizela, Vice-Chancellor of Rhodes University, the celebrated satirist Pieter-Dirk Uys, Fine Music Radio presenter and music impresario Rodney Trudgeon, conductors Bernhard Guellar and Richard Cock, ecological consultant Andrew Cooke from Madagascar, actors Nicholas Ellenbogen, David Muller and 'Doc' Caldwell, financial planner Wouter de Witt, environmentalist Guy Preston (Biosecurity), Tony Leon (Reflections on our Troubled Nation), Chris Taylor (Battle of Muizenberg), Owl John van der Linde (A Twist on Oliver Twiss), wildlife veterinarian Mike Kock, and scientist/entrepreneur Leslie Ansley who, with his wife Paula, has created a premium craft gin flavoured with botanicals derived from elephant dung!

While browsing through the membership roll of the Club I was pleasantly surprised to find that several historic and some recent people whom I have written about or met were members. Sir George Cory, one of the five founding fathers of Rhodes University College, was a member from 1925 until 1935 long before my time. Cory is mentioned in my book, *The Fishy Smiths. A Biography of JLB and Margaret Smith* as he was the professor of chemistry at Rhodes when JLB Smith was first appointed there as a lecturer. Like Cory, who switched from chemistry to history, JLB Smith metamorphosed from a chemist into an ichthyologist, a field in which he achieved world fame. In 1925 Cory resigned from the university and moved to Cape Town where he was the honorary historiographer at the National Archives until his death in 1935.

What is equally interesting is which prominent citizens of Cape Town were *not* members of the Owl Club. As an example, the Iziko South African Museum, a leading cultural and natural history museum founded in the Cape in 1825, has had an illustrious roster of directors yet only two of them, the ornithologist Dr Leonard Gill (no relative of Sir David), director of the museum from 1924 to 1941, and the palaeontologist Michael Cluver have been Owls.

## 4. 'Lunarticks' and Owls

My humble contributions to the Owl Club have included talks on cricket, Einstein, the Cape Town Science Centre, South African inventions, discovery of the coelacanth, JLB Smith, Marjorie Courtenay-Latimer, Lunar Society, words ending in ology, history of the bicycle, Islamic contributions to science, funny episodes in the history of science and the dodo, some of which are presented in expanded form in this book. In my capacity as the Familiar Chat I have also assisted with the recruitment of speakers which was particularly challenging (and rewarding) during the Covid-19 lockdown. Most importantly, I have met, conversed with and been inspired by a remarkable array of accomplished and thought-provoking people and formed lasting friendships that will continue to nurture my curiosity and creativity.

**Discussion**: Human creativity in all fields is dependent on good ideas and the best way to have good ideas, as Edward de Bono has noted, is to have lots of them. The most fertile ground for generating new ideas is that relatively unchartered territory between disciplines and this has, I believe, been a key to the success of both the Lunar Society and the Owl Club. While the Lunar Society fomented an industrial revolution through the dynamic interdisciplinary collaboration of its members the Owl Club's impact on society has been more subtle. The combined wisdom of the Owls has permeated slowly but steadily into society through the articles and books they write, the talks and webinars they give, and through their extensive involvement in their communities.

The Industrial Revolution, facilitated and accelerated by the Lunar Society, was a giant leap forward for 'industrial humans' as it introduced new technologies, created jobs, developed novel products and services, improved public utilities and transport infrastructure and advanced science, but it also did a lot of harm. The exploitationist mind-set that it promoted, and its massive carbon footprint, has wreaked havoc on the natural environment and resulted in unprecedented levels of greenhouse gas emissions and pollution, habitat destruction, the overexploitation of living natural resources, disruption of essential ecological processes, biodiversity loss and climate change. It was inevitable, but also tragic. Interestingly one of the first environmental NGOs to raise the alarm, the Coal Smoke Abatement Society, was founded in 1898 by an English artist Sir William Richmond who was frustrated by the pall cast by coal smoke over the landscapes he painted.

The accomplished economist and founder of the World Economic Forum Dr Klaus Schwab has declared that we are now in the Fourth Industrial Revolution. I am little more than a retired fisherman, although a contemplative one, but I disagree with him. There is little that is 'industrial' about the current revolution which is characterised by gig economies, unparalleled digital connectivity, the broad participation by different age, cultural and socio-economic groups (rather than a narrow elite) in innovation and commerce, the development of clean energy sources and far more enlightened views on sustainable living and biodiversity conservation. It is not characterised by chimneys billowing smoke and top-down technocratic leadership but by nimble, eco-centric, computer-literate techpreneurs who toe the triple bottom line and co-create solutions with collaborators (whom they will probably never meet) in other disciplines. In fact, a more appropriate designation for the current era would be 'anti-industrial' or at best 'post-industrial' as one of its goals is to undo the wrongs wrought by the previous industrial revolutions. The era of the Owl Club is therefore, to a degree, an antithesis of that of the Lunar Society.

As a conservation biologist I have been particularly interested in talks and discussions at the Owl Club on biodiversity conservation, and there have been plenty of them. Douglas Hey, then head of Nature Conservation in the Cape Province, previously expounded on the opportunities and challenges facing conservationists, Ian Player highlighted the goals of the Wilderness Leadership Foundation, John Hanks and Anthony Hall-Martin described the plight of elephants, Andy Siller has spoken about endangered gorillas in Rwanda, Philip Tongue, Walter Mangold, Peter Steyn and Vernon Head have showcased bird conservation efforts, Ian McCallum has elucidated his concept of ecological intelligence and Eric Harley has discussed the rebreeding of the quagga and the role of population genetics in conservation.

There has never been a more important time for discerning and inquisitive men and women to indulge in intelligent conversations and confront head-on the challenges that we face. These challenges are vastly different from those faced by the Lunar Men 240 years ago. The Owl Club does not aspire to be a formal thinktank that promotes a particular ideology (although it could be). Instead, its members are independent thinkers and doers who offer wisdom, common sense and, in general, moderation on a wide range of topics. Their interdisciplinary blue-sky musings can and do trigger novel ideas and insights that are potentially useful to society.

## 4. 'Lunarticks' and Owls

As Owl 'Bags' Baigrie said in a memorable talk in 1969 entitled, 'The Spirit of the Wastepaper Basket' (Rosenthal, 1982),

> 'We, here at the Owl Club do what the world needs so badly – we communicate ideas. May I remind you that no thought exists substantially of itself, but only by its utterance? So we establish tolerance and understanding, by hearing for ourselves and finding out about the interests, joys, achievements, and even the failures, of others. … Brother Owls, in the Owl Club it is always like old times. But the secret is this, that the ideas are usually new.'

Although the press has described the Owl Club as 'this quaint relic of the Victorian age' and 'a cheerful gathering of enlightened Bohemians', the reality is that it is a open-minded institution based on strong traditions that is constantly re-inventing itself. The Club gives its members, in this age of ultra-specialization, opportunities to become better informed about disciplines other than their own and to explore ideas that cut across silos. It might be old but the ideas that its members spawn and discuss are new. Furthermore, these discussions happen in a cheerful ambience that lacks the pomposity that sometimes suppresses creative thinking and ingenuity.

> "Perhaps some of the formalities and rituals are indeed part of a bygone age. But in keeping its members intelligently aware of the issues of the day, the Club is well a part of the 21st century. It moves with the times and is as relevant and vibrant today as it was at the end of the 19th century."
>
> Owl Tony Murray in his introduction to *The Glow of Brotherhood* (2019)

The Owl Club is a remarkable institution that held its 1331st meeting in September 2021. Long may it prosper as its values and ideals of open-mindedness and blue sky thinking are becoming increasing rare in society.

**The author delivering a Wastepaper Basket to the Owl Club in 2021. Sketch by Owl Peter Hyslop.**

# 5.

# Boneshakers to Bloomers: Evolution of the Bicycle

*'When I see an adult on a bicycle,
I do not despair for the future of the human race.'*

Attributed to science fiction writer, HG Wells

INTRODUCTION: Bicycles have fascinated me since I was a child. I rode back-and-forth to school in Cambridge, East London, for 10 years and, with my cousins, pedaled down to Nahoon beach to go swimming. I accompanied a childhood friend, Graham Avery, on adventurous specimen-collecting trips in the bush around East London where he searched for fossils and stone tools and I collected and beetles and skulls. Although I had a bad bike crash as a teenager, when I hit a dog at speed and needed 26 stitches to my face and a new set of front teeth, my enthusiasm for cycling has never waned. Years later I had another accident, this time on the promenade in San Francisco after spending a wonderful day at the Exploratorium science centre. I was racing back to my hotel with bags of educational aids hanging from the handlebars when one swung into the spokes of the front wheel, which sent me catapulting onto the sidewalk. Three attempts in recent years to 'bike jack' me around Cape Town have only strengthened my resolve to promote the bicycle as one of our greatest inventions.

While on my postdoc in London I preferred a bicycle to the Underground, following a congested though scenic route from Russell Square along Oxford Street to Hyde Park, Exhibition Road and the Natural History Museum. Later, on trips abroad, I hired a bicycle whenever possible to explore towns, villages and the countryside. Outside Bangkok Carolynn and I rode along precarious boardwalks through villages perched on stilts in mangrove forests. I explored the canal paths of southern Spain near Valencia and, in Barcelona, visited the *Sagrada Família, Park Güell, Casa Batlló* and other sites where Antoni Gaudi's magnificent art and architecture are on display. Near Arles in France I visited the scenes which Vincent van Gogh immortalised in his paintings *Wheatfield with Crows* and *Lane near Arles*, and near Paje in Zanzibar I sped along the beach at low tide to photograph fishermen setting their traditional fish traps or women beating coconuts to extract fibre for making coir rope. Inland in a crude factory in Mtengani I witnessed the distressing sight of live coral being crushed and burned to produce lime for building. Whether I was attending a conference or holidaying in Canberra, Geneva, Vancouver, San Francisco, Qatar, Copenhagen, Mauritius or Rio de Janiero the bicycle was my favourite ride.

During a three-year contract (2012-2015) in Bahrain I rode the length and breadth of the tiny island country, visiting the Bahrain National Museum, Arad Fort, Barbar Temple and the craft market at Bab Al-Bahrain in Manama City while also venturing southwards into the desert where I came across the First Oil Rig Museum and collected desert roses (gypsum) and quartzite nodules. At the time I was training to compete in the annual Cape Town Cycle Tour but the tallest 'hill' available was a suspension bridge over the Tubli Bay Straits so I rode back and forth across it, much to the dismay of local fishermen. Other than a few 'get-fit' marines from the local American Fifth Fleet naval base I was the only person crazy enough to cycle in the 'heat of the Arabian winter' except for contract labourers from the Indian subcontinent.

In security-conscious Saudi Arabia, while advisor to the director of a new science museum there, I was forced to ride circuits within the campus of the new King Abdullah University of Science & Technology (KAUST) in Thuwal but occasionally ventured out into the desert with the KAUST Cycling Club. My last trip with them ended badly when I broke my hand in an accident just two weeks before the end of my contract. While living in Jeddah I regularly cycled along the picturesque promenade along the shore of the Red Sea shortly before evening prayers on a Friday but always in longs rather than cycling shorts lest I fall foul of the Police for Promotion of Virtue and the Prevention of Vice (who really do exist!).

Bicycles provide a sense of freedom and an escape from internal combustion engines that no other vehicle offers. They are silent, non-polluting, self-propelled and sociable, allowing you to go wherever you like, stop whenever you choose and meet the local people. The fact that you need to put in a bit of effort along the way just adds to the pleasure. It is no surprise to me that the bicycle is now rated as one of the most enduring inventions of all time, but how did it all start?

**What is a bicycle?** The bicycle was the first effective self-propelled wheeled vehicle ever invented yet it only originated about 200 years ago. The wheel first appeared about 5 000 years ago, initially in toys and potter's wheels and, from about 3 500 years ago, in horse-drawn carts with two wheels. Why then did it take so long to invent the bicycle? The reason is that these early carts were stable as the two wheels, side by side, were connected by an axle and formed a stable triangle with the draught animal which allowed them to be used over rough terrain.

The uniqueness of the bicycle is that it has two wheels in line but this idiosyncrasy in design also highlights it biggest flaw - it is inherently unstable at slow speeds and on rough roads. This partly explains why it was invented so recently as early roads were generally sandy or cobbled but there are other reasons, among them the popularity of the horse and horse-drawn carts, the lack of incentive to build an unstable vehicle, conservative dress codes that made it awkward to ride a bike and an inadequate knowledge of materials and mechanics.

**Birth of the bicycle**: The earliest known sketch of a bicycle, reputedly made in 1534 on the back of one of Leonardo da Vinci's drawings, is attributed to Gian Giacomo Caprotti, an apprentice to the master painter. Although German physicist, museum director and authority on the history of the bicycle Professor Hans-Erhard Lessing declared in 1998 that the sketch is a hoax, some historians uphold its validity. I agree with Lessing that it is a fake. A colleague of mine Graeme Murray and I made a full-size replica of this 'bicycle' but it does not work as the front wheel is fixed in line and therefore cannot turn, which makes it unrideable.

In 1649 a compass builder in Nuremberg, Germany, built several crank-and-chain-driven horseless carriages, one of which was bought by the King of Sweden and was a highlight of his coronation. The King of Denmark also bought one which was hand-cranked by two children hidden inside. A drawing from the 1680s depicts a disabled man, Stephan Farfiler, riding a hand-pedaled tricycle to church in Nuremburg, Germany.

The Industrial Revolution from 1760 to 1840, which suddenly equipped the Western world with gears, pulleys, levers, treadles and chains, provided the stimulus, materials and mind-set for the development of the bicycle's mechanics. Machines were developed to tackle every task, and nothing seemed impossible. In 1791 a French nobleman Comte de Sivrac, invented the *celérifère*, a simple wooden beam to which two wheels were affixed inline and which was powered by a rider 'walking' along the ground while seated on the wooden frame. They became popular among the nobility and *célérifère* races were even held down the Champs-Elysées. In 1816 another Frenchman Joseph Niepce, who would later become famous for taking the first photograph in 1826, designed a bicycle comprising an horizontal wooden beam (with a carved animal head on the front end) and two wheels in line, but it was also not steerable.

***Draissiene:*** In 1816 a German Baron Karl von Drais de Sauerbrun improved on this design by developing the hobbyhorse or *draissiene* as an alternate means of transport to the horse to oversee his vast estate. At the time horses were in short supply following the Napoleonic wars and the crop failures and food shortages that followed the 'Year without a Summer' (1816), which resulted in the deaths of thousands of horses. The baron's *draissienne*, which was patented on 12th January 1818, had two wooden wheels in line and a turnable handlebar for steering but no brakes or pedals and was propelled by the rider pushing his feet on the ground, similar to modern kids' 'first bikes'. Graeme Murray and I also made a life-size *draissienne* and found it to be uncomfortable going downhill and impractical going uphill but, over 200 years ago, it did give rise to the first bicycle craze when the Austrian Imperial Military Academy established a *draissiene* riding school in Vienna in 1818. Later fashionable dandies attended bike-riding classes and rode improved, lighter-weight versions, known as dandyhorses, around public parks in London and Paris from the 1820s to the 1850s.

Draissiene *ca* 1816

In 1821 an Englishman Lewis Compertz designed a rack-and-pinion bicycle powered by pushing and pulling with the arms but it gained no traction with the public. Then, in 1839, a Scottish blacksmith Kirkpatrick Macmillan made a novel treadle-powered bicycle which attracted attention when he rode

Kirkpatrick Macmillan treadle-powered bicycle *ca* 1837

over 230 km from Dumfries to Glasgow and back on his contraption. In 1842 he suffered the misfortune of having the first recorded bicycle accident when he knocked over a child in Glasgow and was fined five shillings. Although Macmillan's bike was copied in 1842 by a Frenchman Alexandre Lefèhvre and in 1845 by Gavin Dalzell, a cooper from Lanarkshire, it was not destined to be in the direct line of evolution of the bicycle, although treadle bikes were revived 150 years later in the 1970s.

Velocipede ca 1863

***Velocipede:*** The first commercially successful bicycle, and the beginning of the evolution of the modern bike, was the French *velocipede*, invented in 1863 by Pierre and Ernest Michaux, coach builders in Paris. This bike, also called the boneshaker because of the bumpy ride it gave on metal-rimmed wooden wheels, had cranks fixed directly to the hub of the front wheel, like a modern kid's tricycle. Due to the small size of the front wheel it had a low gear ratio as one turn of the pedals only advanced the bike a distance equal to the circumference of the small front wheel and progress was slow.

The *velocipede* was an instant success at the *Exposition Universelle d'Art et d'Industrie* in Paris in 1867. The Prince Imperial Napoleon III ordered one for himself and also bought twelve smaller versions for his young playmates. Soon there were three- and four-wheeled, tandem, two-abreast and triplet versions of the *velocipede* available and the staff in the Michaux factory increased from two to over 300 in five years. In 1868 James Moore won the first recorded indoor cycle race over 1 200 m on a boneshaker at Parc St Cloud in France and also triumphed in the first road race, from Paris to Rouen, at an average speed of 11.8 km/hr. Interestingly, five of the 200 competitors in this race were women.

The first bicycle to be imported to South Africa (and probably to Africa) was a boneshaker bought by a Mr Griffith of Grahamstown for the astronomer/optician James Carter Galpin in 1869. This elegant machine is still on display in the Observatory Museum (part of the Albany Museum) in Makhanda and has been ridden by the author. In the same year a boneshaker jousting tournament, using lances with blunted tips, was held in London and bicyclists and tricyclists were reported to be prominent if somewhat disruptive during fox hunts in the English countryside! Charles Dickens was a keen cyclist and one of London's most famous cycling clubs, the Pickwick Club founded in 1870, was named after his novel, *The Posthumous Papers of the Pickwick Club* (later *The Pickwick Papers*); each member adopted a name from one of Dickens' novels and had to be addressed by that name at club meetings.

**High-wheeler:** One way to improve the gear ratio of the *velocipede* so as to travel further with one turn of the pedals was to increase the diameter of the front wheel. This lead to the development of the extraordinary high-wheeler or penny-farthing with metal spokes supporting a huge front wheel up to 1.5 m in diameter! The first person to make a commercially successful high-wheeler was James Starley, a self-taught English engineer from Coventry who developed a series of penny-farthings of increasing efficiency and speed in the 1880s. They became so popular, especially with sports cyclists, that they were known as 'ordinaries', i.e., they were the standard bike of the day. High-wheeler races were held regularly from London to Brighton in England and long-distance events became very popular in Europe and North America, sometimes pitting high-wheelers against horses. In 1884 Thomas Stevens of England became the first man to cycle around the world on a penny-farthing, an extraordinary feat. (The first woman to cycle around the world, on a 'safety', was Annie Londonberry of the USA in 1894.)

**High-wheeler *ca* 1880**

In 1885 two members of the Port Elizabeth Bicycle Club, Chas Hallack and Frank Girdlestone, rode (and pushed) their high-wheelers over rough roads and mountain tracks from Port Elizabeth to Cape Town via Knysna, taking 18 days and 12 hours for the epic journey. Outside Cape Town they were joined by a detachment from the Cape Town Bicycle Corps and they all received a hearty welcome on their arrival in Adderley Street. In the early 1890s two of the world's leading competitive cyclists were South Africans, Arthur 'Zimmy' Zimmerman (world champion in 1892) and LS Meintjies (in 1893).

This author's experience with high-wheelers has been mixed. When we mounted an exhibition on 'Cycology: The Science of the Bicycle' at the MTN ScienCentre (now the Cape Town Science Centre) in 2002 I decided that I must learn how to ride the beast so as to understand why it had been so popular. I failed on both counts. I borrowed a high-wheeler from local cycling entrepreneur Chris Willemse and learned how to ride it on the Maritz Brothers fields in Rondebosch. I discovered that riding a high-wheeler is a perilous experience as they are difficult to mount, ride and dismount! If you succeed in mounting and in gaining some momentum it is an exhilarating experience as you rocket along at a frightening speed two metres above the ground on spindly wheels, but this is where the danger lurks. If you hit a bump the tiny back wheel flips up and the rider somersaults over the front wheel to perform what is euphemistically called 'an imperial crowner'. After gaining some confidence high-wheeling around grassy fields I was persuaded to do a film shoot on a tarred road in Cape Town but it ended in disaster! Suffice to say that a bruised and bloodied rider created colourful visuals but subsequently vowed to stick to his 'safety'.

> "This time the Expert took up the position of short-stop, and got a man to shove up behind. We got up a handsome speed, and presently traversed a brick, and I went out over the top of the tiller and landed, head down, on the instructor's back, and saw the machine fluttering in the air between me and the sun. It was well it came down on us, for that broke the fall, and it was not injured.'
>
> Description by Mark Twain of his attempt to ride a high wheeler in *Taming the Bicycle* (1881)

The success of the *velocipede* and high-wheeler led to a period of intense technical and commercial development of bicycle components that resulted not only in the development of the modern 'safety' bicycle but also of the motorbike and motorcar (and, to a degree, the airplane). The most important innovations between 1860 and 1890 included tangential wire spokes, mudguards, differential gears, tubular frames, ball bearings, shaft-driven wheels, hub brakes and chain-and-sprocket drives. These components not only improved the strength and performance of the bicycle but also became essential components of early motorbikes and cars. Another important technological breakthrough took place in 1889 when John Dunlop, a Scottish veterinary surgeon, invented a pneumatic tyre to replace the metal-rimmed wooden wheel and solid rubber tyres that had been used on bicycles until then. The new inflatable tyres provided a more comfortable, efficient and safer ride and were soon adopted by all bicycle makers.

## 5. Evolution of the bicycle

**Safety bicycle:** In 1885 JK Starley, a nephew of James Starley, invented the famous Rover safety bicycle, a revolutionary design that resolved the precariousness issue of the high-wheeler by having two equal-sized wheels with the rider seated on a saddle between them, which gave the bike a lower centre of gravity. The frame

Early safety bike *ca* 1900

comprised a strong triangular design which allowed it to carry more than ten times its own weight, far more than motorised vehicles. Furthermore, the safety bike was propelled, not by pedals acting directly on the hub of the front wheel, but by pedals on a large sprocket (toothed gear wheel) connected to a smaller sprocket on the hub of the rear wheel by a chain, which ultimately became the basic design for almost all modern bikes. By 1886 there were over 70 factories in England making safety bikes. In 1897 Donald Menzies, the Raleigh agent in Cape Town, made the first bicycle in South Africa, a racing safety called the 'Springbuck', the first reference to a *bok* in South African sport (Bruton, 2017).

The success of the Rover kickstarted a boom in cycling in the late 1890s that was experienced throughout the world. The bicycle became firmly established as *the* everyday means of transport, initially for the rich in place of horses but later also for poor folk who 'took to the wheel' when more affordable bikes became available through mass production. Bicycles were also adapted in Asia, Africa and South America for carrying water,

**Early recumbent bicycle**

wood, passengers and foodstuffs and were soon used in many countries by postmen, firefighters, game rangers, soldiers and the police. By 1895 the price of a good horse had reached an all-time low! Pedal-powered machines were also developed for pumping water and air, sharpening knives, grinding corn, mixing fruit drinks, cutting wood, drilling holes, mowing lawns and even turning roundabouts!

**First motorbike and motorcar**: The first functional motorbike was the *Reitwagen* (riding car) developed in Germany by Gottlieb Daimler and Wilhelm Maybach in November 1885. It was essentially a motorized boneshaker with wooden wheels, a wooden frame and a chain drive to the rear wheel hub from an internal combustion engine mounted between the rider's legs. Daimler and Maybach also made one of the first motorcars when on 8th March 1886 they smuggled an American stagecoach into their house, telling neighbours that it was a birthday gift for Mrs Daimler. Maybach supervised the installation of their latest engine, called the 'Grandfather clock engine' as it resembled an upright clock, into the coach and it became the first four-wheeled vehicle to reach 16 km/hr. In 1886 Daimler and Maybach attached their engine to a boat for the first time and in 1890 founded the Daimler-Motoren-Gesellschaft, selling their first car in 1892. After Daimler's death in 1900 the Daimler company and Benz & Cie merged to form Daimler-Benz which lives on today as the premier car company, Mercedes Benz.

*Reitwagen*, the first motorbike
**(1885)**

Independently of Daimler, and just 80 km away in Germany, Karl Benz also produced a *Motorwagen* with his own engine which he patented on 29th January 1886, about 40 days before Daimler and Maybach had assembled their first *Reitwagen*. Benz's car was even more like a motorized bicycle, or rather tricycle, as it had spoked, reinforced bicycle wheels, sprockets and chain and a tiller lever that acted as a steering wheel, brake and clutch. This author has driven a modern replica of this pioneering vehicle which belongs to Daimler-Benz in East London; it does not go very fast but it does provide a smooth and effortless ride. The first 'horseless carriage' in South Africa was a Benz Velo imported by a Pretoria businessman John Percy Hess in 1896. Many early carmakers such as Hillman, Humber, Oldsmobile, Peugeot, Opel and Morris started out as bicycle makers and Henry Ford's first car used bicycle wheels and chains. James Starley, known in England as the 'Father of the Bicycle', established the famous Rover car company that produced an electrically-driven tricycle in 1888, a motorcycle in 1902 and an automobile in 1904, all with some bicycle components. The popularity of safety bicycles also had broader benefits as cyclists campaigned for roads to be improved. The advent of the bicycle literally paved the way for the automobile.

**First airplane**: The bicycle also played a role in the development of the first airplane. Orville and Wilbur Wright, who made the first controlled powered flight at Kitty Hawk, North Carolina, in December 1903, opened a bicycle repair and sales shop, the Wright Cycle Exchange, in 1896 in Dayton, Ohio (it still exists). There they designed, made and sold exquisitely

*Wright Flyer*, the first powered airplane (1903)

engineered 'Van Cleve' and 'St Claire' touring and commuting bikes and used the profits to fund their experiments on airplane design, inspired by the achievements of the pioneer gliders, Sir George Cayley of England and Otto Lilienthal of Germany. Their work was influenced by the belief that an unstable vehicle such as a bicycle or airplane could be controlled and balanced with practice. They tested aerofoils and wing designs by fitting them to a bicycle wheel attached horizontally to the handlebars of a bicycle which was then ridden at speed. They also incorporated lightweight design features from bicycles into their Wright Flyers. Seventy-six years later, in 1979, a bicycle-powered aircraft, the Gossamer Albatross, ridden by Bryan Allen became the first human-propelled machine to fly across the English Channel.

**Efficiency of the bicycle**: The bicycle makes it possible for an individual to travel using the human body as the 'engine'. It is designed to use the most powerful muscles in the body (the thigh muscles, as well as the gluteus muscles in track racing) in an efficient rotary action of the feet while supporting the weight of the body. The cyclist saves energy (relative to a walker) by sitting and does not need to lift his/her rising leg as it is raised by the downward thrust of the other leg. The back muscles are used to support the trunk and the arms holding the handlebars form an efficient triangle of support. Other factors that contribute to the efficiency of the bicycle include minimal friction resistance between the sprockets and the chain and the ability to free wheel downhill without pedalling.

Research has revealed that the bicycle is the most energy-efficient method of transport ever invented or evolved! Scientists have estimated that the rate of energy consumption of a cyclist is about 0.15 calories per gram per kilometre which is five times more efficient than a walking man (0.75 calories) and more efficient even than a swimming salmon (0.4), a galloping horse (0.6), or a flying pigeon (0.9)! A single person driving a medium-sized motorcar is 55 times less efficient than a cyclist! As opposed to a car, 80% of which is unoccupied if there are no passengers, every component of the design of the bicycle is on a human scale as it must relate to the dimensions of its 'engine'. As the 19th century author Louis Baudry de Saunier once said, "The cyclist is a man half made of flesh and half of steel that only our century of science and iron could have spawned."

Other than gravity, wind resistance is the main source of energy loss on a bicycle, using up to 30% of the cyclist's energy. Competitive cyclists wear smooth, streamlined clothing, adopt a crouched riding position and ride bikes that are light and efficiently geared to optimize their performance. The only other significant form of energy loss - rolling resistance of the tyres on the road - is minimized by paved roads and hard, thin wheels.

**The bicycle at war**: Although the bicycle is a benevolent machine it has been used extensively in war as it places little demand on material resources, is silent and provides high mobility with low visibility. Also, unlike horses, they do not need to be fed or watered and do not kick or bite! During the Anglo-Boer War the Boers, who were born into the saddle, initially considered it absurd to use bicycles instead of horses but soon became convinced of their efficacy for dispatch riding and spying. Shortly before the outbreak of this war a champion cyclist Koos Jooste met with Commandant-General Piet Joubert and President Paul Kruger of the Transvaal Republic and convinced them to form the Wielrijders Rapportgangers Corps, which was subsequently led by Danie Theron. They became so effective that in March 1900 Lord Milner labelled Theron as 'the chief thorn in the side of the British' and wanted him dead or alive.

The British also used cyclists during the Anglo-Boer War with, at one stage, three percent of British forces using bicycles. The City of London Imperial Volunteers and two battalions of the Royal Dublin Fusiliers were trained in cycling at Aldershot before they were sent to South Africa. Locally the Cape Cycle Corps was formed in December 1900 with an eventual strength of 500 men led by the indomitable Captain John ('Jack') Rose, a competitive cyclist who held the world one-hour amateur record. Within a month of their formation they were responsible for holding off a Boer attack on Pickaneer's Kloof and they also did excellent work during the Siege of Kimberley.

Fold-up bikes were deployed as well as six-seater, rail-mounted 'war cycles' designed by Donald Menzies and Jack Rose for travelling on railway lines, based on an idea of the Royal Australian Cycle Corps. The war cycles were used for reconnaissance, carrying dispatches, checking railway lines for explosive charges and evacuating the wounded. A faster two-seater war cycle, weighing only 27 kg and capable of 48 km/hr, was later developed by Menzies with a removable steel rim that could be replaced by balloon tyres for normal riding. Examples of these war cycles can be seen in the collections of the Fort Klapperkop Museum near Pretoria and the Iziko South African Museum in Cape Town.

During the Anglo-Boer War cyclists were also used to transport carrier pigeons as a bike trip was less stressful to the birds than a horseback ride. On one occasion Scout Callister of the Cape Cycle Corps achieved fame by riding over 190 km and releasing pigeons from vantage points every time he saw Boer activity. In another interesting innovation kites were released by cyclists for taking photographs of Boer encampments using remote-controlled cameras and, later, for raising aerials for experiments on radio telegraphy. Between the World Wars bicycles again became popular as the industrialized countries suffered from fuel and metal shortages and economic depression. During World War II motorised vehicles took precedence although fold-up bikes and pedal-powered radio transceivers and battery chargers were dropped behind enemy lines with paratroopers.

**Emancipation of women**: Safety bicycles had social benefits for men and women as they freed them from long walks, endless queues and the restrictions of steam train schedules but, or women, the transformation was even more dramatic. The bicycle freed them from household chores and subordination by males and led to them abandoning their ankle-length Victorian dresses and revealing to the world that they also have two legs at a time when glimpsing an ankle was considered outrageous. Bloomers, previously only acceptable in stage musicals such as *The Follies*, were renamed 'rationals' and knickerbockers, culottes and split skirts became the fashion statement of the day. Tight corsets, large hats, prudery and chaperones made way for trousers (which transmogrified from 'unmentionables' to 'slacks'), socks and hair that was set free allowing women cyclists, who became known as 'scorchers' in England, to assert their independence and free will for the first time.

> "I think it [bicycling] has done more to emancipate women than any one thing in the world. I rejoice every time I see a woman ride by on a bike. It gives her a feeling of self-reliance and independence the moment she takes her seat; and away she goes, the picture of untrammelled womanhood."
>
> Susan B. Anthony (1896),
> American social reformer and women's rights activist

An article in the newsletter of the City Cycling and Athletics Club in London published in the early 1900s captured the spirit of the time. 'The Parisian wheel women are going ahead. Long, long ago they discarded skirts for bloomers, and now they have discarded stockings, and ride in socks, which they wear with low shoes. This costume necessitates the display of several inches of bare leg, but the fair creatures do not mind this in the least.' The advent of the bicycle also led to more socializing across class lines and between towns and villages. Women were no longer obliged to marry a local and there is even some scientific evidence that the human gene pool diversified as a result of reduced in-breeding!

**Fluctuating popularity of the bicycle**: In the early 1900s the bicycle craze in Europe and North America suddenly died out due to the invention of the motorbike and the car but bicycles flourished in the developing world as utility vehicles. Simultaneously groups of amateur and professional racers continued to test and improve the bike in sprint and endurance events. In 1903 the *Tour de France* was launched by Henri Desgrange, a journalist working for *L'Auto* newspaper, whose understated description of the first event ('a great cycle race over an interesting route') belied the severity of the brutal 2 400 km, six-stage contest. A women-only Tour de France (*Grande Boucle Féminine Internationale*) was launched in 1984 but was discontinued in 2009.

The tradition of the yellow jersey (*maillot jaune*) worn by the race leader was started in 1919, with yellow being chosen as the sponsoring newspaper (*L'Auto*) was printed on yellow paper. Drug taking was common in the early days of the *Grande Boucle* (big loop) so urine testing was initiated in the 1930s.

On one notorious occasion a male rider who had surreptitiously 'borrowed' a urine sample from a friend was told after the test that he was pregnant! Cyclists from North African countries such as Algeria, Tunisia and Morocco, who trained in the Atlas Mountains, competed in the *Tour de France* from 1913 onwards and were later joined by competitors from Eritrea, Rwanda and Ethiopia, with the first Africa-born cyclist to win a stage in a *Grande Tour* being the Algerian Marcel Molinès in 1938.

Maurice Gavin, winner of the first *Tour de France* in 1903

The first bicycle race for men in South

**New vs old: Unicycle and high-wheeler**

Africa was held in 1881 in Port Elizabeth and the first for 'emancipated women' in Cape Town in 1897. In 1898 the City Club of Cape Town opened a cement track at Green Point where many famous races were held but it was turned into a prisoner of war camp in 1900. From the 1880s onwards a series of highly competitive races was held on a banked track at the Wanderers Club in Johannesburg, with the 25-mile race held in December 1891 being the last to pit high-wheeled 'ordinaries' against new-fangled 'safeties' (a high-wheeler won). In January 1899, before bikes with gears were available, Koos Jooste and Andries de Wet rode on rough, unpaved roads 1 467 km from Pretoria to Cape Town (which they called 'Snoekopolis') in eight days, 22 hours and 49 minutes, averaging about 220 km a day (they did not ride on the Sunday).

**Modern bikes**: A new bicycle boom started in the 1960s when the 'ten-speed' was introduced but the first major innovation in bicycle frame design for 70 years was made in 1962 by the British inventor Alexander Moulton who created a cross-framed, mini-wheeled bike with suspension. Over the next few decades bikes became ever lighter and stronger with the use of space-age materials such as aluminum alloys, carbon fibre, kevlar and titanium and underwent an explosive radiation to cater for the needs of competitive and recreational cyclists. Their designs included Dursley-Pedersens (with a hammock saddle), wooden, bamboo, cardboard and plastic bikes, BMXs, stunt bikes, mountain bikes, downhill racers, krates (low-framed bikes with

**Modern racing bike**

saddles), choppers, unicycles, freight, taxi and rickshaw bikes, prones and recumbents, hand-cranked and treadle bikes, faired bikes with streamlined shells, cruisers, track, triathlete, time trial and cyclocross bikes, various road racers and bikes for picnicking, trekking, touring and camping. In 1998 my son Ryan and I enjoyed a relaxing jaunt observing fishes, dolphins and turtles off Key Biscayne in Florida while riding hydrocycles whose pedals turn small propellors. There are also bikes for multiple riders such as the conference bike (seven passengers), busycle (15) and party bike (17) and even bikes that fly using horizontal helicopter-type blades that spin when the rider pedals. Most recently human-powered ecocars, twikes (human-electric hybrid vehicles designed to carry two passengers and cargo), solar- and fuel cell-powered bikes and hybrid electric bikes are pushing technology to the limit.

Cut-away of faired recumbent racing bicycle

**International cycling boom:** The current international boom in cycling is being driven partly by new events such as cyclocross, downhill racing and trick cycling but also by television coverage of the multi-stage *grande tours* in Europe as well as in South Africa (Cape Epic), Morocco, Zambia, Tanzania, Kenya and Australia. The scenic single-stage Cape Town Cycle Tour, started in 1978, is the largest cycling event in which every finisher receives a time and a medal and was the first event outside Europe to be included in the *Union Cycliste Internationale*'s Golden Bike Series. Most of the 35 000+ participants ride racing safety bikes but some have completed the route on monocycles, high-wheelers and recumbents. My 17 Cape Town Cycle Tours have been ridden on a *dikwiel katonk* (inefficient thick-wheeled contraption) mountain bike and on road racers. During the Covid-19 pandemic in 2020 the live race was cancelled but it was replaced in March 2021 by the Virtual Cape Town Cycle Tour which used the FulGaz Virtual Cycling platform in a race that was divided into three stages to make up the full distance of 109 km. The real race will be restarted on 10th October 2021 without mass starts. The ultimate cycling challenge in Africa is the annual *Tour d'Afrique*, a 11 000-km adventure ride from Cairo to Cape Town that takes about four months to complete. In 2021, due to the pandemic, a 'shortened' version will be held from Arusha in Kenya to Cape Town (6 200 km in 45 riding days).

World production of bicycles exceeded 7 million in 1939, 36 million in 1971, 100 million in 1995 and 132 million in 2019. China is estimated to have about 450 million of the one billion plus bicycles in the world and bike production now outnumbers car manufacture by three to one. Furthermore, the value of the global market for bicycles is estimated to have reached at US$29.2 billion in 2020. The top five producers of bicycles (China, India, European Union, Taiwan and Japan) are responsible for 87% of global production and in Asia alone bikes transport more people than cars worldwide. The most popular bicycle in the world is the Chinese government-made 'Flying Pigeon' with over 500 million made since 1950; sadly, many are discarded and clutter up city dump sites. In Holland I once met an enterprising aquaculturalist who retrieved bicycles that had been abandoned in Amsterdam's canals and suspended them in midwater where they provided a surface for oyster spat to grow to maturity.

**Economic and environmental benefits of the bicycle**: Bicycles offer many economic benefits – they are inexpensive to buy and repair, require minimal infrastructure, parking and road space, reduce traffic congestion, urban noise and chemical pollution, regenerate city centres, stimulate consumer spending, benefit small-scale retailers and are a cost-effective investment. The use of the bicycle is also self-limiting as it allows people to create a new relationship between their life space and their life cycle, between the territory that they traverse and the pace of their living. Bicycles also promote socialisation in an increasingly insular world.

The environmental benefits of the bicycle are equally impressive. They consume far fewer resources (to make and use) than cars, reduce your carbon footprint and promote healthy outdoor lifestyles and enlightened environmental attitudes. To many, this author included, cycling is more than a mode of transport, it is a mind-set based on economy of motion and kindness to the environment. Riding a bicycle does not require the combustion of imported petroleum products that pollute the environment and further enslave us to an unsustainable lifestyle. Increasingly they are being regarded as indicators of how environmentally conscious a community has become.

**Future of the bicycle**: Bicycles are the icon of the global environmental movement and there is every indication that they, unlike most inventions which have a sell-by date, will continue to serve humankind long into the future. They are a classic example of the innovation value chain whereby the spark of an initial idea, catalysed by the availability of new materials and technologies over time and stimulated by an entrepreneurial mindset, creates a continuously improving product that serves an ever-wider set of needs. The bicycle is an unparalleled merger of a toy, utility vehicle and sporting machine yet, while the motorcar is evolving into a bafflingly complex computer-on-wheels, bicycles remain true to their spirit as one of the simplest yet most influential conveyances ever invented. As Frances E Willard stated in an 1895 book, *How I learned to ride the Bicycle*, "I found high moral uses in the bicycle and can commend it as a teacher without pulpit or creed. She who succeeds in gaining the mastery of the bicycle will gain the mastery of life."

At the age of 74 years my relationship with the bicycle will continue whether it is on short shopping trips in Rondebosch, cruising the spectacular coastal roads from Camps Bay to Scarborough, trekking along sandy tracks in the Cape Fold Mountains or cycling in and around foreign capitals once normal air travel resumes. Most recently I have offered to assist Pieter Silberbauer, founder of the first 'bicycle hotel' in Africa, Trail's End in Grabouw, to develop an interactive museum on the history of the bicycle. Watch this space!

As Albert Einstein, a dedicated cyclist, once said, "Life is like riding a bicycle. To keep your balance, you must keep moving". Bikes are unquestionably one of humankind's noblest inventions, up there with the wheel, printing press, radio, television and computer in terms of the impact they have had on our quality of life. They have been a vital part of our past and will unquestionably be a crucial contributor to our future. Not many inventions made over 200 years ago can make that claim.

# 6:

# Who is South Africa's greatest inventor?

*"All creative people want to do the unexpected"*

Hedy Lamarr, Austrian actress and inventor

**INTRODUCTION:** South Africans punch well above their weight in the innovation arena. The many contenders for the honour of being South Africa's greatest inventor are documented in detail in my books *Great South African Inventions* (2010) and *What a Great Idea! Awesome South African Inventions* (2017). A select few are discussed here.

Many traditional knowledge practitioners, women and men, developed novel healthcare, grooming, food and drink products from indigenous plants as well as hunting and fishing gear but sadly they remain anonymous and cannot be acknowledged individually. A Russian immigrant Benjamin Ginsberg first developed the commercial potential of rooibos tea and a Belgian immigrant Ferdinand Chauvier invented the first pump-powered pool cleaner, the Kreepy Krawly. Glynn Jones, an engineer from Johannesburg, developed the super-efficient Jetmaster and Hex fireplaces and the Jones Rotating Cowl, Herbert Sheffel contrived an improved chassis for railway trucks, the Sheffel Bogey, and Lee Dickman conceived the Colindictor, the first automatic telephone answering machine. The first production cars to be developed in South Africa were the brainchild of John Myers (Protea) and Bob van Niekerk and Willie Meissner (Flamingo and Dart), and the first South African electric car, the Joule, was developed by Kobus Meiring and his team.

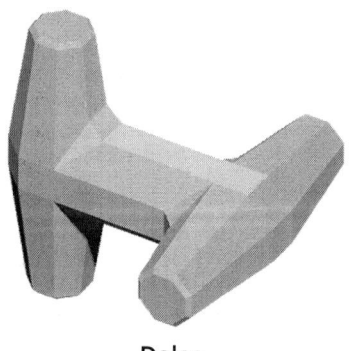

Dolos

Aubrey Krüger and Eric Merrifield invented the iconic Dolos harbour protection device and Jan Policansky

## 6. South Africa's greatest inventor?

developed world class Policansky lever-drag fishing reels. Attie and Uys Jonker produced the highly competitive JS-1 Revelation Glider and Johan Loots created a range of rugged sit-in and sit-on-top Paddleyak kayaks. Mike Beachy Head unveiled the world's first 300 hp V6 diesel outboard engine and the revolutionary Axis Drive Marine Propulsion System and Peter Wallenda first developed parafoil kites. Surajudin Latief invented an innovative Muslim prayer chair, Oliver McLeod-Smith co-invented the Snakeboard, Graeme Murray created an astonishing variety of bicycle innovations and Paul Sinton-Hewitt first suggested the concept of Park Runs. Lucky Netshidzati stunned the world by developing a unique 'Talking Glove' that converts sign language into spoken words. Herman Mashaba has developed a range of hair and skincare products for black people, Viness Pillay and his team at the University of the Witwatersrand have created a range of novel drug delivery systems that have revolutionised the ways in which we take our medications and Professor Kelly Chibale and his team at UCT developed the first one-pill malaria treatment. Graeme Wells created unique Afri-Can guitars and violins and Kit Vaughan at CapeRay has invented the advanced Aceso breast imaging system. On the information technology front Herman Heunis developed MXit, Jacques Blom iStyla, Chris Pinkham Elastic Compute Cloud (EC2), Vinny Langham Clicks2Customers and Adii Pienaar Woocommerce, Woothemes and Receiptful, among many others. To cap it all, Alfred 'Lux' Baloyi invented the *makarapa* sculptured plastic helmet and Freddie 'Saddam' Maakie the much-reviled *vuvuzela*!

**Women inventors**: Did you know that the first 'computers' in South Africa were not machines but women? Around 1910 the Royal Astronomical Observatory in Cape Town employed women calculators called 'computers' whose role was to capture and analyse data collected by the astronomers.

There have been many accomplished women inventors and innovators in South Africa. One of the best-known is Amelia Ball who first concocted Mrs Ball's Chutney, now on sale worldwide, in the 1950s in her kitchen in Fishhoek. Mona Lisa Bentley of Johannesburg famously invented the Bentley Belt which has helped thousands of children to learn to swim and, in 1994, Joan Brown, an ergonomics student, and engineer Ian Brown developed the Keyless Safety Lock to prevent young children from accessing household poisons and medicines. In 2000 Valerie Buhlmann-Strydom of Cape Town invented Easy Turn, a simple but effective device that allows caregivers to turn over heavy patients in hospital beds, which has doubtless been of great benefit during the Covid-19 pandemic. Previously, in 1988, she developed Herb Hair and Herb Coat to promote hair growth in humans and pets using natural herbs.

In 2001 Benedictine Roumega, a single mother from Cape Town, invented Wham Jam, a simple but effective device that improves home security and Debbie Neube developed flavoured Eden All-Natural Peanut Butter. Two innovative women beer brewers have made their mark: Ndumiso Madladla, a chemical engineer who founded the MadMead Brewing Company in Soweto (South Africa's first black-owned microbrewery) where Soweto Gold has been a big hit and Apiwe Nxusani who runs Brewdogs in Johannesburg where she produces the best-selling Vagabond Pale Ale and Punk India Pale. In 2009 Gladys Mawoneke developed an alcohol-free malt beverage Breva that caters for the emerging middle-class market that does not drink alcohol.

A recent invention that has taken off is Reel Gardening, an award-winning innovation that simplifies gardening and makes it fun and effective. The idea was developed by Claire Reid of Johannesburg in 2010 based on a gold medal-winning project at the Expo for Young Scientists. It comprises reels of biodegradable paper strips that are pre-packaged with the seeds of flowering plants, herbs or spices and organic fertilizer, correctly spaced to optimize yield. A similar product, the Super Sprouts Germination Slab developed by Mariette van den Heever of Cape Town is a self-contained mini-garden packed with 'ancient seeds' encased in rich organic soil. Low-tech horticultural innovations made by rural women farmers include the use of crushed ground beetles to control CMR beetles by Rosa Libago from Thohoyandou and of extracts from *Lippia japonica* (also used to make mosquito repellent) by Mutshekwa Rosinah from Limpopo to control insect pests that eat her spinach and tomato plants.

Mashesha stove

In 2015 entrepreneur Louise Williamson from Nelspruit developed the innovative Mashesha Stove for use in rural homes and schools. Her stoves are energy efficient and environmentally friendly as they burn cleaner, use 50% less wood than open fires and produce virtually no smoke. Another useful energy-saving device, Hot Spot, was developed by Sandiswa Qayi from the Eastern Cape. Hot Spot saves electricity by heating water more efficiently as it pushes water from the bottom of the geyser to the top using a thermosiphon. It can be retrofitted over any standard water geyser element.

Kiara Nirghin from Johannesburg is making waves with her invention of a super-absorbent polymer comprising orange peels and avocado oil that stores water in the soil during droughts. Regina Kgatle of Hammerskraal has set her mind on decolonializing computer games and making them relevant to the African experience. She has developed online games with local content that address topics that are relevant to the youth such as bullying, gangsterism and drug abuse. Social activist Anne Githuku-Shongwe uses computer games and mobile technology to help young people re-imagine a better life in Africa. She launched Afroes (African Heroes and Heroines), a digital technology start-up that develops mobile games that encourage African children to look to their own continent for heroes and role models, while also challenging social myths and stereotypes and combating discrimination. In 2013 Anina Mumm and Engela Duvenage launched SciBraai a blog that encourages users to share stories about science and technology and go behind the scenes to find out what makes scientists and inventors tick.

Sue Swain of Knysna is leading the way on the environmental front by pioneering the integration of biomimicry into architecture, engineering and sustainable living and working practices in South Africa. Biomimicry involves the design and production of materials, structures and systems that are modelled on those developed by plants and animals over 3.8 billion years of biological evolution. Amiene van der Merwe started Green Cab, a company owned by women that uses eco-friendly vehicles powered by electricity, low-emission liquid petroleum gas or diesel. Several women have launched women-only hail-and-ride personal transport services in South Africa including Danielle Wright (Chaufher), Joanne Fredericks (Ladies Own Transport) and Thembisile Naomi Mbambo (Pink Cabs). A significant modern innovation, the South African Sustainable Seafood Initiative (SASSI), was developed by Kerry Sink in 2002 in collaboration with WWF-SA. The SASSI app makes it possible

Biowise emblem

for wholesalers, restaurateurs and seafood lovers to make informed choices about which fish and shellfish they buy, serve and eat.

**Discoverers and inventors**: Many South African scientists are both discoverers and inventors. Arnold Theiler was born in Switzerland in 1867 and qualified as a veterinary surgeon in Zurich before emigrating to South Africa in 1891. He initially worked as a farm hand near Pretoria, but tragedy struck three months after his arrival – his left hand was accidently severed while operating a steam-driven chaff cutter. He miraculously survived and, despite the disability, went on to revolutionise veterinary science in South Africa, carrying out thousands of dissections and other delicate procedures with one hand! Through painstaking research he developed methods for the control of rinderpest, horse sickness, biliary fever, gall sickness, wireworm, east coast fever, *lamsiekte*, blue tongue and other farm stock diseases. He was known as 'The Father of Veterinary Science' and became South Africa's Pasteur, Lister, Koch and Jenner rolled into one! Theiler also established the Onderstepoort Veterinary Research Institute and a world-class veterinary faculty at the University of Pretoria. His brilliant son Max followed in his footsteps and won the Nobel Prize in Physiology or Medicine in 1951 for his development of a vaccine for yellow fever (see Nobel laureates chapter).

In the modern era Professor Tebello Nyokong is another example of versatility and enterprise in scientific research and innovation. She was born in the highlands of Lesotho and had to alternate with her brother as a shepherd and school goer. After completing her university studies in England she joined the staff of Rhodes University in 1992 where she is now a distinguished professor. Nyokong has developed many innovative techniques in nanotechnology and photo-dynamic therapy with the latter paving the way for safer cancer detection and treatment without the debilitating side effects of chemotherapy. In 2007 she was rated as one of the top three publishing scientists in South Africa and was named by *IT News Africa* as one of the top 10 most influential women in science and technology in Africa.

Many inventions made in South Africa cannot be attributed to one individual but to teams from research institutes, astronomical observatories, research councils, research foundations, universities, technology hubs, corporations and/or commercial companies especially in the computing, medical, mining, engineering, space and aeronautical industries. They are not discussed here.

# 6. South Africa's greatest inventor

**Foreign-based South Africa-born inventors:** Some South Africa-born inventors have developed almost all their inventions while working abroad, including James Greathead, Henri Johnson and Elon Musk.

James Henry Greathead (1844-1896) is commemorated in the City of London by a towering bronze statue that was unveiled by the Lord Mayor in 1994, yet he is hardly known in South Africa. He was born in Makhanda, the grandson of an 1820 British Settler and educated at St Andrews College, Grahamstown, and Diocesan College, Cape Town.

**James Greathead**

He emigrated to England in 1859 where he became involved in new traction and eventually electric underground railways. At the age of 24 years he successfully tendered for a contract to construct shafts and tunnels for the new London Underground system. He began construction of the Tower Subway in 1869 which, when it opened in 1870, was the first underground tube railway in the world.

Greathead was later the resident engineer for the City & South London Railway, the world's first underground electric railway, and the joint engineer for the Liverpool Overhead Railway, the first overhead electric railway. During this time he worked with many famous British engineers including Sir Benjamin Baker who admired his work and called him 'the practical author of the great London Tube Railway'. His most famous inventions were the Greathead Shield for Underground Tunneling (1869) and the

**Greathead Shield for Underground Tunneling (1869)**

Greathead Grouting Machine (1891) both of which were extensively used on the London, Liverpool and other underground railways in England. The Greathead Shield used a novel cylindrical tunneling shield that not only dug the tunnel and removed the excavated rock but also installed a permanent lining of cast-iron rings to form the tunnel wall. Greathead died of cancer in London at the age of 52 years.

Henri Johnson is an engineer who has used his background in sonar and radar to develop technology including SpeedGun, EDH SpeedBall, RaquetRadar and FlightScope to measure the speed and trajectory of cricket, tennis and golf balls. He initially established his company, the Electronic Development House in Stellenbosch but relocated to Orlando, Florida, USA, in 2008.

Elon Musk

Elon Musk, founder or co-founder of Tesla Motors, SpaceX, OpenAI, Zip2, X.com and SolarCity, has been favourably compared with the Steve Jobs (Apple), Jeff Bezos (Amazon), Bill Gates (Microsoft), Larry Page (Alphabet and Google) and Bob Iger (Disney) in terms of his impact as an innovator. His vision is no less than to change the world, reduce global warming through sustainable energy production and consumption and reduce the risk of human extinction by setting up a colony on Mars! Musk was born in Pretoria in 1971 and educated at Waterkloof House Preparatory School and Pretoria Boys High School. At the age of ten he developed an interest in computing and taught himself computer programming, and at 12 he developed and sold the code for a BASIC-based video game (Blaster).

He emigrated to Canada in 1989 and then to California in 1995 where he enrolled at Stanford University. After just two days he decided to benefit from the internet boom and left Stanford to pursue a business career. He co-founded the web software company Zip2 with his brother Kimbal and sold it four years later to Compaq for US$307 million, which set him on his course. In the USA Musk has established the massive infrastructure and pooled expertise that is necessary for the development of projects as ambitious as Falcon rockets, Dragon spacecraft, liquid fuel- and methane gas-propelled rocket engines, a range of electric cars, vertical take-off and landing supersonic jet aircraft, the Hyperloop high-speed transportation system, solar-powered roof tiles and new-age batteries and battery chargers. He could not have achieved these goals in South Africa at the time and, because he made all his significant inventions in the USA, he is not considered here as a South African inventor.

In January 2021 Elon Musk overtook Jeff Bezos as the richest man in the world (but was deposed again in late February 2021) with a personal worth of US$ 185 billion. Furthermore, one of his companies, Tesla Motors, is now worth US$800 billion and is edging ever closer to the magic US$1 trillion market capitalization mark where it will join an elite clique that includes Alphabet (umbrella of Google), Apple, Microsoft, Amazon and PetroChina. Interestingly, the richest person in the United Kingdom today is the Brexit-backing entrepreneur and inventor Sir James Dyson whose worth exceeds £16.2 billion. Who says that inventing doesn't pay?

**And the finalists are ...:** Taking my research on South African inventions and inventors into account, I nominate the following fifteen individuals (listed in chronological order by birth) as South Africa's greatest inventors on the basis of the multiple impacts and significant downstream repercussions of their inventions (as opposed to their discoveries):

**John George 'Jack' Rose (1876-1973):** Jack Rose (who is also discussed in the chapter on the history of the bicycle) managed to find common ground between his passions for cycling, inventing and military strategising in a remarkable career that spanned three wars and seven decades. Rose joined the Cape Cyclist Corps during the Anglo-Boer War and, with another cycling champion Donald Mentzies, developed a Rail-Mounted Bicycle Reconnaissance Vehicle using pairs of bicycles with flanged wheels riding on railway lines. These 'war cycles' were so successful that over 50 were built and they were extensively used for railway track inspections, espionage and

Jack Rose

dispatch-riding, especially in the Transvaal bushveld. For his personal use Rose mounted a small, air-cooled Ariel car onto one of the war cycles and used it for rapid transport throughout the military operations area. This motorized rail bike created history as it was the first vehicle powered by an internal combustion engine to operate on a military warfront.

War cycle co-invented by Jack Rose

Later, during World War I, Rose together with the South African Pioneers (made up largely of experienced mining and railway engineers) developed a unique Rail-Mounted Motor Tractor with a Reo truck engine and flanged railway bogey wheels as well as small troop-carrying trains powered by Model T Ford engines that ran on narrow gauge railway tracks. During the World War II Rose, then in his 60s, distinguished himself further by coordinating a massive campaign to mobilize the Allied forces in East and North Africa. When he died at the age of 97 years he was celebrated as one of South Africa's most decorated sportsmen, inventors, engineers and military heroes.

**Dr Hendrik Johannes van der Bijl (1887-1948):** In 1913 the South African-born inventor Dr Hendrik van der Bijl, working in the USA, invented an improved thermionic valve that was used for the first transcontinental telephone link in the world and also developed a process for sending photographs by wire. In 1920 he returned to South Africa at the invitation of Prime Minister Jan Smuts to act as his scientific advisor. Within two years he proposed a master plan for the creation of a public utility company that would generate and supply electricity and provide cheap power to industry. Although his idea was ridiculed as 'utopian' and 'unachievable' at the time he succeeded in establishing Eskom in record time. He was not satisfied with this achievement as he believed that the twin pillars on which South

Hendrik van der Bijl

Africa's industrial future should be built are cheap power and locally made steel and proposed the development of a state-aided steel works. This time the experts opined that he was attacking private enterprise and would bankrupt the state but, against all odds, he created a steel-making industry through Iscor that was producing the cheapest steel in the world within five years.

**Sir Basil Ferdinand Jamieson Schonland (1896-1972):** Basil Schonland was schooled at St Andrews College in Makhanda where he was described as a 'rocket-propelled, guided, educational missile' and conducted difficult private physics experiments that had everyone else baffled. In 1911, at the age of 14 years, he passed first in matric in South Africa with distinction in every subject. At Rhodes University College he achieved first class honours in mathematics in 1915 and then, still only 18 years old, entered Caius College at Cambridge University. During World War I he served in the signals section of the Royal Engineers and was soon at the front in France repairing broken field telephones. By the end of the war he had been promoted to the rank of major.

Basil Schonland

Schonland then joined the famed Cavendish Laboratory under Sir (later Baron) Ernest Rutherford where he was a junior member of the team that created the first manmade nuclear reaction. He then accepted a post in the Physics Department at the University of Cape Town where he spent 14 productive years. In 1938 Schonland was appointed director of the Bernard Price Institute of Geophysical Research at Wits University and, at the outbreak of World War II, he was asked to lead a team tasked with developing long-range radar equipment. He attended secret meetings on board a ship off South Africa with Dr EM Marsden, an eminent New Zealand physicist, to learn about progress with radar and then with the backing of a powerful team set about building experimental radar equipment. This top-secret project culminated in the remarkably successful JB1 Radar Transmitter which started life as a brilliant improvisation built from second-hand parts but eventually developed into a formidable 'secret weapon' that saw service in East and North Africa and sometimes outperformed the more elaborate but bulky British radar.

By 1941 Schonland was on loan to the British Army and in 1944 he was appointed as the science and technology advisor to Field-Marshal Lord Montgomery. In 1945 Schonland returned to South Africa at the insistence of the Prime Minister General Jan Smuts to establish the CSIR (of which he became the first president) while also resuming his post as director of the Bernard Price Institute. In 1951 he became the first Chancellor of Rhodes University, retaining this position until 1962, and in 1954 he was appointed as the deputy director and later director of the Atomic Energy Research Establishment at Harwell in England.

**Irvine Bell (1920-2011):** Irvine Bell was fascinated from childhood by tools and machines. After completing his trade apprenticeship as a fitter and turner on a northern KwaZulu-Natal colliery he served for five years in the Army Corps of Engineers during World War II. After the war he settled in Zululand and built a boring machine for extracting water using an old Willy's Jeep engine. He and his wife Eunice then started a farm machinery repair service IA Bell & Company near Empangeni and were soon developing new vehicles to serve the needs of the sugarcane, roadbuilding and forestry industries.

In the early 1960s Bell designed the world's first hydraulic Three-Wheeled Cane Loader, followed by a Three-Wheeled Timber Loader. Later he entered the highly competitive four-wheel-drive dump truck market with his 25A Arctic truck (1984). His flagship Bell B40 articulated dump truck, developed in 1989 with his son Peter has proved to be one of the most durable and fuel-efficient vehicles in its class in the world. Bell also produced the world's largest articulated dump truck, the Bell B50D, in 2002. The Bell company has continued to be developed by Irvine's sons Peter, Gary and Paul and a highly competent staff of over 3 500 people, and now produces a range of articulated dump trucks as well as haulage tractors, forklifts, crawler bulldozers, refuse compactors, road graders and rollers, front-end loaders, tracked hydraulic excavators and motorized conveyors in its factory in Richards Bay.

Bell Trucks has also formed partnerships with some of the world's leading truck and tractor companies including John Deere, Hitachi, Kato and Liebherr and is now a global player in the off-highway bulk cartage, mining, forestry, agriculture and road-building industries. Irvine Bell received the President's Order for Meritorious Service (1987) for 'rendering exceptionally meritorious service in the general public interest' and a Lifetime Achiever Award at the Technology Top 100 Awards in 2009.

## 6. South Africa's greatest inventor

Trevor Wadley with his prototype Radio Tellurometer in 1955

**Trevor Lloyd Wadley (1920-1981):** Trevor Wadley obtained his electrical engineering qualifications from Howard College (part of the University of KwaZulu-Natal) and the University of the Witwatersrand. In December 1940 he joined the army and served during the World War II in the Signal Corps, where he made significant contributions to the design and improvement of radar and radio equipment. After the war, from 1946, he worked for the Telecommunications Research Laboratory in Johannesburg, one of the first two research agencies established by the newly formed CSIR (the other was the Department of Ichthyology at Rhodes University College lead by Professor JLB Smith).

In a stellar career Wadley invented a series of highly innovative devices that carried out a wide range of tasks using radio technology. They included the Panoramic Adaptor, Wadley Loop Receiver, Automatic Ionosonde, Barlow-Wadley Broadband Radio, All-Wave Communications Receiver, Wadley Transistorized Receiver and Rack-Mounted Wadley Receiver, all in the realm of radio transmitters and receivers. But his most significant invention was the Radio Tellurometer, one of the greatest South African inventions which revolutionized surveying worldwide.

**Allan MacCleod Cormack (1924-1998):** Allan Cormack was born in Cape Town and educated at Rondebosch Boys High School and the University of Cape Town where he conducted research on x-ray crystallography. He subsequently worked at the Cavendish Laboratory in England but spent most of his career in the USA. He returned briefly to Groote Schuur Hospital in Cape Town where he carried out research on equipment that was needed for diagnostic radiotherapy. Two of the drawbacks of x-rays are that three-dimensional structures overlap in the two-dimensional image and the different densities of the soft organs scanned cannot easily be determined. Cormack became aware of these difficulties when he was asked to calculate radiation dosages for cancer therapy and found that the methods being used were imprecise. He worked on the problem of measuring and interpreting the absorption of radiation passing through objects from different directions and was then able, in model experiments, to reconstruct an accurate cross-section of an irregularly shaped object such as a body organ.

In 1957 Cormack developed algorithms that use data from x-ray slices of a patient's body to create a 3D image and built a simple, prototype 3D x-ray machine. His reconstructions were the first computerised 3D images ever made although his 'computer' was a simple, desktop calculator. Godfrey Hounsfield of EMI in England took the concept to the market by using Cormack's and other algorithms and, in 1972, developed the first functional CATscanner. In 1979 Cormack and Hounsfield were jointly awarded the Nobel Prize for Medicine or Physiology for their significant invention.

It is widely considered that CATscanners ushered medicine into the space age and they were soon in use in large hospitals worldwide. Like most great inventions the CATscanner was initially developed for a narrow application in medicine but later found wider application in industrial, biological, environmental, space science and astronomy

Allan Cormack (left) and Godfrey Hounsfield, co-inventors of the CATscanner

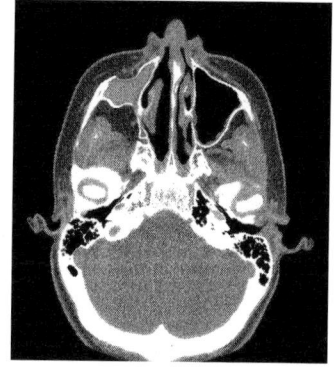

CATscan image of the head

# 6. South Africa's greatest inventor

contexts. Interestingly two other significant South African inventions are derived from CATscanner and x-ray technology – Lodox, a low-dose x-ray machine developed by the De Beers mning company 'to ensure that miners do not use diamonds as a dietary supplement' and Aceso, a 3D breast-imaging system developed by Kit Vaughan of CapeRay in Cape Town.

Aaron Klug

**Sir Aaron Klug (1926-2018):** Electron crystallography is a technique used to determine and model the arrangement of atoms in solids using a transmission electron microscope (TEM). The previously used method, x-ray crystallography, has its limitations as it is difficult to create a realistic image of a crystal. In 1968 Sir Aaron Klug, a Lithuanian-born, South African-educated biophysicist, and David DeRosier demonstrated that they could reconstruct the structure of a crystal in three dimensions using a TEM image. They successfully modelled the tail of bacteriophage T4, a common virus, which was regarded as a major breakthrough in crystallography.

In 1978 Klug was also the first scientist to study inorganic crystals using high-resolution TEM images which led to the modern field of cryo-electron microscopy, a crucial tool for determining the structure of biomolecules that cannot be crystallized. For this research and for his work on virus structures and transfer RNA Klug won the Nobel Prize for Chemistry in 1982. He was also elected a Fellow of the Royal Society of London in 1969 and knighted by Queen Elizabeth II in 1988. In 2005 he was awarded South Africa's Order of Mapungubwe (Gold) for his exceptional achievements in medical science.

**Percy Tucker (1928-2021):** In 1954 theatre impresario Percy Tucker launched Show Service, a manual ticket office in Eloff Street, Johannesburg, that sold tickets for theatre productions. Over the next two decades he became increasingly frustrated by the long queues and decided to develop a computerised ticketing system that would "take the box office to the customer". As he had no computer skills he approached experts and sat with them for six months to set up the first online ticketing system in the world in 1971, 17 years before anyone else achieved that goal. With its high-

Percy Tucker (centre), inventor of Computicket, in the 1960s

end technology and diverse applications Computicket continues to set the standard for online ticketing worldwide and now offers theatre, concert, festival, sport and cinema tickets as well as bus tickets, gift vouchers and classified adverts to customers. It is one of the most recognisable brands in South Africa.

**Dr Selig Percy Amoils (born 1933):** Percy Amoils briefly studied engineering before he switched to medicine but managed to combine the two fields during his career. His major invention was the Cryoprobe, a pencil-shaped device with a frozen tip that is used for eye surgery. When he first demonstrated the device at Oxford University in 1965 he was a young, unknown medical doctor but his invention made him famous overnight as the Cryoprobe dramatically changed the course of cataract and retinal surgery worldwide. The Cryoprobe works by freezing gas that is released from a small nozzle at high pressure in a closed tube. When the probe is inserted into a cut in the eye and the gas is switched on, it freezes to a cataract which can then be removed. In 1983 Amoils used the Cryoprobe to cure British Prime Minister Margaret Thatcher of retinal detachment after laser surgery had failed. He also famously removed a cataract from President Nelson Mandela's left eye the day before he was sworn into office as the first president of

South African coin minted in 2020 to celebrate the 55th anniversary of the invention of the Cyoprobe

a democratic South Africa in May 1994. Miraculously Madiba was able to read his speech at the inauguration ceremony without spectacles!

Amoils also invented the Rotary Epithelial Scrubber to remove epithelial cells from the cornea in preparation for surgery. In 1970 he developed Diamond Vitrectomy Cutters, a family of medical instruments that facilitate fine control of blade depth in eye surgery as well as the Oval Comparator (or Astigmometer) to control astigmatism after cataract surgery. In 1975 he received a Queen's Award for Technological Innovation and the Medal of Honor of the US Academy of Applied Science. In 2006 President Thabo Mbeki awarded him the Order of Mapungubwe (Silver) for 'excellence in the field of ophthalmology and for inspiring his colleagues in the field of science'.

**Kim Pratley (1956-2021):** After the founder of Pratley Putty George Montague 'Monty' Pratley died in 1983 his son Kim took charge of the company after completing his studies in mechanical engineering at the University of the Witwatersrand. Kim proved to be a resourceful and enterprising leader who took the company to new heights. He built on the reputation of the company's most famous product Pratley Putty, the first South African invention to go to the moon (on a NASA Ranger moon module in 1967). Pratley Putty was also used to patch up the Golden Gate Bridge in San Francisco and to repair two sunken ships off South Africa, both of which were subsequently raised and sailed the high seas again. Under Kim's leadership the company produced a range of world class putties, glues, superglues, adhesive tapes, sealants and cable connector boxes. They include Wham! the world's fastest-setting glue, Wondafix one of the strongest-ever adhesives, Eezebond and Pratlock as well as explosive igniters, coolers for deep-level mines and products that clean up nuclear radiation contaminants. The success of these products rests on a solid foundation of research and innovation led by Kim Pratley which has resulted in the company filing over 350 patents worldwide.

Kim Pratley

The Pratley company has also become known for its hair-raising publicity stunts that focus attention on the strength and reliability of its adhesives. On one occasion Kim Pratley stood under a three-ton bulldozer that had been lifted by a crane using two metal plates stuck together using Wondafix and, on another, he stood under a Volkswagen Beetle that had been lifted by a hook attached to a plate just five seconds after the glue had been applied.

**Louis Liebenberg (born 1944):** Anthropologist Louis Liebenberg from Noordhoek in Cape Town is fascinated by the hunting and tracking skills of San hunters and spent eleven years living and hunting with them in the Kalahari Gemsbok National Park in Botswana. In 1996 with Justin Steventon he invented the CyberTracker, a handheld, palmtop computer with a built-in GPS for accurate positioning to document and analyse the knowledge of San hunters. Instead of recording data in text they used icons and pictograms to capture information on sightings, location, habitat, numbers of animals, feeding behaviour and territorial markings. The data could then be downloaded to a laptop and maps could be printed out showing the locations and movements of individual animals and herds. The icon graphic user interface meant that non-literate, but skilled, trackers could collect complex, geo-referenced biodiversity observations in the field and scientists could capture data quickly without having to write text.

The effectiveness of the CyberTracker was further improved by developing a Screen Designer that allows users to design their own electronic field guides, and an Electronic Field Guide which permits the imbedding of descriptions and images to facilitate identifications in the field. Furthermore the Field Map feature tracks the path of the user in real time and the GPS Go app allows the user to define GPS waypoints to guide him/her to specific destinations. CyberTracker has been used for many high-profile research and conservation projects, including monitoring gorilla deaths from Ebola fever in the Congo, and is now widely regarded as one of the the most efficient ways to collect large quantities of field data at a speed and level of detail that was previously impossible. Its software is distributed free through www.cybertracker.org and has been downloaded in over 40 countries.

Cybertracker emblem

Like all great inventions CyberTracker was originally developed for one use but has found application in a variety of other fields

- it is used by ecologists, game rangers and foresters as well as detectives, field educators and disaster-relief volunteers. In addition, a CyberTracker Tracker Certification, now used in Africa, the USA and Europe, has been developed to facilitate the employment of trackers. Louis has written extensively on animal tracking, San hunting and the evolutionary roots of science and is an Associate of Human Evolutionary Biology at Harvard University. In 1998 he won the prestigious Rolex Award for Enterprise for bridging the digital divide by applying modern technology to the understanding and appreciation of indigenous knowledge. CyberTracker is a true product of the Rainbow Nation.

**Sandile Sanele Ngcobo (born 1964):** In 2013 research carried out at the CSIR by Sandile Ngcobo, a PhD candidate at the University of KwaZulu-Natal, under the supervision of Professor Andrew Forbes resulted in the development of the first three-dimensional laser, a feat that many sophisticated laboratories around the world had failed to achieve. The team also included doctoral student Liesl Burger and post-doctoral fellow Dr Igor Litvin. The South African team had a key advantage in that they had one of the few laboratories in the world that specialised in both lasers and holograms. Ngcobo demonstrated that, instead of resorting to expensive optics attached to the front of a laser machine, laser beams could be digitally controlled inside the laser device by programming a liquid crystal display using holograms. In this way the beam is already shaped in the required way when it emerges from the laser cavity and the operator can change the shape of the beam with the touch of a button, which avoids the time-consuming set-up and calibration that is required with conventional methods. The Digital Laser also significantly reduces the cost of laser equipment as users only need one laser to do a multitude of tasks.

According to Dr Igle Gledhill, President of the South African Institute of Physics, "The significance of this breakthrough can be compared with the original invention of the laser about 50 years ago". The then-Minister of Science & Technology Dr Derek Hanekom echoed this sentiment in August 2013 when he stated, "With the 21st century being touted as the century of the photon, with the photonic revolution in which the laser is a key enabling technology, your work has secured a place for South Africa right at the forefront of developments in laser research" (Bruton, 2017).

**Mulalo Doyoyo (born 1970):** The South African entrepreneur and serial inventor Professor Mulalo Doyoyo is a respected researcher in the fields of applied mechanics, ultralight materials, green building and renewable

Cenocell invented by Mulalo Doyoyo

energy who has operated at the interface between academia and industry in South Africa and the USA. Having been born in Limpopo province he was exposed to the hardships that rural people encounter in their daily lives and is passionate about using modern technology to fight poverty, improve the quality of life of rural people and promote green energy and sustainable living. His inventions include Cenocell, a patented 'cementless concrete' produced through a chemical reaction between fly ash and organic and inorganic chemicals that was developed in collaboration with Paul Biju-Duval at the Georgia Institute of Technology. As fly ash is an unwanted pollutant that is a byproduct from coal-fired power stations and mining operations it is cheap and freely available. In addition to its role as an environmentally friendly building material Cenocell is stronger than conventional concrete and therefore needs less reinforcement.

Doyoyo's other inventions include the *Ahifambeni* hydrogen-powered motorbike, Amoriguard, a non-volatile organic paint and skim coating made from industrial and mining waste, and a solar-powered flushing toilet that uses nanofiltration and anaerobic digestion and offers a practical solution in places where the water supply is unreliable. He has also developed two environmentally friendly chemical binders, Solunexz and Glunexz, to reduce pollution by coal dust, and the Ecocast Brick-Making Machine that saves both water and energy. In 2008 Doyoyo established Retecza, a resource-driven technology concept centre that promotes cross-disciplinary industrial research.

**Mark Richard Shuttleworth (born 1973):** Mark Shuttleworth's passion for science and technology first became apparent through his interest in computer games. Later, while studying for a business science

# 6. South Africa's greatest inventor

degree at the University of Cape Town, he became intrigued by the changes that the internet was bringing about in business and society. He initially planned to become a web designer but his interest in internet security prompted him to join an online chat group that discussed coding and digital certification. In 1995 he launched a company from his parent's garage that specialised in internet security verification for electronic commerce which led to the development of the Thawte Internet Security System. Thawte became the first company in the world to produce a full-security encrypted e-commerce web server that was commercially available outside the USA. It was soon recognised by Netscape and Microsoft as a trusted third party and, in 1999, it was acquired by Verisign for R3.5 billion.

**Mark Shuttleworth**

Mark decided to use his windfall wisely not only to create new opportunities for himself but also to invest in the future of science and technology in South Africa. As he believed that entrepreneurs in South Africa have the potential to start businesses that have a global impact he formed HBD Venture Capital ('HBD' refers to the phrase 'Here be dragons' that is often found in uncharted territory on old maps). HBD has since helped many local start-ups in the software, pharmaceutical, electronics and mobile phone services sectors to reach the market. Mark also launched the Shuttleworth Foundation to support social innovation in education in Africa and the Hip2B$^2$ campaign which encourages children to study mathematics and science at school. In 2002, at the age of 28 years, he became the first African in space when he blasted off on Soyuz mission T34 to the International Space Station on a Soyuz-U rocket from the Baikonur Star City Cosmodrome in Russia. After he returned he shared his space experience and enthusiasm for science with over 100 000 learners during a countrywide roadshow in South Africa.

In 2004 Mark founded the Ubuntu Project that creates free desktop and server operating systems. Ubuntu's free open-source software runs from the cloud to smartphones, tablets and laptops and is used by OpenStack, Intel Joule and even the National Gendarmerie in France. As a spin-off the project has created several free tools for software developers such as the Bazaar Control System and Launchpad.net as well as specialized desktop environments including Edubuntu and Kubuntu for schools and industry.

**Siyabulela Lethuxolo Xusa (born 1989):** The youngest 'finalist' is the remarkable 'Siya' Xusa who was born in Mthatha and was hooked on science when, at the age of five, he saw a light aircraft dropping election leaflets over his village. After watching Mark Shuttleworth's space flight in 2002 Siya, a born innovator, began experimenting with rocket fuels in his mother's kitchen. Soon afterwards he built his own rocket but it exploded on the launch pad, which made him even more determined to succeed. His passion turned into a serious science project that culminated in him developing a cheaper and safer rocket fuel. One year later his next-generation rocket broke the national amateur altitude record of nearly 1 000 m!

Siyabulela Xusa

Siya's science project won gold at the National Science Expo and the Dr Derek Gray Memorial Award for the most prestigious project in South Africa. This led to an invitation to the International Youth Science Fair in Sweden in 2006 where he presented his project to the King and Queen of Sweden and attended the Nobel Prize ceremony in Stockholm. He later won two grand awards at the Intel International Science & Engineering Fair in the USA in competition with over 1 500 students from 52 countries, which earned him global recognition and a scholarship to Harvard University. His reputation was further enhanced in 2002 when the NASA-affiliated Lincoln Laboratory named a minor planet '23182 Siyaxuza' in the main asteroid belt near Jupiter after him.

At Harvard Siya focused on developing cheaper solar cells and assessing the commercial viability of solar cell technology as he shares with Elon Musk and Mulalo Doyoyo a passion for using clean technology to solve the world's energy crisis. Despite lucrative offers from abroad he has since returned to his roots in South Africa and launched a high-tech company that focusses on nano-enabled solar technology. In 2010 he was elected a Fellow of the African Leadership Network and in 2011 became a Fellow of the Kairos Society, a global network of top student and global leaders who use entrepreneurship and innovation to address challenging global problems. He has also recently become the youngest member of the African Union-affiliated Africa 2.0 Energy Advisory Panel and was awarded the Order of Mapungubwe (Silver) by President Cyril Ramaphosa in 2017. But Siya also has other talents – he is an accomplished Xhosa praise singer who in 2003 had the honour of performing for former President Nelson Mandela!

**Who is your choice as South Africa's top inventor?**

**And the winner is …:** James Greathead was truly the 'Father of the London Underground', Hendrik van der Bijl and Basil Schonland were master imagineers and Irvine Bell, Kim Pratley, Louis Liebenberg, Mark Shuttleworth, Kerry Sink, Siya Xusa and many others made inventions that had important downstream repercussions but I would narrow the field down to two: Allan Cormack and Trevor Wadley.

Cormack co-invented the CATscanner which is regarded by Philbin (2003) as the 53rd greatest invention of all time (the only one in the top 100 from the southern hemisphere) and is one of only two South Africans (with Sir Aaron Klug) to win a Nobel Prize for an invention. But Cormack only provided proof of principle for the CATscanner and left the construction of a workable clinical device to his co-inventor, Godfrey Hounsfield, an engineer. Furthermore, as the biographer of Cormack Dr Kit Vaughan (2008) has pointed out to me, "Cormack never applied for any patents whereas Hounsfield did, and his employer EMI defended them aggressively!".

My vote goes to Trevor Wadley, whose childhood provides intriguing insights into the development of the mind of a genius. Wadley was born in Durban in 1920, the youngest son of ten children, to Thomas Wadley, a hard-working accountant and erstwhile mayor of Durban, and Florence, a dominatrix who disapproved of his rebelliousness (which probably encouraged it). His kindergarten teacher described him as "Independent and original; conduct fairly good, but he must learn to sit still sometimes" (Wadley von Hirschberg, 2009) and he was later commended for his brilliance in mathematics. Trevor was not interested in sports except on one occasion when he entered a cross-country athletics race and predicted that he would win in record time and that his record would stand for 15 years. His training method involved calculating the time he needed to run each section of the course and then training himself to run at the required pace for each section. He went on to do exactly as he had predicted!

From childhood Wadley was the ultimate tinkerer who liked to take things apart and put them back together again, often in a different configuration. He was also an adventurous, risk-taking youth who delighted in climbing to the top of the giant *Bauhinia* trees in the garden and gazing out over the Umgeni River valley to the chatter of vervet monkeys nearby. In summer the Wadley kids would sleep on the verandah except on nights when a leopard was reported in the vicinity. Trevor was forever carrying out experiments using the simplest apparatus. In 1932 the 12-year-old stood on the verandah vigorously rotating a syrup tin. "Look, Mary, centrifugal force", he said to his younger sister. He once wired the family telephone to the radio in the lounge so that he could listen in on his sister's phone calls to her fiancé!

I had the pleasure of meeting Wadley's sister Mary Wadley von Hirschberg when she visited Cape Town from Swaziland for her 80th birthday celebrations in 2015. She was an accomplished and colourful person in her own right who accompanied her husband on diplomatic postings to Pretoria, London, Vienna and Tokyo and, later in life, served as an attorney in South Sirica. Over a few gin-and-tonics she shared with me memories of her remarkable brother whom she called "an eccentric genius" which she has recorded in her delightful biography, *Trevor Lloyd Wadley. Genius of the Tellurometer* (2009). An interesting point that she made with which I agree is that, when one is trying to ascertain the character traits and true interests that will define a person's future career, one should look at what that person is most passionate about at the age of 12 to 13 years.

At this age Trevor "took infinite pains in pulling to pieces, examining, constructing and poring over any type of machinery and particularly electrical equipment. He also refused to engage himself in anything that did not interest him but pursued relentlessly anything that did. This characteristic he carried through into adult life and he nearly failed the matriculation examination en route!" (Wadley von Hirschberg, 2009). In fact, Wadley only passed matric by dropping Latin which he was doomed to fail and taking up chemistry four months before the exam. To everyone's amazement and through disciplined after-hours study he achieved an A in Chemistry (as well as in Physics and Mathematics)!

In the 1960s, when the then-president of the CSIR Dr Stefan Meiring Naudé asked Mary to explain her brother's genius to him, she emphasised that academic brilliance may be misleading as it could lead an individual into a career for which he/she is temperamentally unsuited, in which case success comes at the expense of happiness. In her brother's case he chose a career for which he was ideally suited and as a result "He lived his working life in a state of euphoria, so happy to have been afforded the means and opportunity to exercise his special talents" (Wadley von Hirschberg, 2009).

Wadley's attendance at Durban High School was followed in 1936 by studies at Howard College where he completed his BSc (Eng) in 1940. He was regarded as a brilliant student who never took notes in lectures but clearly understood every concept and could derive any formula from memory. In examinations he did not confine his answers to the minimum number of questions required but answered every question, sometimes scoring over 100%.

**Wadley's war experiences:** Wadley's war experiences in North Africa with the Special Signal Services are still shrouded in secrecy but we do know that he had a mercurial rise through the ranks from lieutenant to full lieutenant within five months and then to major by the age of 24 years, which no doubt reflects the strategic importance of his work. In March 1941 he was assigned to the Sinai ostensibly as the station commander of a radar station but it is clear from contemporary accounts that his responsibilities were far more important than that. According to a report by JSF Botha, a signaller in the 10th Brigade of the Signal Corps at the time, Wadley "… was never a station commander … for the simple reason that he was never with us in the field long enough. He used to breeze in and out for periods while he adjusted, modified, refined and generally fiddled with the primitive equipment we were using. Then he would be off somewhere else – we never knew where because of the need for secrecy" (letter to Mary Wadley von Hirschfield, 2009).

Wherever he went Wadley carried with him a powerful 'wireless set' (radio) with which he could send secret messages to the Cabinet War Room at Whitehall. He also intercepted enemy radio messages from Germany and Japan (and from submarines) and transferred them to London, which contributed significantly to the Allies' ability to pre-empt enemy moves. The powerful radio in his tent in the desert was also used to broadcast Churchill's inspirational speeches to the troops. Wadley's radio address was, appropriately 'Oppositely London' and he visited the British capital for radar and radio training on at least one occasion in July 1942.

Wadley was intensely engaged with the maintenance, operation and upgrading of field radios and especially the JB1 Radar Transmitters that had been developed by Basil Schonland and his team (including Wadley) at Wits University. Although the JB1s were still in the experimental stage they were so successful that they initially replaced the RAF radar network. Wadley played a vital role in the development of the JB1 and was praised for his 'fertile mind' by Schonland's biographer, Brian Austin (2001).

During the war, at the request of Major ER ('Ted') Cook, commanding officer of a secret section of the South African Corps of Signals, Wadley developed the Panoramic Adaptor (or Spectrum Analyser). This device when affixed to a HF radio allowed the viewer to monitor the full range of radio frequencies being transmitted at one time on a particular waveband, which was of great strategic value. These and other innovations by Wadley helped to ensure the success of aerial raids carried out by the Allies' Desert Air Force, one-third of which comprised South African men and women, who were known as the 'Imperturbables'. Wadley was posted to the 64 Air School in January 1944 and to the Coastal Air Force in March 1944, when he received the award of the 'Africa Star' (and, later, the '1939/45 Star'). He ended the war as Staff Officer Radio, Cape Command, with the reserve officer rank of major.

In 1948 at a University of Cape Town graduation ceremony Field Marshal Jan Smuts, then Chancellor of the university, said to Mary Wadley, "I know your brother! He is a genius - he contributed to the development of radar units in South Africa. At a cost of half a million pounds they did the job of the British units that cost us a million! He is a genius - but do not tell him – it might give him a swollen head" (Wadley von Hirschfield, 2009). Trevor Wadley was only 28 years old at the time and the full flowering of his technical genius was still to come.

**Radio inventions:** After his discharge from the army in February 1946 Wadley was immediately recruited as one of the five founding members of the newly formed Telecommunications Research Laboratory (TRL), now the National Institute for Telecommunications Research, at Wits University. There, as a designer of radio equipment and instrumentation, his inventive genius soon became apparent and his brilliance blossomed in the excitingly free yet challenging atmosphere that Basil Schonland and Frank Hewitt had created. When *Sputnik 1* was launched by the Russians in October 1957 the young team at the TRL developed a tracking system and established the orbit of the satellite within two days!

In 1946 Wadley changed the face of high-performance radio receiver design when he perfected the Wadley Loop Receiver, a unique circuit for cancelling frequency drift which he had developed from the Panoramic Adaptor (Austin, 2007). In 1947 he also developed an improved Automatic Ionosonde, an instrument that measures radio wave reflections from the ionosphere, that part of the atmosphere that contains sufficient free electrons to reflect radio waves. The reflectivity of the ionosphere is what makes the use of shortwave radio for long-distance communication possible as demonstrated initially by Guglielmo Marconi's first trans-Atlantic transmissions in January 1906. As the ionosphere changes constantly in concert with the sun's eleven-year sunspot cycle and less predictably with solar flares and other short-term solar events, it is extremely useful to be able to study, understand and predict its behaviour. Wadley's Ionosonde automatically scanned, measured and displayed the state of the ionosphere and contributed significantly to international data collection in this field until the early 1980s.

In 1949 Wadley also designed a radio receiver that could operate underground in the mines at 335 KHz. Although his design was sound no underground radio communications system is in general use in mines anywhere in the world due in part to the complex geological strata underground (Austin, 2001).

In the work environment Wadley was playful and competitive. Once a technician asked to borrow a piece of equipment; Wadley agreed but insisted that he leave his shoes behind as a 'deposit'. He regularly met with colleagues at Wits University where he was known as a colourful and

Trevor Wadley with a three-wheel electric 'car' that he made in 1962

entertaining personality. In 1962 he bet his colleagues on a Friday that he would arrive at work on Monday in an electric car. He spent the weekend creating a crude but workable three-wheeled, battery-powered vehicle and lived up to his promise. He enjoyed spending tea breaks bouncing his sometimes-eccentric ideas off other staff, always displaying a profound understanding of matters electrical, and often came up with novel solutions to intractable problems through his 'thought experiments', reminiscent of Albert Einstein.

Wadley was often accused of being iconoclastic, i.e., constantly criticizing or attacking cherished beliefs and was very argumentative. If there was no *contretemps* in the offing he would provoke one by assuming a seemingly indefensible position and then proceeding to defend it with great skill (Smith *et al.*, 2008). He was also a betting man, though not a gambler, and almost always won his bets. Punctuality was not his strong point (he was often a couple of hours late for work) and he rebelled against strict office hours and other forms of bureaucracy, but he worked long hours, thinking with an audible whisper. Throughout his life nothing delighted him more than to amaze his audience with the novel, brilliant or outrageous solutions or scenarios that he had conjured. He also enjoyed tackling new challenges. He taught himself to knit and sew, and to play the piano and mouth organ and even made his own electronic organ and telescope!

At the TLR Wadley, now in full flight, developed one of the first practical broadband radios in the world, the Barlow-Wadley Broadband Radio (1947), and then stunned the telecommunications world by

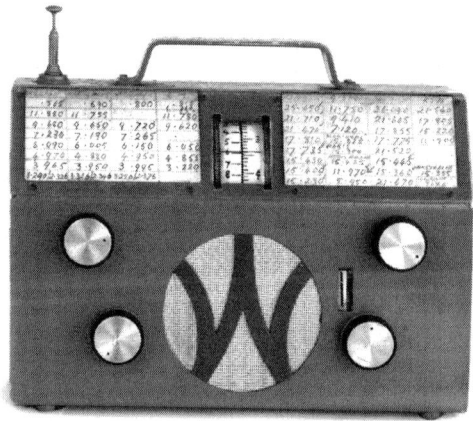

**Prototype of the Barlow-Wadley All-Wave Receiver (1947)**

inventing the first practical broadband crystal-controlled radio, the All-Wave Communications Receiver (1948) based on the Wadley Loop. This revolutionary radio could be accurately and consistently set to any frequency between 0 and 30 MHz and avoided the fading that was typical of most of its competitors. This was achieved by using a single 1 MHz crystal and some clever circuitry. The All-Wave Receiver (RA 17) was made by Racal in Berkshire, England, under licence to the CSIR and subject to the strict supervision of Wadley and became the standard high frequency radio in the British Army, Royal Navy and Royal Air Force for many years. A special model of this radio was designed by Wadley for the South African Post Office.

The international success of Wadley's receivers and their manufacture under licence abroad demonstrated to Basil Schonland how the CSIR should function by fostering a brilliant idea, exploiting its application and then transferring the technology to an industrial company for eventual manufacture.

In 1958 Wadley designed and built a transistorized version of his All-Wave Receiver, the Wadley Transistorized Receiver. This model was simpler and cheaper to build than the valve versions and was a huge seller internationally. One model of this transistorized radio, the Barlow-Wadley XCR30 (1958), was made by Barlows SA and is still one of the most popular receivers used by radio hams worldwide. The Rack-Mounted Wadley Receiver, which Wadley invented in 1960, was made by Racal for the commercial market including telecommunications, military, civil defence and broadcasting applications and was used by the BBC for radio broadcasts between London and South Africa in the 1960s.

**Radio Tellurometer:** Notwithstanding these significant contributions Wadley's greatest invention was still to come. Until the 1950s surveying in South Africa was carried out using Günter's Chains, invar measuring tapes and surveyor's wheels which required baselines to be cleared when measuring the distance between topographical beacons. Although the introduction of the theodolite and the Geodimeter streamlined surveying the early instruments were cumbersome and did not provide the accuracy that surveyors needed. In 1954 Colonel Harry Baumann, director of the Trigonometric Survey of South Africa, requested the CSIR to develop a portable instrument that could measure distances with an accuracy of 1 : 100 000 over 30 miles using the latest radio technology. This request had been made earlier by Schonland, then deputy director of the Harwell Atomic

Energy Research Establishment in England, who stated at a meeting in Johannesburg in 1953 that there was an 'extremely urgent' need to develop an accurate, radio-wave-based measuring instrument (Austin, 2001).

Wadley was the obvious person to take on this challenging task and within two months he had developed a prototype and was conducting field trials, with the first measurements being made near Johannesburg on 14th June 1955. During the early development of the instrument he consulted with Dr Jules Fejer, a similarly gifted young Hungarian mathematician and expert on radio wave propagation at the TRL. Wadley astounded the world by inventing a device that could accurately measure the travel time of a radio wave (or, more accurately, the 'ripples' in a radio wave) by comparing the phase of a transmitted signal with that received back from a second instrument. He called it the Tellurometer ('earth measurer'), a name he derived from the Latin word *tellus* for 'earth', although other names such as Scelometer, Telourimeter and Horimeter were also considered at the time. The entire process from the formulation of the user requirement to the delivery of a patented and functional operating system embodying a totally new concept with new capabilities took slightly more than two years, a remarkable achievement especially during peace time (Smith *et al.*, 2008).

Although the Tellurometer operated on basic principles the design of a practical instrument to exploit these principles was far from simple. Wadley knew that the accuracy of distance measurement using radar was not sufficient for survey purposes so he devised a method of superimposing a low frequency wave on the signal using a highly precise method of phase comparison to determine time and therefore distance. By using microwaves he produced an instrument that could be operated in broad daylight and measure distances to an accuracy of 3: 100 000 (thus exceeding Baumann's specification) over a range from 3 to 60 km. Furthermore, the Tellurometer was relatively light and portable and could be used in all weather conditions, day and night and in rugged terrain without the need to clear a baseline. Wadley's prototype Tellurometer (MRA1) operated at 3 GHz with a wavelength of 10 cm but later models (such as the MRA; 1956) operated at 10 GHz and 3 cm and were even more accurate. Over time Tellurometers were not only used from land bases but also from helicopters, fixed-wing airplanes, ships and artificial islands.

Land surveying was revolutionised by the Tellurometer. Wadley demonstrated it to great acclaim in South Africa, England, continental Europe, the USA and elsewhere and in 1957 presented a paper, 'The

## 6. South Africa's greatest inventor

**Trevor Wadley using a whirling hygrometer to measure humidity while demonstrating the Tellurometer in England in 1957**

Tellurometer system of distance measurement' to the Royal Geographical Society in London and received a standing ovation. He so impressed the British Prime Minister Harold McMillan that he is said to have wondered out aloud why British scientists had not come up with something similar! Wadley also stunned his hosts in England by showing that the distance between two points on Salisbury Plain, which was used as a baseline for British surveying, had been wrongly measured by 1.5 m! He was even retained for a while by the US military to apply the principle of radar to detect the movement of troops and vehicles in densely afforested areas; this research is classified and is still secret. Although Wadley stood only 154 cm tall and was somewhat squat he was a towering figure in the world of electronic surveying.

The Tellurometer also speeded up surveying work. On 9th November 1958, the *New York Times* reported that a survey of Manhattan Island which would normally take four men five days to complete had been done by two men in 3½ hours using the Tellurometer. In 1958 the US Federal Bureau of Publications estimated that highway surveys using the Tellurometer were so quick that a cost saving of 40% would be achieved compared to the traditional methods of triangulation and traverse surveys. Tellurometers were thus responsible for reducing the cost of building roads, railways and harbours worldwide during a post-war period when this infrastructure was desperately needed. Wadley later learned that US and Canadian scientists

# *Tellumat*

**Tellumat logo**

---

had spent millions of dollars trying to develop their own 'earth measurers' but without success.

Tellurometers were initially manufactured by two companies in Cape Town, Tellurometer (Pty) Ltd and Phillips Denbigh (Pty) Ltd, and then by Tellumat. The current status of Tellumat is illustrated by the fact that the current non-executive chairman and director of the company is the CEO of the CSIR Dr Sibusiso Sibisi. Over 20 000 instruments were exported worldwide, the first major electronics export from South Africa, and they earned the country foreign earnings in excess of R300 million (in 1960s terms). By 1958 Tellurometers were in use in over 60 countries surveying, *inter alia*, road and railway networks, harbours, bridges and forests in South Africa, Rhodesia (Zimbabwe), Nigeria, Belgian Congo (Democratic Republic of Congo), Ethiopia, Kenya, Greenland, Ireland, Belgium, Norway, Sweden, Denmark, Finland, Switzerland, Germany, Poland, Falkland Islands, Indonesia, Bahamas, Australia, Canada and the USA,

**Tellurometer in use in Antarctica**

canals in Panama, the new Carlton Centre development in Johannesburg, offshore navigational beacons in Saudi Arabia and the Cabora Bassa dam in Mozambique. They were also used to conduct hydrographic surveys and mapping in Antarctica and Malaysia and to facilitate the maintenance of the Defence Early Warning system (DEW line) which protected the USA and Canada from possible bombardment by ballistic missiles during the Cold War (Buderi, 2004). In 1964 Wadley took early retirement from the CSIR and joined the board of Racal-SMD (Pty) Ltd.

The definition and defence of a strong patent for the Tellurometer, enthusiastic and highly innovative research by Wadley and his successors, and an active programme of upgrading and introducing new models by the commercial manufacturers established South Africa's position for many years as the leading supplier of electronic measuring equipment to the Western world and illustrated the benefits that accrue from the successful application of a brilliant idea and being first to do so. Through the Tellurometer (and Wadley's pioneering radios) South Africa became internationally renowned for the extraordinary development of its electronics industry which created an infrastructure of skilled employment opportunities that continues today. A predominantly mining and agricultural country at Africa's tip was propelled into the age of information technology (Austin, 2007; Wadley von Hirschberg, 2009).

**Accolades and awards:** What was truly remarkable about Wadley was his exceptional ability to imagine ingenious solutions that overcame seemingly intractable problems and then to convert these ideas, quickly and efficiently, into practical devices, a rare combination in one individual. Dr LE Howlett, Director of Applied Physics at the National Research Council in Ottawa, Canada, commented in 1957 with respect to the Tellurometer, "Very great credit is due to Mr Wadley for his originality and the rare practical commonsense with which he has carried out his initial development. He has worked with beautiful simplicity and freedom from the ever-present curse of over-elaboration". Earlier, in 1956, Howlett had commented with respect to Wadley, "It is a matter of some considerable satisfaction to find it proved once again that in scientific research one man with a good idea can still prove himself superior to teams of researchers organized to find an idea" (Smith *et al.*, 2008).

Wadley received many international awards for his inventions including the prestigious Frank B Brown Medal from the Franklin Institute in Philadelphia in 1970 which his sister Mary received on his behalf. The recipient of this award in the previous year had been the architect and designer Frank Lloyd Wright and co-recipients in 1970 included the co-inventor of the aqualung Jacques-Yves Cousteau. Wadley was also awarded honorary doctorates by the universities of the Witwatersrand (DSc [Eng.]; 1959) and Cape Town (DSc; 1976). In 1960 he received the Gold Medal of the South African Institute of Electrical Engineers and in 1976 the National Award of the Associated Scientific and Technical Societies of South Africa. In February 1979 the South African Post Office issued a 15-cent stamp commemorating the 25$^{th}$ anniversary of the invention of the Tellurometer but they had been pipped by Belgium which had issued a 10-franc stamp featuring a Tellurometer MRA2 in 1971 to mark the 10$^{th}$ anniversary of the Antarctic Treaty.

**Family life and death:** Wadley married twice, first to his childhood sweetheart Madge Senfftleben in 1941, with whom he had two sons, and second to Gill du Plessis in 1961 with whom he had three daughters. In later life he and Gill lived in a smallholding at Warner Beach where they kept cows and chickens and ran a roadside stall selling fresh vegetables from their garden. What an odd sight it must have been seeing one of South Africa's most brilliant minds selling cabbages by the roadside! Sadly Trevor Wadley died of colon cancer at the age of only 61 years at his home in Warner Beach on 21$^{st}$ May 1981. The inscription on his tombstone in the Stellawood Cemetery in Durban reads simply, 'Trevor Lloyd Wadley. Philosopher, Scientist, Beloved Family Man. 5.2.1920 – 21.5.1981'.

Stamp issued by the South African Post Office to commemorate the 25$^{th}$ anniversary of the invention of the Tellurometer

## 6. South Africa's greatest inventor

**Infra-Red Tellurometer:** Wadley's Radio Tellurometer was eventually superceded in 1970 by the Infra-Red Tellurometer, also invented by a South African Hobbe Dirk ('Dick') Hölscher from the TLR in Johannesburg, who had contributed to the development of the Radio Tellurometer and had accompanied Wadley on demonstration trips to the USA. The Infra-Red Tellurometer has an astonishing accuracy of 1 mm in 1 km but is only effective over short ranges (50 m to 2 km) and is mainly used in harbour and urban contexts. Today Tellurometers have largely been replaced by laser beams which can measure the distance to the moon with an accuracy of centimetres and by satellite-based Global Positioning Systems (GPS) which can tell us our exact position on Earth. Since 1970 a variant of the MRA7 Radio Tellurometer has been used in vertical lift shafts in mines to measure the position of the lift cage and miniaturized Infra-Red Tellurometers are still widely used in short-range surveying instruments.

**Conclusion:** Although Wadley's inventions have largely been superceded by more modern technology his 'innovation blitz' during the 1940s, 50s and 60s changed the world for the good and showcased South African technology to the world. He was our Edison, Tesla and Marconi combined and is worthy of the accolade as 'South Africa's greatest inventor'.

# 7.
# The comical art of naming new species

*'One man's fish is another man's poisson'*

Mark Gatiss, in *The Vesuvius Club* (2005)

**INTRODUCTION:** Although there are strict and somewhat archaic rules about naming new species of animals (and plants), taxonomists (scientists who classify organisms) sometimes have fun coining unusual and even humorous common and scientific names for the new species that they have discovered. Most of the examples that I give below are from my field of specialisation, ichthyology, the study of fishes, but there are many others. The Rules of Zoological Nomenclature state that an animal species must have a binomial (two-word) scientific name comprising a generic name, such as *Homo*, and a specific name, such as *sapiens*, which is, of course, our name, *Homo sapiens*. It means 'wise man', an appellation that we rarely live up to. An example of a descriptive scientific name from the fish world is *Hydrocynus vittatus*, the 'striped water dog', an appropriate moniker for the African tigerfish, a voracious predator on other fish. Other descriptive names of fishes include *Pseudocrenilabrus philander*, a notorious adulterer, and the jumping bean fish, *Xiphopops acanthops*, whose name is almost onomatopoeic!

**Naming fishes**: South Africa's marine fish have some interesting common names, including the measles flounder, pineapple fish, giraffe seahorse, lookdown fish, prodigal son, old woman, gorgeous gussy, harry hotlips, sergeant major, chocolate dip, smiling goby, evil-eyed blaasop and puzzled toadfish. We also have an international coterie of marine fishes called the Englishman, Roman, German, Scotsman, Dane, Fransmadam, Zulu damsel and Arab blenny.

New species of fishes and other animals are often named after people. A famous example is the coelacanth, *Latimeria chalumnae*, named by JLB Smith after Marjorie Courtenay-Latimer who kept the first specimen for science and the river off which it was caught. Smith also initially named the second coelacanth specimen known to science, caught in the Comoros in December 1952, *Malania anjounae*, after the then-South African Prime Minister, Dr DF Malan who had provided a military airplane to fetch the fish and the island where it was caught, but it was later found to be the same as *L. chalumnae*.

In an article published in *Grocott's Daily Mail* on 19th June 1956 JLB Smith wrote, "It is one of the privileges enjoyed by the research biologist (naming new species after people) to be able to commemorate in this way valuable support of scientific work". He named a new species of goby *Trimia naudei* after the then-President of the CSIR Dr SM Naudé, another new species after Dr Cecil van Bonde, Director of Sea Fisheries and the magenta splitfin *Nauria addisi* (now *Luzonichthys addisi*) after Sir William Addis, governor of the Seychelles where he had special permission to collect fishes. In fact, at least 56 of the 392 new species of fish that JLB Smith described as new were named after people, including *Chlidichthys johnvoelkeri* (John Voelker was the chairman of the Trustees of the Sea Fishes of Southern Africa Book Fund), *Gobiichthys lemayi* (Hugh le May was a major sponsor of Smith's research), *Sideria schonlandi* (Basil Schonland was a President of the CSIR which funded his research at the time) and *Tharbacus vanecki* (HJ van Eck was also a President of the CSIR). Smith, who was always on the lookout for funds for his research, certainly knew on which side his bread was buttered!

Even though he was a serious scientist Smith was sometimes quite flippant when he named new species. On Wamizi Island in Mozambique in August 1951 he reported, "Besides many rarities, I found two fishes new to science that morning. One was a beautiful, small creature with a red snout that reminded us of a man who imbibed too freely ... I found it to be of a new genus as well and christened the fish *Wamizichthys bibulus*, part in reference to Wamizi, where we first found it, and partly from its bibulous appearance" (Smith, 1956).

Smith named four new fish species after his wife Margaret: *Canthigaster smithae* and *Chlidichthys smithae* from Mauritius, *Trachurus margaretae* from Durban and *Pseudocheilinus margaretae* from Aldabra Island. In his description of *P. margaretae* he wrote, "This exceptionally beautiful creature is named as a small tribute to my wife, whose contribution to all phases of our work is probably greater than my own." He also named a new species, the kaalpens goby *Bathygobius william* (now *Monishia william*) from Xora River mouth (where the Smith family often spent their winter holidays) after their son William.

Scientists often pay respect to one another by naming new species after them. Doug Hoese from the Australian Museum named the smoothscale goby *Hetereleotris margaretae* from Sodwana Bay after Margaret Smith and William Smith-Vaniz from the Academy of Natural Sciences in Philadelphia named the halfscaled jawfish *Opisthognathus margaretae* after her. George Coulter, previously of the Department of Ichthyology & Fisheries Science at Rhodes University, named a new cichlid fish from Lake Tanganyika *Simochromis margaretae* after Margaret. Margaret also had a new species of nematode worm from a cow's intestine named after her – we never let her forget that one! This author's only claim to fame in this regard is that he had a tiny beetle *Notoxus brutoni* that he had collected on the shores of Lake Ngami in Botswana named after him by the entomologist Dr Bob van Hille of Rhodes University. It is a tiny, inconsequential animal of little ecological importance!

Several new species of marine fishes were named after JLB Smith including the conger eel *Bathymyrus smithi*, shortfin pipefish *Choeroichthys smithi*, flounder, *Engyprosopon smithi*, the cuskeel *Ophidion smithi* and the mini-clingfish *Pherallodus smithi*. Margaret Smith named 14 new species of fishes the most significant of which was the rare six-gill stingray *Hexatrygon bickelli*, a new species, genus, family and suborder of fishes that she described with Phil Heemstra. That is the equivalent, in the mammal world, of finding the first primate (the taxonomic group that includes lemurs, monkeys and apes)! They named the stingray after the Eastern Cape journalist and angler Dave Bickell who found the first specimen and provided great support for their work.

Some new species are named after a peculiar feature of a fish. Examples include the stargazer mountain catfish *Amphilius uranoscopus* (sky watcher) which has upturned eyes on top of its head and the eastern bottlenose *Mormyrus longirostris* which has a long snout. Fish may even be named after their taste – the Cornish jack *Mormyrus deliciosus* produces a tasty fillet, or after an aspect of their behaviour, such as the east coast lungfish *Protopterus amphibius*, an air-breather that can live in or out of water

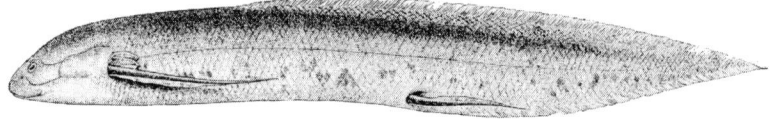

**African lungfish *Protopterus annectens***

or *Clarias cavernicola*, a cave-dweller. The name of the African lungfish *Protopterus annectens* has an interesting origin. It was named '*annectens*' (connecting) as lungfishes at the time of their first discovery were also known from South America, so it was seen as 'connecting two continents' (they have since been found in Australasia). Their generic name *Protopterus* (before proper fins) refers to their atypical, threadlike fins. The name of the first lungfish to be discovered, the South American *Lepidosiren paradoxa*, reflects the confusion that this unusual fish created in taxonomic circles.

Many fish are named after the place where they were first caught such as the sharptooth catfish *Clarias gariepinus* which was first collected in 1811 by William Burchell from the Orange River, known to locals then as the 'Gariep'. But the most interesting names are those derived from the people who first collected them (or others associated with their discovery). The Cape whitefish *Barbus andrewi* was named by one Director of the South African Museum (SAM) Keppel Barnard after a previous Director of the same institution, Andrew Smith. Rex Jubb, in turn, named a barb *Enteromius barnardi* after Barnard. The rainbow bream *Sargochromis carlottae* was named by George Boulenger after the wife of the collector, Charlotte. The collector was WL Sclater, also a former director of the SAM in Cape Town, who had the rock-catfish *Austroglanis sclateri* named after him. Yet another director of the SAM Leonard Gill had the Clanwilliam rock-catfish *Austroglanis gilli* named for him! It all sounds rather inbred!

Some new species are named after a scientist whom the discoverer admires. The western bottlenose *Mormyrus lacerda* was named by the naturalist Count Francis de Castelnau, French consul in the Cape in the 1860s, after the Portuguese explorer and astronomer Francisco Jose Maria de Lacerda. He also named the threespot tilapia *Oreochromis andersonii* after the Swede Charles John Anderson who explored Namibia in the middle 19th century where he endured terrible hardships and eventually died on the Cunene River. The banded tilapia *Tilapia sparrmanii* was named by Andrew Smith, the first director of the SAM after the Swedish naturalist Anders Sparrman who sailed around the world with Captain Cook and visited South Africa in the 1770s.

New species of fishes are not always named after the first person to find them; sometimes they are named after the relatives of someone associated with their discovery. The snake catfish *Clarias theodorae* was named in 1897 by Max Weber after his wife's niece, the Dutch artist Theodora Jacoba Sleeswijk and was, as far as we can establish, the first African fish to be named after a woman. The Zambezi dwarf stonebasher *Pollimyrus marianne* was named after Marianne Elfriede Kramer, mother of a famous researcher on electricity generation in mormyrid fishes, Berndt Kramer. The spottail barb *Enteromius afrovernayi* honoured Arthur S Vernay, an American who sponsored the 1930 Vernay-Lang Kalahari Expedition during which fishes and other animals were collected.

The well-known South African-born British ichthyologist Humphry Greenwood named a new species of barb found by Graham Bell-Cross, a previous Deputy Director of the East London Museum, *Barbus neefi*, as Graham's nickname was '*neef*' (Afrikaans for nephew). Graham had previously named a new species of deepcheek bream *Sargochromis greenwoodi* in honour of Humphry. As this species was originally placed in the genus *Haplochromis*, commonly known as 'happies', the common name for it was Greenwood's happy, which he usually was. Other cheerful southern African freshwater fishes have included the slender happy (every woman's dream) and the rainbow happy. The latter would have been the perfect national fish for South Africa but it only occurs further north in the Okavango and Upper Zambezi river systems. Some fish have been named after institutions rather than people – Vic Springer from the National Museum of Natural History in Washington, DC, called a blenny *Mimoblennius rusi* after the 'Rhodes University Smith Institute' (RUSI) and JLB Smith named a grenadier *Macruroplus ori* after the Oceanographic Research Institute (ORI) in Durban.

Fish have even been named after mythical characters. Shark expert Len Compagno, who worked at the Ichthyology Institute (now SAIAB) in Makhanda and at the SAM for several years, named a New Zealand shark *Gollum attenuates* after the anti-hero in JRR Tolkien's *Lord of the Rings* trilogy. He also named a requiem shark *Iago omanensis* after the villain in Shakespeare's *Othello* as it was so difficult to classify. The German research

*Iago omanensis*

submersible *Jago*, which has been used extensively in coelacanth research, was named after *Iago omanensis*.

Interestingly the common name of a group of popular African freshwater fishes, tilapia, is derived from the Tswana word *thlape* meaning 'fish'. The genus *Sandelia*, which includes the endangered Eastern Cape rocky *Sandelia bainsii* and the Cape kurper *Sandelia capensis*, is named after the charismatic 19th century Gaika chief Mgolombane Sandile who lead the Xhosa armies in several frontier battles. The name *Sandelia bainsii* was given by de Castelnau to honour the 19th century South African engineer Andrew Geddes Bain.

At the official launch of Paul Skelton's *Complete Guide to the Freshwater Fishes of Southern Africa* in October 1993 I gave a short speech that played on the names of our freshwater fishes:

> "The Thin-faced Largemouth was a really Scaly character who suffered from *Hypseleotris* and *Varichorhinus* veins. His mother, who was a real old Trout, had forced him to marry a Red-breasted Southern Deepbody who wore long Herrings and was well-Eeled. She was a Fiery Redfin with a Suckermouth, whom he affectionately called his 'Wimpelstert-suierbekkie'. Her friend, *Liza* Makriel, was a Largemouth Sucker who *Tinca*'d in other people's business and was a bit of a Climbing Perch who would have made Greenwood Happy. Her father, a Long-bearded Stonebasher, was *Nkupe*-rating after a Swordtail had struck him in the *Clarias*. The rest of the family was of *Myxus*-blood and included Slimjannie, Cornish Jack and Rudolph the Rednose Labeo."

**Naming other animals**: Looking beyond fishes, a plethora of new animal species have been named after celebrities. The famous broadcaster and conservationist Sir David Attenborough has had an extinct marsupial lion *Microleo attenboroughi* named after him. A beetle with bulbous legs resembling bulging biceps is called *Agra schwarzeneggeri*, after body builder turned California governor Arnold Schwarzenegger, and an extinct trilobite with an hourglass-shaped shell *Norasaphus monroeae* was named for Marilyn Monroe. A marine worm that has 'dreadlocked' tentacles is called *Bobmarleya gadensis*, an extravagantly coloured spider *Heteropoda davidbowie* and a horsefly with a golden bottom carries the name *Scaptica beyonceae*. The late Australian zoologist Steve Irwin, whose favourite expression was 'crickey', had a critter named after him called *Crikey steveirwini* and there is even a tree frog called *Hyloscirtus princecharlesi*! The late Nelson

 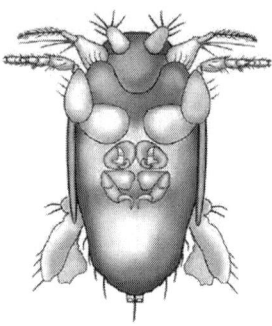

*Agra schwarzeneggeri*

Mandela had crustaceans, spiders, a varieties of strelitzia (Mandela's Gold) and chrysanthemums (Madiba), an extinct woodpecker (*Australopicus nelsonmandelai*) and a sea slug named after him. In 1973, while Mandela was in prison on Robben Island, researchers at the University of Leeds named a newly 'discovered' nuclear particle for him but it turned out that their equipment was faulty so the 'discovery' was discounted!

New species of animals have also been named after Adolf Hitler, Johnny Depp, Shakira, Angelina Jolie, Freddy Mercury, Charlie Chaplin, John Cleese, Ringo Starr, Shaka Zulu, Bill Gates, Barack and Michelle Obama, Andre Brink, Michael Jackson, Donald Trump, Elvis Presley, Mick Jagger, Kate Winslet, Ellen DeGeneres, Lady Gaga, Dick Cheney and George W Bush, some in praise of them, others as insults! Appropriately, a new beetle 'dressed in black' found near Folsom Prison in California was named for Johnny Cash. There is no prize for guessing who has had the most species named after him – Charles Darwin, of course, with over 120 and counting!

 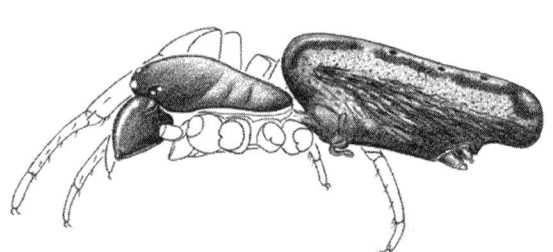

*Crikey steveirwini*  *Singafrotypa mandela*

# 8.

# How well do you know your ologys?

*"He gets an ology and he says he's failed ...*
*you get an ology, you're a scientist"*

Maureen Lipman to her grandson after he had failed most of his exams but passed sociology, in a famous British Telecom advert in the 1980s

INTRODUCTION: As a museologist I inevitably collect stuff. My collections include postage stamps and other philatelic items depicting the coelacanth, coelacanth arts and crafts, different language editions of JLB Smith's famous book *Old Fourlegs. The Story of the Coelacanth*, an unusual assemblage of over 400 different clothes pegs from around the world (including wood, bone, stone, metal and plastic pegs) and many examples of South African inventions. My most unusual collection is a catalogue of words, specifically words ending in ology, which means 'knowledge of' or 'study of', or 'logy' (figures of speech or characteristics of writing). To qualify, these 'ology' or 'logy' words must be part of the English language in that they have appeared in English dictionaries, books, technical journals or other publications or are or have been part of public discourse.

Only one-word ologys are listed so two-word terms such as Algebraic topology, Evolutionary biology or Social anthropology are not included even though they are independent disciplines. Alternate spellings are not counted separately as many ologys such as Oenology (Œnology, Enology, Oinology; wine and wine making), Paedology (Pedology, Paidology, Paidolology; children) and Semeiology (Semiology, Semology, Semeilogy; signs and symbols in language) are spelled in several different ways. An ology word may signify a genuine field of research or hobby or it may be used to add grandeur or the impression of scientific rigour to humble or even pseudoscientific pursuits. Sometimes an ology word is created to emphasise that a particular product was developed through a scientific process, such as the wines produced by a winery called Vinologist in South Africa. Throughout this essay I will honour the ologys with a capital initial letter even though they are not proper nouns.

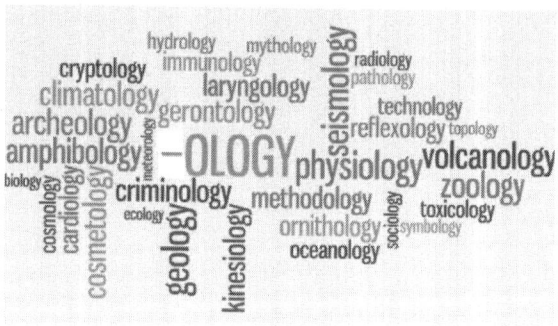

Wikipedia lists 1 056 different ologys but my list, which I started compiling in my second year at Rhodes University 54 years ago, now totals 1 692 words ending in ology and 75 in logy, a total of 1 767, from Abiology to Zymotechnology. Examples from the logy sub-list include Amphilogy (use of confusing words), Battology (needless repetition of words), Cacology (poor choice of words), Dittology (having a two-fold meaning), Leptology (tedious discourse on trifling things) and Pseudology (telling lies; first used in 1658), all of which I hope to avoid in this essay. My collection of ology words, on which I will focus here, is curated, updated and extended almost monthly and is a source of animated discussion and debate among friends and relatives. To add to this folly I also have a collection of 292 words ending in -phobia and 83 ending in -mania!

**Origin and diversity of ology words**: The earliest known use of an ology word is Meteorology (study of the weather, not meteorites, which is Astrolithology), which was first used in 340 BC as *Meteorologie* by Aristotle; it entered the English language in 1620. Another early ology is Geology (*ca* 735 AD, as *Geologie* by St Bede), initially to distinguish earthly from godly matters; it only acquired its modern meaning in 1866. Others include Tropology (figurative use of language; 1519), Aetiology (causes of diseases; 1555), Physiology (internal processes and functions of living organisms; 1564), Tautology (saying the same thing twice in different words; 1579), and the first ology to gain widespread use in the English language, Anthropology (diversity and evolution of human cultures; 1593).

Other early ologys include Demonology (demons and evil spirits; 1597), Theology (religious doctrines and beliefs; 1603), Archaeology (human antiquities; 1607), Pathology (diseases; 1611), Philology (origins and meanings of words; 1614), Technology (originally uses of the mechanical arts and applied sciences; 1615), Orthology (originally correct use of words;

1619; now research on genes that have been separated by a speciation event), Ichthyology (fishes; 1646), Cryptology (making and deciphering codes; 1655), Micrology (microscopes and microscopic organisms; 1656, 66 years after the invention of the microscope by Hans Lippershey in 1590), Phytology (plants; 1658), Mythology (myths and legends; 1659), Phraseology (choice of words; 1664), Osteology (skeletons and bones; 1670), Neurology (nervous system, 1681), Mineralogy (rocks; 1690) and Psychology (structure and function of the human mind; 1693). It amazes me how early some fields of specialization acquired specific names.

Some old ologys that are no longer in use include Idiatrology (medical science, now replaced by a plethora of ologys, see below), Lemology (infectious diseases), Mesology and Hexicology (interactions of organisms and their environments, now Ecology), Piscatology (previously fishes, now art of fishing; replaced by Ichthyology), Taxology (classification of organisms, now merely Taxonomy), Thalattobiology (life in the sea, now marine biology and other fields) and Theriology (wild beasts).

Some recently introduced ologys include Toyology (toys; first used by the retailer Toys-R-Us in 2008), Blogology (impact of blogs on society), Robopsychology (use of robots to treat medical disorders), SHE-ology (promotion of girls' and women's rights and opportunities), Mobilology and Mociology (impact of cell phones on lifestyles), Astrophysiology (stresses of space travel on human physiology) and Boffinology (popular history of scientists, science and technology, first coined by Justin Pollard in 2010). Other recent ologys include SETIology (search for extraterrestrial intelligence), Cyborganthropology (addition of electronic and robotic parts to humans), Webliology (websites), Webtechnology (history, structure, function and impact of the World Wide Web), Wikiology (impact of the internet on modern society) and Yahoology (Yahoo and its applications on the internet).

**Derivation of ology words**: While it is easy to fathom the fields of study of some ologys, such as Clownology, Coelacanthology, Egyptology, Fairyology, Ghostology, Hamburgerology (a fast food course offered by McDonalds), Hieroglyphology, Mapology, Martyrology, Mirthology, Ozonology, Treeology, Trilobitology,

Pieology, the study of pizza

UFOlogy and Woodpeckerology, many require some knowledge of Latin or Greek to determine their meaning. For example, Teratology (malformations early in human life) is derived from the Greek noun *teras* meaning 'monster' but in modern parlance also refers to the breeding of novelties in aquaculture. Other ologys whose identities are revealed by their classical roots include Agathokakology (good and evil), Anemology (wind), Brontology (thunder), Campanology (bell ringing), Cynology (dogs), Felinology (cats), Hypnology and Somnology (sleep), Kalology (beauty), Logyology (study of studying), Ludology (games and play), Nephrology (clouds), Odontology (teeth), Ombrology (rain), Plangonology (dolls), Seriology (silk production and silkworms), Tachology (speed), Toxology (archery), Trachology (wheels and rotary motion), Trichology (hair), Tsiology (tea), Vexillology (flags as national symbols) and Xylology (wood). Areology, the study of the planet Mars, is derived from the name of the Greek god Ares whom the Romans called Mars, the god of war. Perhaps you, the reader, could work out the fields of study of the following ologys: Autobuyology, Periodontology, Phytotoxicology, Retrovirology, Saccharology, Sanitechnology and Taratology?

The fields of study of some ologys appear to be obvious but are sometimes obscure. Batology is not about bats but brambles and blackberries (the study of bats is Cheiropterology), Plutology is not the study of the dwarf planet Pluto but of wealth, Kidology is about kidding and not kids, Hipponosology is not about hippopotamus' noses but the diseases of horses, Pimpology concerns pimping and not pimples and Rhinology is about noses and not rhinoceroses! Likewise, Googology is not the study of the impact of Google (Googleology) but about giant numbers; a Googol, after which Google is named, is 1 followed by 100 zeroes).

The shortest ology is simply Ology (1786), a knowledge of all things or any science or theory, and the second shortest is that bygone hobby of little boys and girls, Oölogy (1831), the collection and study of bird's eggs. A more modern six-letter ology is Nology, about the impact of the internet and other digital communications. Seven-letter ologys include Naology (church and temple architecture), Noology (origin of ideas) and Otology (anatomy, functioning and diseases of the ears). The longest ology is not Ophthalmootorhinolaryngology (eyes, ears, nose and throat; a mere 28 letters) as suggested by Google, but Ethnotraumapsychopharmacology (impact of drugs on behaviour, stress management and the mind; 29 letters). Another long one is Psychoneuroendocro-immunology (impact of the mind and the nervous and endocrine systems on immunity; 28 letters). You will have noticed that all three of these long terms are from the medical field. In fact, medicine, together with the pharmaceutical and veterinary sciences, has by far the largest number of ologys compared to other sciences (413, about one-quarter of all ologys, and counting), with the environmental sciences (especially biology), physical sciences (especially astronomy) and social sciences coming next.

It seems that medics have a need to distinguish themselves from other medics, and other mere mortals, using fancy terms. They even have a term Omphalology for a medical field of study whose specialty has not as yet been defined! After all, why be an eye, ear, nose and throat specialist when you can be an ophthalmootorhinolaryngologist! In other fields, why be a housewife when you can be an Oikologist or a farmer if you are really an Agro-ecologist? A large number of ologys begin with medical or pharmaceutical prefixes such as gastro-, haemo-, immuno-, neuro-, onco-, osteo-, patho-, pharmaco-, physio- and psycho- and more recently radio- and tele-. Common ology prefixes in other fields include aero-, agro-, anthropo-, astro-, bio-, cosmo-, cyto-, dendro-, eco-, electro-, ethno-, exo-, geo-, hydro-, macro-, micro-, nano-, palaeo-, phyto-, socio-, theo- and zoo-.

Further examples of medical ologys which indicate the extreme specializations of some medical practitioners include Aetiology (origins and causes of diseases), Andrology (male health), Aphasiology (diagnosis and treatment of linguistic problems arising from brain damage), Cardiotocology (impact of childbirth on the heart), Chronopathology (patterns of disease manifestation), Cyesiology (pregnancy and midwifery), Haemopathology (diseases of the blood), Psychopharmacology (effects of drugs on mental processes and behaviour), Psycho-oncology (impact of cancer on mental health) and Telenosology (diagnosis of diseases through the exchange of information by telephone or the internet). There is even a term, Pedology, for the use of mud as a therapeutic agent in medical treatments. But it is not only the elite in society who can claim to be ologists. As a kid in East London I once watched a railwayman, covered in soot, tapping the metal wheels of a slow-moving steam train with a metal hammer to detect cracks and other defects; I later learned that he is a Tapologist.

As science is concerned with precise measurements and predictions many ologys cover these fields. Dendrochronologists and Dendro-pyrochronologists date past events and wildfires respectively using tree rings, Lepidologists age fishes using scale rings, Skeletochronologists gauge the age of birds and mammals using bone rings and Tephrochronologists calculate the dates of volcanic eruptions and predict future eruptions by studying layers of deposited ash and rocks. What is the field of study that determines the age of long-lived marine animals such as giant clams using deposits of radioactive material from past nuclear blasts laid down in their tissues? Perhaps radiochronology? That ology was born in the last decade.

**Dubious ologys:** Some dubious fields of study have also acquired the honour of ologydom including the pseudoscience of Astrology (divining information about human affairs and terrestrial events by studying the movements and relative positions of celestial bodies) which is countered by Sporalogy, a parody of Astrology invented by the Finnish professor of astronomy Nils Mustelin in the 1980s. Sporalogy 'predicts' a person's future on the basis of the positions of trams in Helsinki at the moment of a person's birth, which parodies Astrology's claim of being able to do the same using the positions of the planets and constellations. Mustelin points out, tongue in cheek, that trams are much nearer to the person in question and are likely to have a much greater effect than distant planets and stars. Scientology, originally developed by L Ron Hubbard as a religious knowledge system with the aim of helping people to reach their full potential, has transformed into a pseudoscience but is countered by the field of Sci(fi)entology. The captain of the *HMS Beagle*, Vice-Admiral Robert Fitzroy, used the pseudoscience of Phrenology (studying bumps on the head and the shape of the nose to characterise a person's personality) to determine the suitability of Charles Darwin as a naturalist on his famous round-the-world cruise (1831-1836). Although Darwin's nose did worry him he agreed to take him onboard.

Other laughable pseudo-fields of study include Bosstrology (using astrology to predict and explain your boss's character traits), Rumpology (fortune telling by the examination of rumps and buttocks) and Xenoarchaeology (fictional science of studying the physical remains of alien organisms). Some ologys refer to seemingly trivial fields of interest: Spinology

LIFEOLOGY

Lifeology, a platform that brings scientists, artists and storytellers together examines the skill of spinning drumsticks, Mixology the art of inventing, preparing and serving mixed drinks such as cocktails and mocktails, Lipsology is the study of lip prints, Promenadology and Strollology examine the benefits of strolling and Roundology is the long-lost art of getting-round-to-it!

**Ologys with different meanings**: Some ologys have two different meanings: Anthology is the study of flowers or a collection of short stories, Algology concerns pain but also algae, Bluseology is the study of blues music but also the name of Elton John's first band, Boomology is either the study of the ebb and flow of commercial businesses or the impact of rock music on society, Buttonology is about buttons or basic training in computer literacy, Carcinology about crabs or cancer, Gelatology about laughter or Italian ice cream, Glossology is the study of the tongue or of linguistics, Herpetology is about amphibians and reptiles or herpes, Hibernology about Irish history or winter hibernation in animals and Orchidology is about orchids or testes. Others include Jeanology (denim jeans or the impact of American consumerist culture on other societies), Pyrology about fire or the effect of fevers and Tobaccology concerns tobacco and nicotine products but is also the code name of a secret mediaeval philosophical society, the Pednosophers. In some cases the same field of study has several different names: the study of children is either Tecrology or Pedology, of ants Myrmecology or Formicology, of cheeses Ecchinology, Cheesology or Fromology, of Chinese culture Sinology or Chinology, and the study of saints is either Sanctilogy or Hagiology depending on your point of view.

Some ologys have their origins in the names of books or fictional characters such as Ripperology (Jack the Ripper and his unsolved murders) and Potterology (impact of the Harry Potter books on children's literacy). Buyology was used by Martin Lindström in 2008 as the title of his book on neuromarketing, Cyberpsychology was introduced by the psychologist Kent Norman in an eponymous book in 2008 and Universology was first used as the title of a book on the universe by Japanese astronaut Mamoru Mohri, executive director of Miraikan, the National Museum of Emerging Science and Innovation in Tokyo. Redology is named after the famous Chinese novel *Dream of the Red Chamber* and Panicology was first used as the title of a book on irrational fears in modern society. Other ologys are named for experts in different fields, such as Buffettology (musings of the investor Warren Buffett) and Zimology (study of traditional African jazz), named after the Port Elizabeth-born jazz supremo Zimasile ('Zim') Ngqawana who founded the Zimology Institute and released an album entitled *Zimology*.

Some ologys refer to a passion for collecting objects rather than a professional interest. They include Argyrothecology (money boxes), Deltiology (picture postcards), Discology (vinyl records), Labology (beer and wine bottle labels), Machirology (knives), Numismatology (coins and medals), Panelology (comic books), Pegology (clothes pegs), Rivology (walking sticks), Tegestology (beer mats) and Timbrology (postage stamps).

**Ologys on our specializations**: Many ologys demonstrate the incredibly wide range of human knowledge and endeavour but also our sometimes-absurd levels of specialization. They include Aerodontology (effects of high altitude flying on teeth), Archaeomusicology (ancient origins of music and musical instruments), Astro-archaeology (relationship between the design, location and orientation of ancient megalithic monuments at the time that they were built to celestial events), Beierlology (stalkers), Biogerotechnology (products and services that slow ageing in humans), Biomusicology (biological origins of music), Bootyology (booty and other ill begotten gains), Curvology (symbolism of women's curves), Ecotheology (interrelationships of religion and nature), Enigmatology (puzzles), Entreprenology (creation and extraction of value from an environment), Ethno-entomology (relationships of humans and insects), Ethnomusicology (music of different cultures), Ethno-ornithology (relations of humans and birds), Exobiology (likely nature of life on other planets), Gizmology (gizmos and gadgets), Graphopathology (hand writing as a symptom of mental or emotional disorders) and Hedonology (human pleasure).

Other specialized ologys include Hypertrichology (facial hair), Legology (educational value of LEGO building blocks and toys in general), Mirthology (jokes and humour), Molinology (windmills, watermills and animal engines), Nessology (Loch Ness monster), Neuro-archaeology (brain and intelligence of prehistoric people), Ophiotoxicology (snake venoms), Ornitho-musicology (songs and music of birds), Palaeo-oncology (incidence of cancer in prehistoric people), Petrogeology (geology of oil and gas exploration and exploitation),

Pherology (human-carrying capacity of planet Earth), Plebiocology (lives of ordinary people and their common desires), Proto-anthropology

## 8. How well do you know your ologys?

(humans prior to the invention of writing), Spytechnology (technology of espionage), Theo-astrology (interrelationship between religion and astrology), Timpanology (drums and other percussion instruments), Xenodochieiology (hotels and inns), Xenoglossology (foreign languages), Zoomusicology (music produced by non-human animals) and, last on the list, Zymotechnology (art and technology of fermentation).

**Conclusions**: For my sins, as an anti-specialist, I have had the privilege of studying and publishing in the fields of at least 30 ologys, some obvious such as Ichthyology, Coelacanthology, Ecology, Herpetology, Mammalogy, Oceanology, Ornithology and Timbrology, others less so including Cryptozoology (rare, newly extinct or mythical animals), Islamology (Islamic culture), Peluchology (use of plush toys to teach children about biodiversity), Cycology (bicycles), Dodology (dodos), Neomuseology (using innovative interactive techniques in museum displays), Ecomuseology (developing museums to create a greater awareness of sustainable living practices) and, of course, Ologyology (study of ologys). A colleague Tengku Nasaria, director of the Petrosains Science Centre in Kuala Lumpur, Malaysia, has introduced the term Wonderology, the art of inducing wonder through the use of innovative teaching techniques in science centres, which I have also attempted to do.

Which are my favourite ology words? That is a difficult question but they would include Ferro-equinology (iron horses *aka* railway locomotives), Kookology (kooks and other eccentric people), Bubbleology (use of bubbles for understanding and teaching science), Greeniology (reducing your carbon footprint), Garbageology (household rubbish as a marker of social status), Frogology (frog racing), Beology (the art of being yourself) and Omnology (everything).

Finally, I pity those poor souls who labour through life and only become a judge, surgeon, musician, geographer, politician, astronomer or mathematician, without an ology to their name. It is clear to me that anyone worth his/her salt must surely be an expert in one or other ology. It has even been suggested that someone without an ology behind their name might come from the shallow end of the gene pool!

# 9.
# Finding Old Fourlegs: Act 1

*'All the world's a stage,
And all the men and women merely players;
They have their exits and their entrances;
And one man in his time plays many parts.'*

William Shakespeare, *As You Like It*, Act 2, Scene 7

**HOW IT BEGAN:** Every great scientific discovery is populated by compelling characters and associated with intrigues, scandals and occasional unsavoury incidents, like a real-life play. In fact, by their very nature, truly epochal breakthroughs will inevitably be controversial as they tend to upset current thinking and overturn accepted wisdom. Examples that come to mind include Galileo Galilei and the heliocentric universe, Antoine Lavoisier and the phlogiston theory, Samuel Wilberforce, Thomas Huxley, Charles Darwin and the theory of evolution, the Bone Wars between rival paleontologists Edward Cope and Othniel Marsh, the Piltdown hoax, Pierre Teilhard de Chardin and Sir Arthur Conan Doyle, and Rosalind Franklin and the structure of DNA. The coelacanth saga is no exception although the complexity of its plot and subplots may exceed that of most other great discoveries. Like a classical drama the story unfolds with an explanation of the events that preceded the main plot, the introduction of the primary 'conflict' and discussion of the main characters' back stories, followed by rising action, a climax, falling action and a denouement.

Full details of the coelacanth story can be found in JLB Smith's *Old Fourlegs. The Story of the Coelacanth* (1956), in my books *When I was a Fish. Tales of an Ichthyologist* (2015), *The Annotated Old Fourlegs. The Updated Story of the Coelacanth* (2018), *The Amazing Coelacanth* (2018), *The Fishy Smiths. A Biography of JLB and Margaret Smith* (2018) and *Curator and Crusader. The Life and Work of Marjorie Courtenay-Latimer* (2019), and in many other publications (see Reference List). Some of the fascinating back stories and subplots are discussed here.

**Discovery of coelacanth fossils:** The scene was set by an enigmatic Swiss-American palaeontologist Dr Louis Agassiz who discovered the first coelacanth fossils, which he named *Coelacanthus granulatus*, in England in 1836. 'Coelacanthus' means 'hollow spine' and refers to the hollowed-out spines in the tail fin, a diagnostic feature. He wrongly classified the coelacanth with the placoderms and other ancient fishes, an error that was later corrected by the British palaeontologist Sir Thomas Huxley who placed them with the lobe-finned fishes. Although Agassiz would later establish the celebrated Museum of Comparative Zoology at Harvard University in 1879, achieve fame as an evolutionary biologist and establish the new field of glaciology he never accepted Darwin's theory of evolution by natural selection and believed that species were 'ideas in the mind of God'. Furthermore, he was a creationist and his writings suggest that he was also a 'scientific racist' who believed that animals including humans were created in 'special provinces' with distinct populations and that these populations were endowed with unequal attributes. Although Agassiz never supported slavery his views emboldened proponents of bondage.

**Louis Agassiz**

Agassiz's legacy has recently been re-assessed and some of the places and institutions named after him have been renamed. During the 1906 earthquake in San Francisco his statue toppled from the façade of Stanford University's zoology building and landed headfirst in the concrete below; some faculty suggested that it should stay that way. Interestingly, other leading scientists who did not believe in Darwinian evolution include the famous palaeontologist Sir Richard Owen who established the Natural History Museum in London in 1881 and coined the term 'dinosaur', and South Africa's Robert Broom who claimed in his book, *The Coming of Man. Was it Accident or Design?* (1933) that 'spiritual agencies' had guided evolution as animals and plants were too complex to have arisen by chance.

**JLB Smith:** The second main character to strut onto stage was Professor JLB Smith who described and named the first living coelacanth known

JLB Smith

to science, *Latimeria chalumnae*, in 1939. Like Agassiz he was a larger-than-life character who made an impact in several fields. Smith completed his MSc in Chemistry at Victoria College (now Stellenbosch University) in 1918 and his PhD (on mustard gases) at Cambridge University in England. In 1923 he joined the staff of Rhodes University College as a lecturer in organic chemistry and carried out pioneering research on the essential oils of indigenous plants. His favourite hobby, after bee keeping, was angling and he regularly travelled to the coast to catch fishes. Few people who are aware of Smith's achievements in ichthyology realize that he was also a world-class organic chemist (and excellent lecturer) before he became immersed in the world of fishes. He was elected a Fellow of the Royal Society of South Africa at the relatively young age of 39 years and won that society's prestigious Marloth Medal (with one of his students) in 1946 for his work in organic chemistry.

JLB's slender frame, boyish looks and brush haircut sometimes lead people to underestimate him and comments such as, 'What! Is this skinny little fellow your expert?', reveal that they had difficulty reconciling his fragile figure with his towering intellect. In December 1952, as Smith prepared to fly to the Comoros to fetch the second coelacanth, the burly Air Force officers looked askance at this scrawny character who had persuaded the prime minister to provide a military airplane to fetch the fish in the Comoros but Smith's assertive manner and fanatical attention to detail soon made them realize that there was more to him than met the eye.

Smith transferred his analytical approach to chemistry to his research on fishes. In the three chemistry textbooks that he wrote he devised keys to the identification of unknown chemical substances using progressive tests with known reagents. As an ichthyologist he devised the first numerical keys to the classification of fishes using fin spine and ray counts. Furthermore, Smith's knowledge of chemistry and geology and his broad understanding of science enabled him to understand the intricacies of, for example, carbon isotopes, radiocarbon dating, the evolution of the planet and of life on Earth and continental drift, better than most specialist ichthyologists at the time which allowed him to reach insights that were beyond the ken of specialized fish scientists. His idea that radioactivity might have triggered the transformation of inorganic matter into organic life (Smith, 1956) was well ahead of its time and may still prove to be true.

Smith also had a keen appreciation of the importance of that crucial evolutionary leap by backboned animals from water onto land about 320 million years ago in which the coelacanth and its relatives may have played a crucial role. The most recent research has revealed that lungfishes, the nearest living relatives of coelacanths, are closer to the direct line of origin of the four-legged animals (tetrapods; amphibians, reptiles, birds and mammals, including humans) than coelacanths. Furthermore, Smith's observation that 'the trend in evolution was normally towards a smaller size' (1956) anticipated the Lilliput Effect proposed by Adam Urbanek in 1993 that predicts a decrease in body size in animal species that have survived a major extinction, which further highlights his excellent understanding of evolutionary processes.

JLB Smith suffered many hardships and survived several near-death experiences in his youth and early adulthood. As a youth he had a serious bicycle accident which caused internal injuries that plagued him for the rest of his life and may have destroyed one of his kidneys. When World War I broke out Smith enlisted but was initially returned to school as too young to campaign. He was determined to contribute to the war effort, so he arranged to travel to England and join the Royal Flying Corps (RFC) in late 1915 but was thwarted in this endeavour. Instead, he became a foot-slogging infantryman jammed "with thousands of other half-trained men of all ages" into a transport ship, "fed mainly on bread, tinned rabbit, and tea" and shipped off to Mombasa to join the East Africa campaign. There he contracted malaria, dysentery, the acute rheumatic enlargement of various organs and other diseases and nearly died in a Kenyan hospital (Smith, 1956). If he had joined the RFC, where the survival rate was low, or had succumbed to one of the tropical diseases ichthyology in South Africa and the coelacanth saga would have followed a different course.

The professor of chemistry when Smith joined Rhodes University College was the redoubtable Dr (later Sir) George Cory, one of the five founding fathers of the institution. Initially Cory and Smith worked well together but Smith's increasing interest in angling and other factors eventually drove a wedge between them and Cory did not recommend Smith for the vacant professorship when he took early retirement in 1925. This was one of several factors including the discovery of the first coelacanth that caused

Smith to switch careers from chemistry to ichthyology. Like Smith, Cory would become better known for his late-career contributions to another field, South African history.

Considerable intrigue revolves around JLB Smith's marriage to his first wife, Henriette Pienaar. She was from a prominent Western Cape Afrikaner family known as the 'Rugby Pienaars' as her eldest brother Theo was the captain of the first Springbok rugby team to tour Australia and New Zealand and several of her other brothers were also first-class players. Smith did not make a good impression on the conservative Pienaars. Although they respected his academic achievements, they did not find him a likeable character and objected to his eccentric behaviour. For example, he would walk into the historic *Ou Pastorie* (old rectory) almost naked 'with only a type of nappy on like Mahatma Ghandi' (Bruton, 2018) and, at the time, was following a special diet of pawpaws.

Henriette was JLB's wife for 14 years, bore him three children and sustained him during his formative years as a scientist. She was a homebody so his increasing involvement in rugged angling trips strained their relationship which eventually reached breaking point when she took their third child Shirley to her parents' home in Somerset West, never to return. Their divorce in 1937 was a disgrace on the Pienaar family who refused to have anything further to do with Smith even after he achieved world fame. He, in turn, never mentioned Henriette again in his extensive writings. During my 18 years in the ichthyology establishment in Grahamstown no-one ever broached the subject of JLB's first marriage, and I only found out about it when I did research for my book, *The Fishy Smiths. A Biography of JLB and Margaret Smith* (2018).

Less than a year after his divorce from Henriette JLB Smith married Margaret Mary McDonald, a promising young science student who had been in his chemistry class. In a TV documentary screened in 1976 Margaret recalled that she had been nervous at the wedding as she was not sure how she would be able to live with this 'great brain' whom some claimed had less than five years to live. But she also realised that he would give her complete freedom to become the tomboy that she had always wanted to be. At the age of 21 years (he was 40) Margaret was catapulted into a ready-made family and quickly adapted to her role as stepmother and research assistant. She would later comment, "A wife can be independent or indispensable, not both. I chose to be indispensable" (Bruton, 1982). She took a keen interest in his work in both chemistry and ichthyology and they soon formed a formidable partnership that lasted until his death in 1968, enduring unbelievable hardships but also enjoying many triumphs together.

**Hendrik Goosen and Marjorie Courtenay-Latimer**: The next two characters to enter the fray were Hendrik Goosen, skipper of the steam trawler *Nerine* and the young curator of the East London Museum Marjorie Courtenay-Latimer. The capture by Goosen of the first living coelacanth known to science in relatively shallow water (about 73 m) on the gently sloping, sandy continental shelf off the Chalumna River mouth near East London was an astonishing fluke as we now know that this species prefers deeper water (150-400 m) and steep, rocky offshore canyons and caves. Marjorie was summoned to Buffalo Harbour to examine the catch and found "the most beautiful fish I have ever seen" among a ragtag pile of sharks and rays (Bruton, 2019). She forced her taxi driver to take the "stinking fish" back to the museum and then struggled in vain to preserve the 56-kg specimen intact, a low-key beginning to one of the greatest scientific adventures. Many years later, in 1987, while I was assisting Professor Hans Fricke with a TV documentary on the coelacanth, we interviewed a man in East London who claimed to be *the* taxi driver. He turned out to be a fake as our research later revealed that he was the nephew of the long-deceased cabbie! Such is the evil lure of fame.

Postcard depicting the first East London Museum, Marjorie, the trawler *Nerine* and the first coelacanth

Marjorie had had no formal training in science but she was a knowledgeable, self-taught naturalist who had developed into an authority on plants, fossils and birds during a carefree childhood in the bush (Bruton, 2019). When she examined the coelacanth her naturalist instincts made her realize that "the strange blue fish" was something special, although she did not know what it was. The chairman of her museum board Dr James Bruce-Bays, a medical doctor, whom she described as "a very sarcastic old gentleman", poured scorn on her discovery, stating that "it's just a rockcod" and "all your geese are swans" (Bruton, 2019). Undaunted, Marjorie sent a letter and drawing of the strange fish to JLB Smith, the only scientist in the Eastern Cape at the time with a knowledge of marine fishes, but had to wait 44 agonizing days before he was able to visit East London and confirm that the fish was very unusual and worth keeping.

Keppel Barnard

**Keppel Barnard**: The only other scientist knowledgeable on fishes in South Africa at the time with whom Smith could communicate was Dr Keppel Barnard, director of the South African Museum (now the Iziko South African Museum) from 1946 to 1956. Barnard typified the dying breed of polymath zoologists of the 19th and early 20th centuries as he conducted research on marine and freshwater fishes, cave animals, insects and crustaceans. He was initially cynical of Smith's claim that the fish caught off East London was a coelacanth but assisted Smith in his research by sending him books on fossil fishes. Later in life Barnard, a shy and retiring man, took exception to the high public profile that Smith enjoyed.

**Smith's detractors**: JLB's high profile upset other scientists as well including staff in the Zoology Department at Rhodes University and Professor John Day at the University of Cape Town. Later, Smith's habit of using posters, leaflets and financial rewards to locate a second coelacanth also raised a few conservative eyebrows, especially in his funding organization, the CSIR. While Smith professed to hate publicity he also courted it and was a master at manipulating the media, which he saw as a vehicle for publicising the importance of science. He realised that celebrity is a fickle mistress but was streetwise enough to know that the coelacanth could focus attention on his work, and on ichthyology in South Africa, and attract funding. I followed JLB's example when I became director of the Ichthyology Institute in Grahamstown in 1982 and presented a paper at a South African Museums Association conference entitled, 'The coelacanth is a mammal'. I did not, of course, propose that it is a lactating, fur-bearing animal as opposed to a fish but that it could be 'milked' for marketing and fund-raising purposes.

One of the many people who tried to knock JLB off his perch was Leonard Thesen, a neighbour in Knysna where the Smiths had a beach cottage. To trivialize Smith's coelacanth discovery Thesen painted a picture of the fish on a curtain which he hung in his cottage, claiming that it was based on a specimen washed up on the beach in the 1920s. It was later found that the painting was based on an image from Smith's 1956 book, *Old Fourlegs*. During this author's career as an ichthyologist he received several unsupported claims that coelacanths had been washed up at various locations in the Eastern and Southern Cape, South Africa, including Gonubie, Robberg Peninsula, Knysna and elsewhere. We were once sent a photograph of a coelacanth which a Zimbabwean angler claimed he had caught near Beira in Mozambique. We were able to refute his claim by examining the pattern of white spots on the specimen (which is unique for each individual coelacanth) and matching it with that of a coelacanth that had previously been caught off Grande Comore and flown to Zimbabwe in a deepfreeze airplane. He had posed the photo in Harare!

**Identification and description of the coelacanth**: JLB Smith received Marjorie's first description and drawing of the coelacanth while on holiday in Knysna. As he had a good knowledge of fossil fish and was aware of the main diagnostic characters of extinct coelacanths he quickly realized that

**Marjorie's iconic sketch of the first coelacanth**

her description fitted this group of fishes, but, surely, he was contemplating the impossible? He knew that the fossil record of the coelacanths ended 65 million years ago so his identification would be regarded as preposterous by most scientists. He needed to be sure of his identification for his sake as well as Marjorie's. He later wrote of his first sight of the coelacanth in the East London Museum,

> "Although I had come prepared, that first sight hit me like a white-hot blast and made me feel shaky and queer, my body tingled. I stood as if stricken to stone. Yes, there was not a shadow of doubt, scale by scale, bone by bone, fin by fin, it was a true Coelacanth."
>
> JLB Smith, *Old Fourlegs. The Story of the Coelacanth* (1956)

Marjorie's troubles with her board chairman continued. After Bruce-Bays realised that the fish was valuable he decided to sell it to the British Museum. Marjorie, supported by a very indignant JLB Smith, threatened to resign if he did so and Bruce-Bays relented. A second secretive attempt by him to sell the fish a few months later was also thwarted. The decision to keep the specimen in South Africa was a wise one as it put the East London Museum and South African ichthyology on the world map. Subsequently all the hundreds of thousands of fishes that Smith collected in Africa stayed in South Africa in defiance of the colonial trend to send specimens to well-established museums in Europe.

Calendar featuring Marjorie Courtenay-Latimer issued in 2021 to celebrate the centenary of the East London Museum Society, with a 3D printed coelacanth model

Margaret Smith had a baptism of fire as the wife of an ichthyologist as the first coelacanth was discovered only eight months after their marriage in 1938. Over the next four months, working after hours so that he could fulfill his obligations in the Chemistry Department, JLB worked furiously on the description of the fish which was laid out on the dining room table at home, with Margaret on hand to assist him at all hours. In June 1939, thirteen months into their marriage, the manuscript was finally submitted for publication and four days later their son William was born! In his Foreword to my book, *The Annotated Old Fourlegs. The* Updated *Story of the Coelacanth*, William Smith wrote, "I was told that during the months preceding my arrival it was all hands to 'the fish' so when I was born there were no baby clothes ... Fortunately I was not born with scales as some had predicted!"

Notwithstanding the arduous conditions under which they had worked Smith's description of the fish was later described by the eminent American ichthyologist Dr Carl Hubbs as "perhaps the most meticulously detailed account ever accorded a fish specimen" (Hubbs, 1968) and was only superseded by the detailed accounts published 20 years later by a team of French anatomists. Despite Smith's extraordinary achievements he continued to work under primitive laboratory conditions until his death; his laboratory from 1946 to 1968 was a corrugated iron building constructed during the Anglo-Boer War (1899-1902). The new Ichthyology institute building, originally named the JLB Smith Institute of Ichthyology (now the South African Institute for Aquatic Biodiversity; SAIAB), was opened in 1977 nine years after his death.

JLB Smith named the fish *Latimeria chalumnae* and Marjorie's name was further immortalized as he placed it in a new family Latimeriidae which was later included in the new suborder *Latimerioidei*. He was criticised for naming the fish after Marjorie as she had failed to save the soft organs, which are not preserved in the fossil record, of the living fish. This accusation infuriated JLB (and Marjorie) as she had gone to enormous trouble trying to preserve the fish even

**Marjorie Courtenay-Latimer as a young woman**

though her museum was ill-equipped for handling large, wet specimens. In time JLB and Marjorie's achievement came to be regarded as one of the greatest biological discoveries of the 20th century, a classic 'black swan' event that was a total surprise and led to significant 'downstream' consequences.

**Errol White**: Dr Errol White, the pre-eminent fish palaeontologist in Britain at the time, was a thorn in Smith's side as he ridiculed the efforts of the amateur, colonial ichthyologist. White also unethically pre-empted Smith's detailed description of *L. chalumnae* by publishing a popular article, 'One of the most amazing events in the realm of natural history in the twentieth century', in the *London Illustrated News* on 11th March 1939, using content he had snatched from Smith's brief announcement of the discovery in the journal *Nature*. This incident infuriated Smith as did White's prediction that the coelacanth is a deep-water fish which led to futile expeditions (clothed in secrecy at the time) by Jacques-Yves Cousteau and Jacques Millot searching for 'old fourlegs' in the abyssal depths off East Africa in 1953 and again in 1963.

Whereas White was a palaeontologist who studied extinct fishes Smith was an ichthyologist and angler who was familiar with the habitat preferences and functional anatomy of living fishes. His prediction that the coelacanth lives in relatively shallow water (compared to the average depth of the oceans; 3 688 m) proved to be correct. JLB also accurately predicted that coelacanths are ambush predators that inhabit rocky reefs in remote parts of East Africa that are fished by traditional fishermen.

Smith was even right in predicting that most coelacanths are likely to be found south of Cabo Delgado in northern Mozambique where the Mozambique current splits and runs southwards. All the coelacanths from South Africa, Mozambique, the Comoros and Madagascar are from more southerly latitudes with only those in Tanzania and Kenya from further north. Smith (1956) also predicted, based on his examination of its anatomy, that the coelacanth "moves quietly about reefs", "catches its food by stealth and cunning", lives in habitats "with rough rocky bottom washed by a strong current", that it is "not speedy" but is a "pouncer" (ambush predator) and "would easily take a baited hook" (Smith, 1956). He was right on every count.

**Fish collecting in East Africa**: After the fuss of the first coelacanth was over JLB set his mind on capturing a second specimen so that he could examine and describe the whole fish, including its soft anatomy. His predictions led him to cast his eye to the north up the east coast of Africa and he and Margaret mounted a series of epic fish-collecting expeditions along these remote and unexplored shores from the mid-1940s to the 1950s. While at Cambridge University Smith had studied explosives and was able to make his own small 'bombs' for collecting fish during the East Africa expeditions. He even envisaged catching a coelacanth in this way as he stated in his famous radio broadcast from Durban in December 1952 that he hoped "one day to see a Coelacanth's belly breaking the foam after a blast" (Smith, 1956).

Smith exploded his bombs in midwater so as to stun fishes but avoid damage to coral reefs. He collected thousands of fishes in this way arguing that no other method (or combination of methods) would have been as efficient given the primitive circumstances and remote locations in which they had to work. In 1952, in an article published in the *South African Angler*, he even bragged that he had collected "10,000 fishes in 10 days"! One wonders, though, what impression he created on local fishermen. Today blast fishing, used indiscriminately and with little regard for its long-term impact on coral reefs, is together with the use of large mesh gillnets the most destructive form of fishing in the Western Indian Ocean. As explosives can now be made from readily available ingredients (diesel fuel, nitrogen-based soil fertilizer and powdered urea foam mixed with small amounts of gasoline and gunpowder) their use is increasing despite their being banned.

While he did exercise caution while using explosives Smith's lifelong struggle with ill-health caused him to develop a cavalier attitude towards his own mortality and he and Margaret had some hair-raising experiences. Once he dropped a bomb then rowed away but a wave washed them back over the site when it exploded. They were blown out of the water (and luckily the 20 kg of unexploded gelignite in the boat did not ignite) and escaped unhurt! In a caption to a photograph taken during their 1951 expedition to Mozambique JLB wrote, "Silhouette Island, where bombing fish caused an island to disappear", which suggests that he did cause some habitat damage.

Smith's use of rewards to entice fisherman to catch coelacanths was also controversial but eventually proved to be effective. He initially offered £10 then £100 which the French doubled to £200. In the 1950s the American government offered US$5 000 for a specimen and in 1975 the Steinhart

Aquarium in San Francisco upped the ante to an all-expenses-paid, two-week round trip to Mecca for a lucky Muslim fisherman!

The nickname of the coelacanth 'old fourlegs' was coined by a journalist and not by Smith. It derives from the posture depicted in the first mounting of the fish by an amateur taxidermist (Robert Center) which shows the two pairs of lobed fins pointing straight downwards which they never do in life. We now know that the coelacanth does not walk on its lobed fins but hovers above the bottom with its paired fins performing exquisite sculling movements. The bony support of the paired fins does, however, suggest that a smaller ancestor of the modern coelacanth with a more flattened body shape might have waddled around in shallow water on its lobed fins but they never developed the other accoutrements for life on dry land such as lungs and strong skeletons.

The discovery of the first coelacanth, the subsequent hunt for a second specimen and JLB's career pivot from chemistry to ichthyology changed his personality from an easy-going person to a stern and relatively unfriendly character. To a large extent he did not conform to the norms of society – he was an academic who liked fishing and a chemist who studied fishes – but his manic work ethic allowed him to pursue his twin passions without dropping standards. He admitted that his attitude later in life "did not always create the most cordial relations" and that he was driven to achieve more and more ambitious goals such as the production of the monumental *Sea Fishes of Southern Africa* book, first published in 1949. His goal to produce a book on the marine fishes of the Western Indian Ocean was never achieved. This project would only come to fruition some 70 years later in 2021 when the South African Institute for Aquatic Biodiversity (SAIAB) in Makhanda published a five-volume treatise on the *Coastal Fishes of the Western Indian Ocean* that took 20 years to produce and was co-authored by 101 scientists from 19 countries.

According to Denys Davis, an illustrator who worked with the Smiths in the 1940s, "He cut music out of his life completely as it stirred emotions, and this was wasteful.' This must have been hard on Margaret as she loved music and singing. After JLB died in 1968 Margaret immediately re-involved herself in choirs and music especially after her beloved sister Flora, a retired music teacher, joined her in Grahamstown. In my biography *The Fishy Smiths* I described Margaret's transformation, "It was almost as if she had undergone a metamorphosis from a working caterpillar into a vibrantly colourful butterfly that could spread its wings and fly".

## 9. Finding Old Fourlegs: Act 1

Several people who have read my books on the coelacanth, the Smiths and Marjorie Courtenay-Latimer have enquired whether JLB and Marjorie had a love affair. After all, they started collaborating in 1933 (five years before JLB married Margaret) and continued to be in close contact until Smith died 35 years later. I have found no evidence to substantiate this rumour. Marjorie and JLB had a close and respectful relationship that transcended ichthyology and included their hobbies and pastimes, but it never went further than that. Margaret and Marjorie had a slightly strained relationship in the beginning but they soon became good friends, attended conferences and functions together and regularly exchanged letters. Marjorie was occasionally jealous of the amount of publicity that the Ichthyology Institute received from the coelacanth discovery, but she took it in her stride. JLB's dedication of his book *Old Fourlegs* to Marjorie with the words, "This book is dedicated to Miss M. Courtenay-Latimer, one of South Africa's most able women", further cemented their friendship.

The fish-collecting expeditions that JLB Smith organised up the East African coast in 1947 and 1948 when he was over 50 years old would not have been contemplated by an ordinary man. They covered vast areas of relatively unknown and hostile terrain where, even today 70 plus years later transport logistics and communications are problematic. They lived and worked under harsh and primitive conditions and were imperiled by tropical storms, rough seas, bush fires, giant snakes, man-eating lions, moray eels, stonefish and other dangers yet he and Margaret collected, illustrated, preserved, identified and labelled vast collections of fishes that were transported back to Grahamstown. In 1948 they even managed to proofread the manuscript of the *Sea Fishes of Southern Africa* book in the field! As recently as 2014 man-eating lions (and hyaenas) were reported to be preying on humans in northern Mozambique (Wessels, 2014).

**Margaret and JLB Smith illustrating and examining fish in Shimoni, Mozambique, in 1952**

Notwithstanding their hardships in the field when the Smiths reached large ports such as Mombasa, Dar-es-Salaam and Lourenco Marques (now

Maputo) they were able to travel on the splendid, lavender-hulled Union-Castle liners. On these trips they would have looked and behaved very differently from the other holidaying passengers dressed in their rugged khaki clothes and pith helmets and preferring to eat tinned pilchards and peas in their cabin rather than partaking of the epicurean menu in the dining room. JLB would also have had no appetite for the ship's flippant onboard entertainments (tombola, tennisette, quoits, dancing and card games) and would have busied himself studying his notes, updating his diary and proofreading his papers.

**Eric Hunt**: When JLB and Margaret Smith visited Stone Town in Zanzibar in September 1952 while returning from a fish-collecting expedition in Kenya they met Eric Hunt, the skipper of a trading schooner *N'duwaro* (billfish in kiSwahili). The Smiths were most fortunate to have met him as he was an extraordinary man in his own right. He was born in London and schooled at Eton but forsook high society England for the adventurous life of an East African trader. Hunt was a trained engineer and a dashing, handsome man, apparently a dead ringer for Errol Flynn. During World War II he enlisted with the Royal Engineers and served in East Africa where he received citations for bravery (Stobbs, 1996). With progressively larger vessels he traded tea, coffee, spices (especially ylang-ylang), cloths and cloves between Dar-es-Salaam, Zanzibar, the Comoros and Madagascar. He was also a keen angler and aquarist with a particular interest in freshwater killifish and was deeply concerned about the destruction of marine life off East Africa decades before environmental activism became popular.

When the Smiths showed Hunt the coelacanth reward poster he was intensely interested, questioned them in depth and promised to find a specimen for them. As they parted, he joked, "When I get a coelacanth, I'll send you a cable". They all burst out laughing but Margaret did later remark to JLB, "... that man is all there. I think we can rely on his judgement if ever he gets a coelacanth; he is sound" (Smith, 1956). Tragically, Hunt died four years later when his second schooner *Hairiako* struck Geyser Bank between Madagascar and the Comoros during a storm and he and 21 of his crew and passengers were lost at sea. Hunt was on his final trip before retiring and joining his young wife in Madagascar where he planned to run an aquarium fish farming and trading company.

## 9. Finding Old Fourlegs: Act 1

**The second coelacanth**: When the Smiths arrived in Durban at the end of their 1952 expedition they were astounded to find a telegram from Eric Hunt, 'HAVE FIVE FOOT SPECIMEN COELACANTH INJECTED FORMALIN HERE KILLED 20TH ADVISE REPLY HUNT DZAOUDZI', sent from Dzaoudzi island off Mayotte in the Comoros. Most mortals after an exhausting expedition to the tropics would have settled for a delayed trip to fetch the fish but JLB's sense of ownership of the specimen that he had sought for 14 years compelled him to look northwards again and what followed was surely one of the most bizarre episodes in the history of ichthyology. In a nutshell, over Christmas JLB Smith telephoned some of South Africa's most high-ranking politicians and military leaders asking them to, as a journalist later wrote, "to provide a military airplane to fetch a dead fish for a mad scientist from a foreign country". Eventually, in desperation as he knew that *his* fish was rotting in the tropical heat, he telephoned the prime minister Dr DF Malan late at night at his holiday home in Strand and emphasised to him that the collection of the fish was "an issue of national importance" (in later years South Africa deemed it to be so as it published postage stamps depicting the coelacanth and even minted a gold coin bearing its image). In doing so Smith laid his reputation on the line; if Hunt's identification had been wrong he would have made a fool of himself and Malan.

Malan, a deeply religious Calvinist who did not believe in evolution,

Eric Hunt (right front) with the second coelacanth on his schooner in the Comoros in December 1952

was fortunately also an angler (and a savvy politician) who had a copy of Smith's *Sea Fishes of Southern Africa* book at his bedside. He realised that the man who wrote that monumental tome would not phone him for trivial reasons and patiently listened to his plea. He then said, "Your story is remarkable, and I can see at once that this is a matter of great importance. It is too late to do anything tonight, but first thing in the morning I shall get through to my Minister of Defence to ask him to allocate a suitable aeroplane to take you where you need to go". Smith later recalled, "As I put down the receiver, I felt dazed, like a man reprieved on the very scaffold, like somebody suddenly jerked from the hollows of hell to a high hill-top in heaven" (Smith, 1956), a slight exaggeration. Within two days he was flying north in Douglas C-47 Dakota 6832 (later nicknamed the 'Flying Fishcart' by the *Pretoria News*) with six crew *en route* to the Comoros to fetch his fish.

After refueling in Lourenço Marques and overnighting in Lumbo in northern Mozambique they flew across the Mozambique Channel and, with trepidation, approached the small airstrip at Pamanzi, not knowing whether they would be able to land there as the strip had been built by South African troops during World War II but its condition was unknown. However, a Dakota with a competent crew is a versatile aircraft and they

JLB Smith with the second coelacanth in the Comoros in December 1952

landed without mishap. After meeting Hunt and the local governor JLB confirmed the identity of the fish which they quickly loaded onboard lest the French change their mind and took off again, so executing one of the most audacious fish-jackings in history!

Despite the harrowing circumstances the SAAF crew, somewhat bemused by this unorthodox use of a military airplane, played a trick on Smith soon after take-off. They pretended to receive an urgent radio message, "... a squadron of French fighter planes left Diego Suarez before we took off from Dzaoudzi with orders to intercept us and to compel us to turn back to Madagascar". Smith quickly replied, "I don't know how you chaps feel about this, but I'm not going back. I don't believe they would dare shoot us down if we refused to turn, but I would be prepared to chance that rather than turn back." Captain Peter Letley burst out laughing and the tension was eased (Smith, 1956; Bruton, 2018).

JLB Smith cannot be accused of fear of failure on this epic trip. His view was that nothing was more important than his search for the coelacanth and that no barrier was insurmountable in the fulfillment of this quest. Other scientists gasped in awe at the sheer audacity of his gambit but also secretly admired his chutzpah. When they reached Durban an exhausted Smith gave a 40-minute interview to the media that was broadcast (live and

JLB and Margaret Smith with the crew of Dakota 6832 and the coelacanth in its coffin in December 1952

recorded) worldwide. It was a dramatic speech delivered in painstaking detail without notes that captured the spirit of his maniacal determination and exposed many listeners for the first time to the agony and ecstasy of making a great scientific discovery. During the interview JLB Smith wept as he recalled the dramatic events that had unfolded and his audience wept with him. Decades later people who had heard the interview told me that they (or their parents or grandparents) had never been so moved by a scientific event. I occasionally play excerpts from the recording of this interview in my talks on the coelacanth and they never fail to elicit an emotional response from the audience. Other than Professor Christiaan Barnard's first human heart transplant in 1967 I doubt that any scientific event in South African history has had such an emotional impact on society.

**DF Malan**: The next day they flew to Grahamstown, picked up Margaret and William (in contravention of military regulations) and flew to Cape Town as Smith was determined that Dr Malan should be the first to see the precious fish. *En route* they broke yet another military rule when they flew low over Knysna and dropped a parcel containing a message for JLB's son Robert updating him on their news! When Smith showed the fish to the prime minister he exclaimed, "My, it is ugly. Do you mean to say that we once looked like that?", to which Smith deadpanned, "H'm. I have seen people that are uglier." The next morning they took off early for Grahamstown but first flew low over the Strand and 'bombed' the prime minister's estate with copies of the early morning papers! In Grahamstown they were met by an enthusiastic crowd that included Marjorie Courtenay-Latimer. "She kissed me before them all, and nobody was embarrassed, not even myself", Smith recalled. Eventually the exhausted crew was free to fly back to Pretoria just in time for their Old Year's Eve party on 31st December 1952.

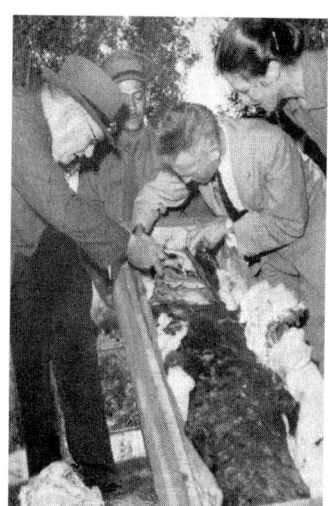

JLB Smith showing the second coelacanth to Prime Minister DF Malan

The total duration of the return flight was 34 hours and five minutes airtime, covering over 7 400 km across three countries. Dakota 6832 continued to do sterling service for many years thereafter and in December 1992 flew to Grahamstown and back with three of the original crew to commemorate the 40th anniversary of the epic flight. After her retirement 6832 became a permanent display at the Ysterplaat Air Force Base in Cape Town, resplendent in her original livery.

In 1956 Smith visited the prime minister again and presented him with a copy of his book, *Old Fourlegs. The Story of the Coelacanth*. There was an interesting sequel to this subplot 30 years later when Margaret Smith, now widowed and director of the JLB Smith Institute of Ichthyology, and I visited the then-Minister of National Education (later President) FW de Klerk at the Union Buildings in Pretoria. We gave him a copy of the 1986 edition of *Smiths' Sea Fishes* and, in thanking us, he laughingly said, "I will keep it next to my bed in case I receive an urgent call from you". He clearly knew about the incident 34 years earlier; such is the power of the coelacanth story!

**Jan Smuts**: In his book *Old Fourlegs* Smith praises DF Malan but criticizes the more science-orientated politician Field Marshal Jan Smuts who refused to support his work. In 1945 Smith had asked Smuts, then prime minister, to provide a military airplane to collect fishes killed by a submarine disturbance off Walvis Bay in South West Africa (now Namibia) but Smuts refused to even meet him. Smuts appeared to have little interest in marine life as his natural history passion was terrestrial plants especially grasses. He participated in several plant-collecting expeditions organised by Kew Gardens in the 1920s and 1930s in South and East Africa and had two new species of plants *Digitaria smutsii* and *Pteronia smutsii* named after him. In contrast Malan was an ordained minister of the Dutch Reformed Church and a politician who championed Afrikaner nationalism and laid the foundation for the repressive apartheid regime in South Africa.

**Jan Smuts**

**Franco Prosperi**: The fish-jacking caused considerable tension with the French authorities who promptly banned foreign scientists from collecting or researching coelacanths in the Comoros (and Madagascar) with the result that 26 of the next 29 specimens caught went to French museums. The foreign embargo was broken in 1954 when, in a bizarre subplot, a team of Italian divers *Spedizione Zoologica Italiana* lead by Franco Prosperi secretly dived off Mayotte and subsequently claimed that they had photographed a live coelacanth "in its dim and aerie habitat" at a depth of 15 m. They published illustrated articles on their spectacular scoop in the popular media in England, France and Italy but their image turned out to be a fake as they had photographed a crude, homemade inflatable coelacanth hovering over a shallow coral reef! To an informed observer everything about the image was wrong but this farcical exercise probably hardened French attitudes towards foreign research on the coelacanth in the Comoros.

**Jacques Millot**: The first live coelacanth to be seen by a scientist was a 142-cm-long female caught off Mutsamudu on Anjouan island in November 1954 and towed back to the harbour by a fisherman in his dugout canoe. It was kept alive for about 20 hours in a flooded boat where it was briefly observed by the French scientist Dr Jacques Millot (1954) from Madagascar. He recorded its unusual swimming movements and its aversion to bright light and also noticed that the fish, although a predator, was docile and unaggressive and did not try to escape. The subdued behaviour of coelacanths would later allow submersible divers to study the fish up close and even remove scales and insert tags during the course of their research.

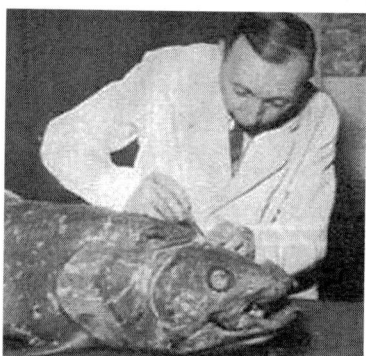

**Jacques Millot examining a dead coelacanth in France**

# 10.

# Finding Old Fourlegs: Act 2

*'And then the justice,*
*In fair round belly with good capon lined,*
*With eyes severe and beard of formal cut,*
*Full of wise saws and modern instances;*
*And so he plays his part.'*

William Shakespeare, *As You Like It*, Act 2, Scene 7

**COELACANTH DETENTE:** The second act in the coelacanth saga follows a more peaceful trajectory as the charismatic Smith departs the scene, nationalist imperatives are cast aside and internationally collaborative research teams take centre stage. The initial period of hostility between the French and their rivals was followed by an era of 'coelacanth detente' during which the French and Comorian authorities donated coelacanth specimens to museums and other institutions and the coelacanth became a 'peace maker'.

The truce was initiated in July 1956 when Jacques Millot and Jean Anthony of the Muséum National d'Histoire Naturelle de Paris donated the 15th coelacanth caught to their archrival the British Museum (Natural History) in London. In 1976 the Comoros donated a specimen to Algeria and in 1982 (and again in 1983, 1984 and 1997) coelacanth diplomacy was extended to China. Since then the French and Comorian authorities have generously donated coelacanths to museums, corporations and individuals in the USA, Japan, Kuwait, South Korea and South Africa. The most unusual gift received by the United Nations on its 40th birthday in 1985 was probably a coelacanth in a beautiful carved mahogany cabinet.

In 1967 President Charles de Gaulle presented a coelacanth to a Japanese media mogul M Shorikim who had made substantial cultural contributions to the Comoros. When President François Mitterrand visited the Comoros in July 1990 he was presented with a large female specimen which now has pride of place in the *Grande Gallerie de l'Evolution* in the Muséum National d'Histoire Naturelle in Paris. Some of the diplomatic donations did not go according to plan. In April 1985 Comorian President Ahmed Abdallah sent a specimen to a construction company in Japan but it was confiscated by Japanese customs 'due to inappropriate export permit' and subsequently claimed by the Ibaraki Museum. Another specimen was highjacked by Tokyo customs officials in 2009 and now languishes on permanent display in the Customs House in East Kyoto Minato-ku!

In 1991 President Djohar donated an 80-kg female coelacanth to the then-South African Minister of Foreign Affairs Roelof 'Pik' Botha to thank him for his role in developing cordial relations between South Africa and the Comoros. Pik accepted the gift with his usual grace and promptly donated it to the Ichthyology Institute in Grahamstown. When Eugene Balon from Canada, Humphry Greenwood from the British Museum (Natural History) and I dissected this frozen specimen in December 1991 we found that it contained 67 eggs. We also drank some of the notochord fluid in case the unproven rumour that it is a life-prolonging elixir proved to be true!

Old Fourlegs. The Story of the Coelacanth by JLB Smith, first published in English in 1956

**Writing *Old Fourlegs*:** After the discovery of the second coelacanth in 1952 JLB Smith was persuaded by his wife Margaret to write up his involvement in the coelacanth saga. In 1953 JLB had nearly drowned while angling when he was washed off the rocks by a freak wave in the Narrows at Knysna. Margaret admonished him and insisted that he should put pen to paper immediately before it was too late. He reluctantly agreed to do so and penned the manuscript in longhand during long spells floating on Knysna Lagoon in his aluminum skip *Blikkie* in the company of his terrier *Marlin*, the only circumstances in which he could find peace and quiet.

The memoir *Old Fourlegs. The Story of the Coelacanth* was published by Longman, Green & Co. in London in June 1956 and became an international best seller eventually selling over 800 000 copies in five English editions (plus braille) and ten foreign language editions including German (1957), French (1960), Russian (1962), Estonian (1964), Afrikaans (1965), Czech (1969), Slovak (1970), Dutch (1973), Latvian (1977) and Japanese (1981). The Russian edition (1962) had an extraordinary first print run of 100 000 copies, the Latvian edition (1977) 65 000 copies and the Estonian (1964) 30 000 copies (Bruton, 2018b)! JLB benefitted privately from the sales of *Old Fourlegs* and bought a prime piece of real estate, the Western Heads at Knysna, with the profits. Some say that he bought this property as it included his favourite fishing spot JLB's Rock in the Knysna Narrows.

*Old Fourlegs* was not just a chronicle of discovery but also an analysis of the intimate thoughts of a practicing scientist and of the acute frustrations that he experienced in achieving his goals. It reveals a great deal about JLB's character, resourcefulness and determination and his passion verging on *terribilità* to find an intact second specimen of the coelacanth. The book gave many laypeople their first insight into the workings of a scientist's mind and introduced them to the arcane disciplines of ichthyology and palaeontology. It is also an extraordinary example of how JLB Smith was able to share his enthusiasm for science with non-scientists.

I have often wondered whether 'old fourlegs' would have become such a stimulating icon of science if it had been discovered by an unsung anatomist beavering away in, say, Antananarivo, who published his findings in technical journals. Would it have been doomed to the same obscurity as the poor lungfish which is as important in the pantheon of evolution as the coelacanth but is hardly known to anyone? The coelacanth needed someone with the audacity, dare one say arrogance, of JLB Smith to bring it to the world's attention and use it as a flag-bearer to tell the story of evolution. The timing of the first discovery (1938) helped as it created excitement and an escape from reality in a dreary world that was about to be engulfed by World War II. When the second specimen was found in 1952 the world was a different place, full of optimism and with an increased enthusiasm for science. The mood was jubilant and people were primed to share in the success of JLB's 14-year search. Moreover, as Keith Thompson (1991) at the California Academy of Sciences would later write, "At a time when science seemed to have all the answers and threatened to take all the mystery out of life, *Latimeria* made zoology romantic again and science the realm of real people, like fishermen and small-town museum curators."

**Death of JLB Smith**: JLB Smith drew the curtain on his career in January 1968 when he committed suicide at the age of 71 years. A chemist to the end he poisoned himself with a lethal dose of cyanide. One of his two suicide notes read, "For some years I have suffered from severe mental depression ... the sight of one eye has almost gone ... back pressure is proving troublesome ... I live in perpetual fear of becoming bedridden and helpless ... I prefer to take this way out, probably only a brief anticipation of nature' (MM Smith, *pers. comm.*, 1986). He had strode the South African and international stage with authority and aplomb and made a lasting impact on ichthyology in Africa and beyond. Despite his frail body he was a man of incredible energy, drive and enthusiasm who lived several lifetimes in one. In the end not even his formidable will power could reverse the ravages of time. As Romeo said on his deathbed, "Come, bitter conduct, come, unsavoury guide! The dashing rocks thy sea-sick weary bark!"

**Coelacanth conservation**: The coelacanth is known in the Comoros as *gombessa* which means 'taboo', perhaps a reference to its oily flesh and foul taste (although they are occasionally eaten in Madagascar and Tanzania) but more likely due to its sacred status. In April 1987 we established the Coelacanth Conservation Council/*Conseil pour la Conservation du Coelacanthe* (CCC) in Moroni, Grande Comore, to coordinate coelacanth research and conservation efforts worldwide. In the 1990s Hans Fricke and this author, together with Jerry Hamlyn from the Explorer's Club in New York and many other scientists and conservationists, mounted an international campaign to conserve and de-commercialize the coelacanth. Eventually *Latimeria chalumnae* (and later the Indonesian coelacanth *L. menadoensis*) was elevated to Appendix I of CITES (which means that it cannot be traded for commercial gain) and to Endangered on the IUCN Red Data List.

But international policies alone will not save the coelacanth – its ultimate survival depends on the attitudes and actions of the people in the countries where it lives. In this regard the Comoros is taking the lead. In 1989 a decree by President Said Djohar introduced control over trade in coelacanths and stipulated that all specimens caught must be sold to the government and in 1994 he signed the convention (CITES) that forbids trade in coelacanths. In November 1995 an Association for the Protection of Gombessa was established in Dzahadjou, Grande Comore, and a Coelacanth Centre was set up. In March 2011 a second Coelacanth Centre built by Comorians was opened in Itzounzou on Grande Comore.

In Tanzania the Tanga Coelacanth Marine Park was established near Pangani in 2009 but conservation measures are not strictly enforced there as both gillnetting and blasting take place within the park and over 60 coelacanths, including pregnant females, have been killed in the past decade. Madagascar has recently established several marine protected areas and recommendations were made in 2021 for a strictly protected marine sanctuary to be proclaimed around the Onilahy canyon near Anakao where the highest concentrations of coelacanths are known to occur (Cooke *et al.*, 2021; Bruton *et al.*, 2021).

The discovery of a coelacanth colony living off the coast of northern Zululand in South Africa in October 2000 by Peter Timm and co-divers contributed to the decision to rebrand the Greater St Lucia Wetland Park as the iSimangaliso Wetland Park and the declaration of this sanctuary as a World Heritage Site. It also resulted in the establishment of the multi-disciplinary African Coelacanth Ecosystem Programme (ACEP) which is now in its third phase and has done remarkable work on the fish and its habitats along the south-east and east coasts of Africa under the overall leadership of the South African Institute for Aquatic Biodiversity (SAIAB) in Makhanda.

A live coelacanth was seen by mixed-gas divers off the south coast of KwaZulu-Natal in November 2019, over 300 km south of the iSimangaliso Wetland Park, which indicates that they are more widespread in South Africa than previously thought and suggests that the first specimen caught off East London may not have been a stray.

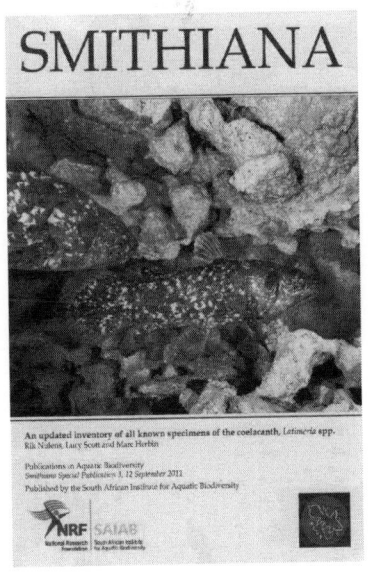

Coelacanth inventory produced by Rik Nulens, Lucy Scott and Marc Herbin in 2011

**Rik Nulens:** A remarkable aspect of coelacanth documentation and conservation is that a detailed inventory of every specimen known to have been caught has been kept initially by French, American and Japanese scientists then by this author and his staff and, since 1994 by coelacanth guru Rik Nulens in Maaiseik, Belgium, with collaborators Lucy Scott in South Africa and Marc Herbin in France. The latest inventory lists 337 *L. chalumnae* and *L. menadoensis* specimens caught since 1938 in seven countries (Nulens, 2020). Not even very rare and critically endangered species such as the California condor, Sumatran rhinoceros, mountain gorilla, leatherback turtle, pangolin and Yangtze finless porpoise have been inventorised to the same level of detail.

Rik is another example of someone from a different profession (engineering) who became fascinated by the coelacanth and is now an authority on the capture history of the fish. Like Hans Fricke who was inspired to study coelacanths after reading the German edition *Vergangenheit steigt aus dem Meer* of Smith's *Old Fourlegs* book, Rik was stimulated to update the coelacanth inventory (and bibliography) by reading the Dutch edition, *Vis op de Loop*.

Bob Denard

**Bob Denard**: Bob Denard, the legendary French soldier of fortune, was the villain in the coelacanth drama. He was involved in four coup attempts in the Comoros and from 1978 to 1989 headed President Ahmed Abdallah's 500-strong *Guarde Presidentiale* (GP), at the same time developing a lucrative business empire that included land, hotels and the GP itself. He married seven times, had ten children, converted to Islam, adopted the name Said Mustapha Mhadjou and eventually became a citizen of the Comoros. During one of our expeditions in the 1990s he was effectively in charge of the archipelago when we met him in his office to discuss logistics which were problematic at that time. He was seated on a throne-like chair in the dark while we sat on brightly lit, low chairs surrounded by the GP.

Denard radiated intrigue and connivance but he did help us. To my eternal shame we asked him for assistance and protection while moving around the islands as we were experiencing difficulties with security and transport. He was very obliging and we soon had priority access to flights and GP escorts wherever we went. On one expedition with Hans Fricke some of our camera equipment was stolen. We alerted the GP who visited the local village elder and instructed him to ensure that, within an hour, the equipment should be returned. When we went back to the village we found not only our missing equipment but also a large cache of other cameras and paraphernalia that had been stolen from other visitors.

On another coelacanth expedition we had an audience with President Said Mohamed Djohar in his palace. The meeting was intended to be a serious one about marine conservation but the decorum was broken when one of our expedition members, Canadian palaeontologist Richard Cloutier, sat on a precious carved mahogany chair which shattered into pieces. The president was initially horrified but then burst into raucous laughter, which effectively ended the meeting. Later we presented him with a drawing of the incident by our expedition artist Roy Reynolds. Another less happy event occurred at Hayahaya Airport when an overzealous customs official, against our vigorous objections, sniffed and tasted a packet of formalin powder (which we use to preserve fishes), thinking that it was an illegal drug. Predictably he collapsed in a heap and had to be rushed to hospital.

**Hans Fricke, Jürgen Schauer and Karen Hissmann:** The next player to appear on stage, in 1969 and again in 1975, was a youthful German marine biologist Hans Fricke who initially searched for coelacanths off Madagascar and the Comoros using SCUBA, but without success. At the end of the second trip, he said to his wife, "Next time I come here, I'm coming with a submarine". Nobody believed him but his perseverance paid off when he returned to the Comoros in 1986 with the *Geo* research submersible and

**A youthful Hans Fricke on an expedition in the Sahara Desert in 1963**

carried out a series of dives to a depth of 200 m off Grande Comoro.

In January 1987 submersible pilot Jürgen Schauer located the first live coelacanth off the Comoros and, over the next 21 years, using the manned submersibles *Geo* and *Jago* (capable of diving to 400 m) Hans and his team carried out an extraordinary 350 plus dives and documented the distribution, abundance and behaviour of 141 individual fishes off Grande Comore. One 27-kg male caught by an 80-year-old Comorian fisherman Ali Mmadi, now on display in the Fukushima Aquarium in Japan, was recognised by its pattern of dots as a specimen photographed by Hans and his team off Grande Comore on ten occasions in 1995, over 10 years earlier! This is the only specimen that the Jagonauts documented live in its natural environment that was subsequently caught by a fisherman. In 1999 Jürgen Schauer and Karen Hissmann were the first to see and film Indonesian coelacanths alive in their natural environment in a cave at a depth of 155 m off Sulawesi Island (Karen Hissmann, *pers. commun.*, April 2021). The Japanese coelacanth researcher Masa Iwata from the Fukushima Aquarium also needs to be recognised for the significant contributions he has made to our understanding of the biology of Indonesian coelacanths.

Hans Fricke, coelacanth researcher extraordinaire

Hans and his Jagonauts have also contributed strongly to the ACEP project by identifying 32 fishes in collaboration with Kerry Sink of the South African National Biodiversity Institute in the iSimangaliso Wetland Park, extracting scales for tissue analysis and carrying out behavioural studies. By 2018 researchers in iSimangaliso had encountered coelacanths 108 times of which 46 observations were made by experienced mixed-gas diver Peter Timm, who tragically died in an unrelated diving accident in 2014. Another famous diver who has contributed to the ACEP project is the French underwater photographer Laurent Ballesta from *Andromeda Océanologie* who has taken spectacular photographs of the coelacanth in iSimangaliso.

In 2009 Hans and his team continued their coelacanth research off Grande Comore using *Octopus*, the megayacht owned by the late co-founder of Microsoft Paul Allen, as their mothership and sophisticated remote-operated vehicles that could dive deeper than the maximum diving depth of *Jago*. They found that ecological conditions in water deeper than about 500 m off Grande Comoro were unsuitable for coelacanths but their research was cut short by the threat of piracy. Other millionaires who have contributed to the coelacanth project include the American publisher Herbert Axelrod, founder of *Tropical Fish Hobbyist* who supported research by John Maisey on fossil coelacanths in Brazil and 'Shogun Noodle', the diminutive Japanese industrialist and inventor of instant noodles who has sponsored some of Fricke's recent research.

Hans, Jürgen and Karen have also been deeply involved in coelacanth conservation, sometimes in unorthodox ways. In September 1989 a Japanese team attempted to catch a live coelacanth for the Toba Aquarium using traps deployed from the mothership *Pacific Oak*. Karen and Jürgen made a nuisance of themselves by secretly diving in the *Jago* and placing a note 'Let them be where they are' in the trap using the mechanical arm of the submersible. When the Japanese retrieved the trap and read the note they upped anchor and left the Comoros. Their attempts to catch a live coelacanth had been unsuccessful but they did obtain valuable video footage of live coelacanths at 184 m using remote-operated cameras. No live coelacanths have as yet been caught and displayed in a public aquarium anywhere in the world.

Today the main direct threats to coelacanths in the Comoros, Madagascar and along the east coast of Africa are large-mesh gillnets set in deep water to catch sharks for the shark-fin trade and explosives. Explosives are regularly used off the coast of Tanzania and occasionally off the Comoros. In 2008, while diving at a depth of 170 m off Grande Comore in the *Jago*, Hans and Jürgen felt the jolt of dynamite depth charges being exploded by fishermen! If the typical pattern of 'development' of Western Indian Ocean fisheries is followed - traditional fishing methods followed by gillnets and dynamiting - then coelacanths are severely endangered.

On 17th November 2008 I had the rare privilege of diving with live coelacanths in the *Jago* off the south coast of Grande Comore near Singani. The submersible, piloted by Jürgen Schauer, gingerly entered a cave inhabited by coelacanths at a depth of 198 m and we spent an unforgettable hour with the ancient fishes in their watery lair. "They were serenely tranquil, almost spiritual, and stared blankly back at us across space and time. I was transported back to a primeval age when they had wandered the seas and freshwaters before the dinosaurs had evolved. ... Somehow their looming presence was foreboding to me. They are the ultimate survivors, yet we are concerned about their conservation. How arrogant we are!" (Bruton, 2015).

**Jean-Louis Geraud**: The 'court jester' in the coelacanth drama is the amiable French dive-instructor Jean-Louis Geraud, founder of *Gombessa* Plongée (Coelacanth diving) in Moroni, Grande Comore. This popular fellow, brimful of *joie de vivre* and *bonhomie*, was able to talk bluntly to high level officials about environmental degradation in the Comoros and stress the importance of coelacanth conservation, of which he was a passionate advocate. During several expeditions from the JLB Smith Institute of Ichthyology (now SAIAB) to the Comoros in the 1980s he helped us to track down dead specimens in a variety of unlikely places, such as hotel lobbies, restaurants, private houses and the presidential palace, as well as in museums, research institutions, dive schools, military camps, food storage companies, taxidermy workshops and fisheries schools. One of the taxidermists we met, S Bakari in Mutsamudu on Anjouan island, showed us the coelacanths that he had crudely mounted, stuffed with coconut coir and supported internally with wooden planks. Sadly, in the process, he had discarded valuable internal organs and even eggs and juveniles. Jean-Louis kept his own dried *coelacanthe de compagnie* under his bed! He has observed and photographed several tethered live coelacanths that had been brought to the surface by Comorian fishermen and accompanied this author on a memorable scuba dive off the west coast of Grande Comore near Hahaya where we saw giant hollow sponges the size of jacuzzis.

**Mark Erdman and Arnaz Mehta**: On 18th September 1997 a young American marine biologist Mark Erdman and his wife Arnaz Mehta, who were researching coral reefs in Indonesia, decided to visit the local fish market in Manado, North Sulawesi, with friends. To their astonishment a fisherman nonchalantly wheeled a coelacanth past them on a wooden cart! Mark had read Smith's book *Old Fourlegs* and was aware that coelacanths had only been reported from the Western Indian Ocean, so he found it difficult to comprehend that he had made such a sensational discovery under such everyday circumstances. He was subsequently caught up in a web of politics, intrigues and turf battles which committed him to secrecy for over a year and even had the honour of describing the new species snatched from him by local scientists who named it *Latimeria menadoensis*.

Mark wrote to me a year later, on 24th September 1998, "I simply couldn't believe that we were viewing something which was unknown to science. ... it is a humbling and exciting reminder that humans have by no means conquered the oceans ...". The discovery of *raja laut* (King of the Sea), as it is known in Indonesia, was a sensation as it meant that *Latimeria* is far more widespread than we had first thought and even hinted that they may occur over undersea volcanic fields in the Pacific Ocean.

On 25th September 1998 I replied to Mark, "I do agree, though, that the coelacanth tends to bring out the worst in people; I have experienced this many times, sometimes in fairly nasty ways ... Welcome to the weird and wonderful world of cryptozoologists and coelacanthophiles. It's a bug that never let's go – enjoy the ride!". In July 1998 Mark and Arnaz's fortunes changed when they had a live coelacanth brought to them by a fisherman and were able to swim with it for over an hour in the tropical waters off North Sulawesi.

**Robert Gess**: The remarkable association of the coelacanth with the Eastern Cape in South Africa has continued with the discovery by Robert Gess of the Albany Museum of juvenile coelacanth fossils in road cuttings near Makhanda. The extinct fishes, which he has named *Serenichthys kowiensis*, are over 300 million years old (the oldest coelacanths known from Africa) and represent the oldest find of a nursery of prehistoric coelacanths anywhere. One of Gess' fossil sites (Waterloo Farm) is under the hills over which I once walked with JLB Smith.

Robert Gess

*Serenichthys kowiensis*, a new species of extinct estuarine coelacanth described by Robert Gess

**Fate of coelacanth specimens**: Coelacanth specimens around the world have suffered many different fates. The Aga Kahn IV, alias Prince Shah Karim Al Hussaini, had a smoked specimen on display in his yacht before he donated it to the Palaeontology Institute at the University of Pavia in Italy. A large 90-kg female caught off Madagascar, which most museums would give their eye teeth for, is on display in the Fort Dauphin Town Hall in Madagascar. Another 35-kg specimen caught by migratory *Vezo* fishermen off Madagascar was used as bait except for 200 scales that were saved for science. A coelacanth caught off Tanzania in 2009 never reached a museum as it was eaten by the fishermen! Further examples of the destiny of coelacanth specimens are given in my autobiography, *When I was a Fish* (Bruton, 2015).

**The coelacanth as lead actor**: But the most extraordinary actor in the coelacanth drama is, of course, the fish itself. Butterflyfishes may be more beautiful, stonefishes more dangerous, blobfishes uglier and sailfishes faster but the coelacanth has a unique mystique. Their anatomy is a potpourri of traits associated with bony fishes, cartilaginous fishes and even four-legged animals. Like bony fish they have a bony skull, bony supports of the lobed fins, scales, a symmetrical tail and a lateral line but like sharks they have a cartilaginous, tube-like notochord (and no articulated, bony backbone), heart chambers in the form of a tube (straight in coelacanths, S-shaped in sharks), a rostral organ on the snout that is sensitive to electrical currents, spiral valve in the intestine, rectal gland, fatty liver, teeth in clusters, elevated salt levels in the blood and high urea levels in the tissues.

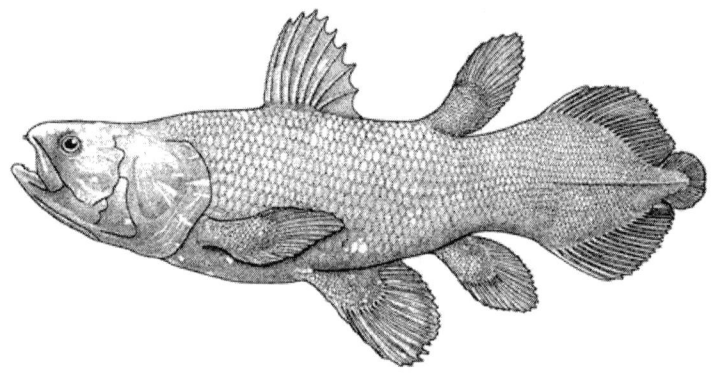

*Latimeria chalumnae*

Coelacanths resemble four-legged animals in the structure of their inner ears (with birds) and slow metabolic rate (with some amphibians). They uniquely have very elongate scales that overlap threefold, two pairs of lobed fins and six other fins, an extra lobe in the tail fin and the ability to increase their gape by lifting their upper jaw (made of cartilage) *and* lowering their bottom jaw. They even have nano-sized optical 'lenses' on their scales that improve the visibility of their white spots, a novel way of harvesting light in the ocean's twilight zone. Like crocodiles, bats, kangaroos, kiwis and some cats they have a reflective layer (*tapetum lucidum*) behind the retina of their eyes that reflects visible light back through the retina, making it easier for them to see in dim light. The swimming action of coelacanths is most unusual with the alternate sculling motions of the paired fins resembling those of the ocean sunfish.

Other traits of the coelacanth are puzzling and make one wonder how they have survived this long. They have the lowest known blood haemoglobin count of any fish and a low gill surface area, so they have a limited ability to absorb oxygen and must live in cool, well-oxygenated water. To continue the superlatives, they also have the lowest resting oxygen consumption of any fish and the slowest metabolic rate of any vertebrate! Consequently, even though they are predators, they swim slowly, drifting with the ocean currents like underwater albatrosses, although they are capable of lightning-quick ambushes to catch prey. Far from being degenerate they are highly sophisticated fishes despite their ancient lineage.

Coelacanths are large, with females of *L. chalumnae* reaching 179 cm and 98 kg and males 165 cm and 65 kg. Some extinct species in the genus *Mawsonia* (from Texas, of course) grew to three times that size. The only other Indian Ocean reef fishes of comparable size are sharks, Napoleon wrasses, a few large groupers, kingfishes and wreckfish. The latest research suggests that coelacanths may live for over 100 years, the longest for any tropical fish and exceeded only by Greenland sharks and some other polar fishes. The largest population of coelacanths known is that off Grande Comore, which Fricke and his team have estimated to number between 300 and 500 adults, but the population sizes in canyons off Tanzania, the northern coast of Mozambique and the south-west coast of Madagascar, all of which offer promising habitats, are unknown.

The Indian Ocean coelacanth is iridescent blue in colour with distinct pink-white spots while the Indonesian coelacanth is brown with white patches and brilliant gold flecks over its upper body and fins, although these colour differences may reflect ambient light conditions. Coelacanths gather in caves or under overhangs in groups of up to 16 fishes during the day and hunt at night for fishes and squid. They are able to hunt in the dark as they use mainly non-visual senses such as electro-reception to locate their prey. During these nocturnal hunting forays they perform peculiar headstands, perhaps to detect the electrical fields of prey hidden in the sand with their electro-sensitive organs. Hans and Jürgen have even been able to induce coelacanths to do headstands by emitting weak electric currents from the *Jago* submersible!

**The ultimate survivors:** Another remarkable feature of the coelacanth group of fishes is their longevity. Their fossil record stretches back at least 420 million years, longer than that of any other bony animal. They evolved more than 170 million years before the dinosaurs and co-existed with the 'terrible lizards' for 185 million years, then out-survived them. We even have evidence that dinosaurs ate coelacanths as a specimen of the extinct North African fish-eating dinosaur, *Spinosaurus aegyptiacus*, was found to have a fossil coelacanth in its stomach! Coelacanths are the ultimate survivors, having persisted through four major extinction events, epochal changes in marine and freshwater ecosystems, massive sea level changes and the formation and breaking up of vast continents and ocean basins.

They were also the first animals to evolve many advanced anatomical features such as bony skulls, jaws and teeth, armoured scales and lobed fins but, uniquely during their evolution, they have lost their upper jawbone, the maxilla. Although their external form is conservative and has hardly changed they have an advanced breeding strategy (see below)

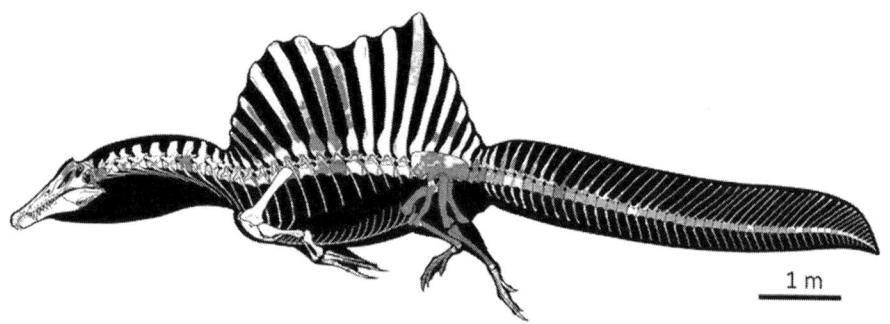

*Spinosaurus aegyptiacus*

and a sophisticated physiology that has allowed them to survive in widely changing environmental conditions. Charles Darwin coined the term 'living fossil' to describe living species that have changed so little over time that they resemble extinct animals. The coelacanths, together with the lampshells, horseshoe crabs, nautiluses, velvet worms, sharks, hagfishes and platypuses, are probably the best examples. Furthermore, coelacanths and lungfishes are closest in the evolutionary lineage to that ultimate trailblazer, the first backboned animal to venture onto land. Recent research has shown that the lungfishes are the closest surviving relatives of the tetrapods although the modern species are highly modified compared to the original tetrapod ancestor.

**Sex and the coelacanth:** No drama is complete without sex and the spicy coelacanth saga is no different. The one prediction by JLB Smith on the biology of the coelacanth that proved to be incorrect was his statement that "the coelacanth doubtless sheds its eggs inside a special case, quite possibly like those produced by some sharks and rays. Who will be the first to find one?" (Smith, 1956). In one of the biggest *faux pas* of his career Smith turned down the gift of a 65-kg female specimen by the Comorian government in 1962, saying, "My coelacanth work is done". On his recommendation the specimen was sent to the American Museum of Natural History in New York where it languished in the collection for 14 years before it was dissected by James Atz and colleagues. To their astonishment they found that the fish was pregnant with five late-term, unborn embryos, the first evidence that it is not an egg layer like most fishes but a live bearer (about 5% of fishes). The discovery was announced by Atz (1976) in an article entitled, '*Latimeria* babies are born, not hatched' and forced scientists to rethink their views on the 'primitive' fish.

In fact a British palaeontologist Professor DMS Watson had been the first scientist to reveal that coelacanths are live bearers. In 1926 he found clear impressions of two unborn embryos inside the fossil of a Triassic coelacanth which means that coelacanths were giving birth to 'babies' at least 40 million years before the mammals first evolved! Also, as far as we know, the coelacanth is the only live bearer in which the male does not have an intromittent organ for internal fertilization – the female extrudes her oviduct to receive the sperm.

Numerous coelacanths with eggs and pups have been caught and studied including a 179-cm, 98-kg female caught off Pebane in Mozambique in

August 1991, the largest living coelacanth known, which contained 26 late-term pups measuring 308 to 358 mm in length and weighing between 410 and 502 gm each. Most coelacanth produce less than 70 eggs but an astonishing 197 eggs in three different size groups were found in a female caught off Anjouan in 1955 which indicates that they produce eggs in batches like some sharks and rays. A 79.2-kg female caught off Tanzania in 2010 contained 25 fertilized eggs although others were probably lost during capture. A coelacanth's eggs, which are the size of an orange, are the largest of any fish and they give birth to the largest-known juveniles. The number of juveniles produced by each female over her lifetime, estimated at about 140 by Fricke, is the lowest breeding rate of any fish and more akin to the reproductive strategy of reptiles. As far as being actors in the evolutionary drama is concerned the coelacanths, like the peripatus, axolotl, lamprey, platypus and kangaroo have torn up the script and fashioned their own storyline.

The fact that has astounded most scientists about the coelacanth's breeding strategy is its incubation period – the time from the fertilization of the eggs to the birth of the juveniles – which is an astounding 36 months, 14 months longer than that of the next longest incubation period (22 months in the African elephant). The slow rate at which coelacanth young grow is illustrated by the fact that they weigh about 500 gm at birth whereas new-born African elephants weigh about 120 kg! One mystery that has not as yet been solved is whether coelacanths guard their newborn young. Based on my knowledge of live-bearing fishes I predict that they do as parental care makes evolutionary sense for a fish that has invested so

Pregnant coelacanth caught off Mozambique in 1991

much energy in each of a few, large young. There is, however, no proof of this theory as we have never seen recently born juveniles in the vicinity of their parents.

A controversy has also broken out as to whether coelacanth embryos might have a placenta-like organ, comprising a vascular labyrinth between the walls of the juvenile's yolksac and the mother's uterus, through which nutrients and gases could be exchanged. Such an organ occurs in some other live-bearing fishes such as sharks and surfperches and in live-bearing amphibians, reptiles and mammals. One group of ichthyologists supports the idea of 'placental vivipary' (live bearing with a placenta) in coelacanths but others don't. The jury is still out and the debate continues which is good as coelacanth research has thrived on controversies! Of course, the term 'live bearer' is a misnomer as eggs that are laid outside the mother's body are just as alive as newborn juveniles.

**The naked coelacanth:** A further debate has raged about the desirability of displaying and studying live coelacanths in a public aquarium. The initial failure of very well-equipped expeditions from Japan, the USA and Belgium to catch a live specimen, and an international outcry led by Fricke, this author and many others, resolved the problem temporarily but the issue is still being discussed. The fact that many coelacanths are being caught and brought live to the surface (although often damaged or weakened) off Tanzania and that coelacanths can, almost at will, be encountered at close range by mixed-gas divers and manned or unmanned research submersibles in the iSimangaliso Wetland Park makes it easier to envisage how a live capture could be achieved but there is considerable opposition to the idea. To succeed it would need to be an internationally collaborative effort that focuses on research and conservation rather than

**Juvenile coelacanth**

on the commercial benefits of displaying the fish to the public. One of the challenges would be to decide which aquarium would be the lucky host as a worst-case scenario would be for there to be a rush on coelacanths, which would further threaten their survival.

The general opinion among coelacanth biologists is that they *may* be relatively easy to keep and feed in captivity but that specimens should not be caught for this purpose. Rather, coelacanths that are already being caught as a bycatch of the sharkfin industry off Tanzania or Madagascar should be considered for captive display. Another alternative which this author is pursuing in Cape Town is to display realistic life-size, 3D-printed robotic coelacanths in aquaria to satisfy the public's yearning to observe the strange 'fish' without endangering it.

**Cultural history:** Unlike most fishes but in common with many mammals and extrovert reptiles (birds) coelacanths have a rich cultural history in terms of the multitudinous ways in which they have interfaced with humans not only as food or museum specimens but also as symbols. In fine art, poetry, prose, performing arts, movies, crafts and even politics they have symbolized surprise, rarity, longevity, tenacity, immortality, survival and primitiveness. Like other fishes they have also symbolized virility, fecundity and elusiveness.

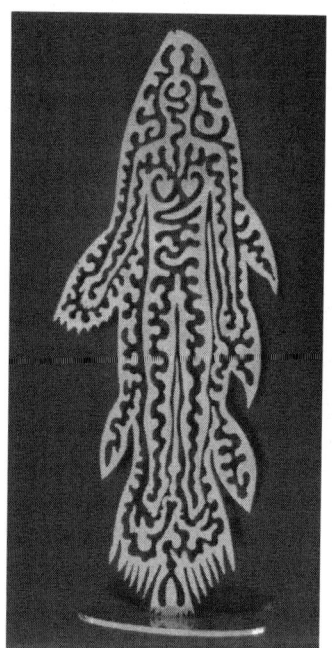

Stainless steel sculpture of the coelacanth by Uwe Pfaff

Perhaps the most remarkable coelacanth artwork is a linocut by South African artist Hylton Mann called *Oncanthusphere* in which coelacanths symbolise our excessive dependence on fossil fuels. It depicts three businessmen assembled around a coelacanth egg, the source of life, while the fish spirals off into a desolate landscape and extinction. Cape Town artist Uwe Pfaff has produced a series of exquisite sculptures of the coelacanth in stainless steel and a street artist in Woodstock, Cape Town, has created a magnificent mural on the ancient

## 10. Finding Old Fourlegs: Act 2

fish. Outdoor sculptures of coelacanths in soapstone and metal are common in the Comoros and South Africa and a coelacanth-shaped wind vane can even be found atop a clock tower in Ulaan Baatar in Mongolia! In the Albany District around Makhanda, where the coelacanth is part of the collective DNA, craftworks and jewelry have been made using materials as varied as wood, soapstone, ceramics, steel wire, recycled aluminum cans, ring-can openers, colourful fabrics and even used tea bags and elephant dung!

Coelacanth pendant

On our 1987 coelacanth expedition to the Comoros Eugene and Christine Balon and I produced a *gyotaku* (Japanese fish print) of a coelacanth at the fisheries school in Mutsamudu on Anjouan island. We brushed paint onto the fish then draped a piece of white linen over it and rubbed it until every detail of its scales, fins and body had been transferred onto the cloth, which was then peeled off. To this day my *gyotaku* hangs in my lounge in Cape Town. Some visual artworks depicting the coelacanth predate the discovery of the living fish. Two silver artefacts from Bilbao and Toledo in Spain which accurately depict coelacanths have been examined by Fricke and dated by expert silversmiths to the 17$^{th}$ or 18$^{th}$ centuries. The unusual shape of the fish probably inspired the artists but where did they obtain their models?

Logo of Coelacanth Brewing in Norfolk, Virginia

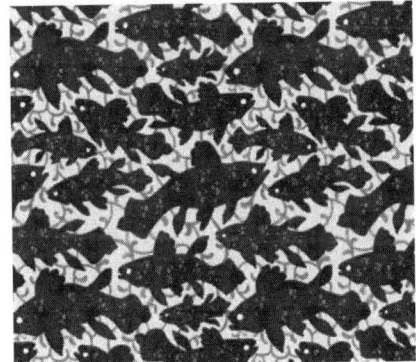

Coelacanth fabric design

**Coelacanth stamp issued by the Comoros in 1954**

In everyday language the word 'coelacanth' is used, somewhat unfairly, to refer to people who are old-fashioned and conservative or have risen Phoenix-like from the dead. Winston Churchill once referred to a backbencher in the House of Lords, who had remained silent for nearly 20 years and then got up to make a great speech, as "that coelacanth of a man". Churchill himself was referred to as a "political coelacanth, a prehistoric monster fished up from the depths of the past", and Harold Macmillan was described as "the coelacanth of modern British politics – the last of the old and the first of the new" (Bruton, 2018b).

Coelacanths have appeared on the postage stamps of at least 22 countries including 16 where they do not occur. The Comoros has issued 12 different sets of coelacanth stamps as well as coins and notes depicting the iconic fish.

In 1989, at our instigation, the South African Post Office issued a set of four coelacanth stamps to commemorate the 50th anniversary of the description of *L. chalumnae*.

**Five-franc coin depicting the coelacanth issued by the Comoros in 1984**

## 10. Finding Old Fourlegs: Act 2

Coelacanth stamps issued by South Africa in 1989

**The denouement**: Coelacanths are an important part of our heritage not for their nutritional or commercial value but because of their evolutionary, scientific, ethical and cultural significance. They also have inspirational and intellectual value and occupy a unique place in the human consciousness. Despite all the controversy, comedy and tragedy that has been associated with their discovery they have opened our eyes to the wonders of the natural world and emphasised how little we know about the oceans and why it is so important to conserve them. The chance discovery of the coelacanth also suggests that there might be other large, undetected marine animals in the oceans. In 1976 a huge 5.5-m-long megamouth shark was accidently caught in the sea anchor of an American naval ship near Hawaii. Despite intensive searches less than 100 specimens of this rare plankton-eating shark have since been sighted or caught.

In common with other endangered species such as the giant panda, great white shark, Przewalski's horse and California condor coelacanths have many of the attributes of animals that are vulnerable to extinction such as rarity, large size, slow gene flow, low dispersal rates, genetic bottlenecking, infrequent breeding, few offspring and a high level of specialization. There is therefore widespread concern that human interventions in the ocean may end their 420-million-year-long reign. Ultimately the coelacanth story is about survival – the survival of an ancient lineage of fishes but also that of the more recent and very fragile human experiment. Our existence and that of all other organisms on the planet depends on the welfare of the oceans.

The coelacanth is a beacon of hope in an otherwise bleak world (from an environmental point of view). By allowing us to peek into the past it has empowered us to glimpse the future. This perspective has made us realize that we need to undergo a fundamental change in our mind-set and behaviour if we are to become more responsible custodians of the biosphere. Judging by our recent record I suspect that, long after *Homo sapiens* has shuffled off this mortal coil, coelacanths will continue their lonely vigil. The denouement of the coelacanth saga, that part of the plot when the threads come together, is simple: let them be where they are. What we need to do, as the evolutionary misfits, is to minimise our damage to the biosphere, and especially the oceans, and leave space for 'old fourlegs' to continue living in its age-old way.

*'Last scene of all,*
*That ends this strange eventful history,*
*Is second childishness and mere oblivion;*
*Sans teeth, sans eyes, sans taste, sans everything.'*

William Shakespeare, *As You Like It*, Act 2, Scene 7

# 11.

# Lessons from the dodo

*'Her visage darts forth melancholy, as sensible of Nature's injurie in framing so great a body to be guided with complimentall wings, so small and impotent, that they serve only to prove her bird.'*

Sir Thomas Herbert (1634)

**INTRODUCTION:** I cannot claim to be a genuine dodo researcher but I have gone to considerable lengths to familiarise myself with this iconic bird and its habitat during several holidays in Mauritius and visits to museums in South Africa, Mauritius, Tanzania and England. In Mauritius I tramped along the seashore and lowlands where dodos once lived, visited the caves where they sheltered from storms and waded out to the islet of ïl d'Ambre off the north-east coast of the main island where they were last seen by a sailor Volkert Evertsz in 1662, over 350 years ago. In the south-east I visited the wetland *Mare aux Songes* near Mahébourg where schoolteacher/naturalist George Clark discovered hundreds of perfectly preserved dodo bones in 1865. Further research at this site by international

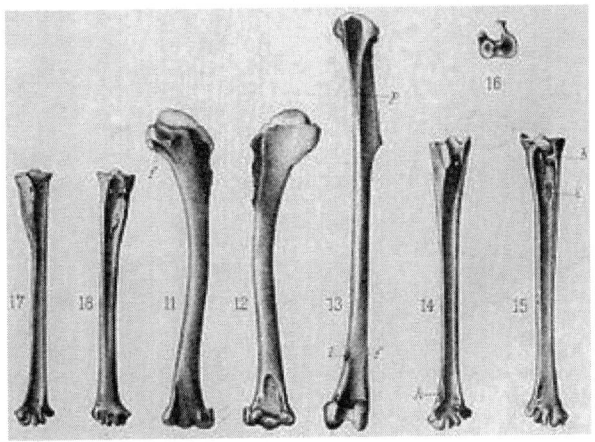

**Dodo bones from *Mare aux Songes***

teams in 1891/2, 1920s, 1993 and 2005/6 has resulted in the discovery of additional dodo bones as well as the remains of an extinct giant tortoise.

Although I did not dig for dodo bones when I visited *Mare aux Songes*, which is now covered with reeds and lined by eucalyptus, voyager palm and rose pepper trees, signs of the earlier excavations were obvious and it was thrilling to be in a place where dodos once lived. Today one goes there not to see dodos but to experience the breathless silence which they have left behind. The *Songes* deposits were laid down in a similar way to the much older (4.5-million-year-old) fossil deposits at the West Coast Fossil Park north of Cape Town, i.e., in a depression where flooding rivers deposited suspended material *en route* to the sea. The dodo deposits are too recent to have been fossilized and have been preserved by the high alkalinity and anoxic properties of the wetland muds.

In the Mauritius Natural History Museum in La Chaussée, Port Louis, I examined the dodo displays and part of their collection of over 3 000 dodo bones, and the art and craft markets around *Grand Baie* provided copious proof of the fact that the dodo is still very much alive in the minds of people today. In particular I was entranced by the evocative paintings of the dodo by the Mauritian artist, Vaco Baissac, which capture the essence of its exotic mystique and tragic loss.

During 1976/77 I spent my postdoctoral year in London doing research in the Fish Section of the Natural History Museum. While there I was shown a dodo foot in the bird collection but it had gone missing when I visited the museum again 20 years later (I don't have it!). In the distressingly decrepit Natural History Museum in Stone Town, Zanzibar, my wife Carolynn and I discovered an almost complete dodo skeleton on display behind a barricade of sturdy iron bars. The curator told us that several attempts had been made to steal the bones as they are reputed to have life-prolonging properties. I have also scoured the international literature to find out as much as possible about the natural history of the bird and the reasons why it went extinct.

But what is the relevance of the long-gone dodo today? Surely there are more pressing issues to discuss? I argue that the dodo's extinction teaches us valuable lessons that we can use in addressing one of the most urgent challenges facing humankind today, the extinction of the planet's biodiversity. Furthermore, it is interesting to speculate why, among the many millions of animals that have gone extinct in past millennia, several thousand at the hands of humans, we have forgotten most of them but not the dodo (or *Tyrannosaurus rex*). Why is this? I suggest that the dodo's demeanor, its interesting natural and cultural history and its appearance in a popular children's book are the reasons why we still remember it.

**Early research:** Our knowledge of the natural history, behaviour and appearance of the dodo comes mainly from the diaries of sailors and early settlers in Mauritius, the Western Indian Ocean island that is its only known home. Although Mauritius and nearby Rodrigues Islands are first mentioned in sailor's logs from the 10$^{th}$ century onwards and were regularly visited by Chinese and Arab mariners from the 13$^{th}$ century, they left no record of the dodo. The

**Sailing ship and the dodo**

first record, a very brief mention by Dutch sailors, was in 1598. In 1601 a Commander Harmenszoon recorded in his ship's log, "… these birds can be caught on the Island of Mauritius in vast quantities, for they are unable to fly, and provide good sustenance and refreshment". Within three years of their discovery they had become a standard food source for mariners. The last reliable record of the dodo is in 1662, at the time when Jan van Riebeeck was building his fort in Cape Town. It took humans only 64 years to kill it off and it was one of the first large animal species whose extinction can be blamed directly on humans and their pets.

Another notable extinction brought about by humans in a short time was that of Steller's sea cow, a giant relative of the dugong that lived in the northern Pacific Ocean. It was first reported by sailors in 1741 but by 1768,

**Postage stamp of Steller's sea cow, hunted to extinction by 1768**

Early sailors hunting dodos and giant tortoises in Mauritius

only 27 years later, it had been hunted to extinction. I once saw a skeleton of this remarkable animal in the Museum of Natural History in Brussels – it is an awesome creature that deserved a better fate. Furthermore, it is not only rare species that have gone extinct at the hands of humans. There were hundreds of millions of passenger pigeons in North America before European colonists arrived in the 16$^{th}$ century but hunting and habitat destruction decimated their numbers and the last one, the celebrated 'Martha', died in Cincinnati Zoo in 1914.

**Dodo ancestors:** The closest extinct relative of the dodo is the solitaire *Pezophaps solitaria*, a flightless bird from Rodriguez Island, 617 km to the east of Mauritius. The solitaire was discovered in 1634, 36 years after the dodo, and died out in 1761, 127 years later and 99 years after the dodo. It was taller and swifter than the dodo, had a lighter beak, larger wings and longer legs and tail and lived in relatively inaccessible mountainous terrain, forests and caves, hence its more prolonged survival.

Research published by Oxford University scientists in 2002 on DNA from the dried leg of a dodo in the university's museum revealed that the dodo and solitaire diverged about 26 million years ago, yet the island of Mauritius is only about 8 million years old. They must therefore have lived on one of the many islands that previously existed in the western Indian Ocean but have now sunk beneath the ocean surface. Rodrigues Island, where solitaires lived, is even younger (four million years). There are rumours that a 'white dodo' occurred on Reunion island, but no hard evidence has been found to substantiate this claim.

# 11. Lessons from the dodo

**Early seafearers' reports:** A report written by an unknown sailor in 1631 referred to the dodos as 'mayors':

> "These Mayors are superb and proud as they displayed themselves to us with a stiff and stern face and wide open mouth, very jaunty and audacious of gait, and would scarcely move a foot before us, their war weapon was the mouth, with which they could bite fiercely, their food was raw fruit, they were also not well-adorned, but were abundantly covered with fat, and so many of them were brought aboard, to the delight of all of us."

Two scientists Hugh Strickland and Alexander Melville who reviewed the literature on the dodo in the mid-19th century reported:

> "Their wings are too short and feeble for flight, the plumage loose and decomposed and the general aspect suggestive of gigantic immaturity ... so rapid and complete was their extinction that the vague descriptions given of them by early navigators were long regarded as fabulous or exaggerated .... and the birds became associated with the griffin and phoenix of mythological antiquity."
>
> Hugh Edwin Strickland and Alexander Gordon Melville,
> *The Dodo and its Kindred* (1848)

People began to wonder whether the dodo ever existed.

Many other species also went extinct on Mauritius and Rodriguez at that time including giant saddleback tortoises, domed tortoises, geckos, green parrots, parakeets, a night heron, flightless ibis and various pigeons, starlings and owls. (The giant tortoises that occur on Mauritius today were imported from Aldabra.)

We know a great deal about the dodo not only from the logs of sailors but also from the notes of scholars who studied them as dodos, dead and alive, were distributed widely around the world and several dozen landed up in museums and private menageries in the southern and northern hemispheres. When Admiral Jacob Cornelius van Neck left Mauritius in 1599 to return to Amsterdam after a tour of duty he took a live dodo with him and a second

**Early painting of the dodo by Edouard Poppig**

bird arrived in Amsterdam six months later on another ship. One of these dodos was sent to Prague where it became part of the fabled menagerie of Emperor Rudolf II of the Hapsburg Empire, surviving until at least 1610.

In June 1628 Emmanuel Altham, who had accompanied the English diplomat Sir Thomas Herbert to Mauritius, sent several presents back to his family in Essex, England, on board the *William* under Captain Perce; they included a 'jarr of ginger', some 'beades' and a live dodo! On 25th July 1647 the following note was addressed to Willem Verstegen, Head of the Dutch Embassy in Japan,

> "Dear Willem, Please accept the following gifts for the Honourable Shogun. We do not have any exotic animals at the present time, so please accept, as token of our affection, a white deer … as well as a doddaers bird from the island of Mauritius."

**Real dodo eggs and specimens**: The East London Museum claims to have a genuine dodo egg, possibly the only one in existence, but this claim has never been proven as a lawsuit between the family of the donor and the museum prohibits any invasive research on the egg. The egg was gifted to Lavinia Bean, great aunt of the first curator of the museum, Marjorie Courtenay-Latimer, in January 1847 by Captain von Syker who traded between South Africa and Mauritius and Lavinia gave it to Marjorie in 1935. Some scientists have suggested that the egg is that of an ostrich but Marjorie, who had an excellent knowledge of bird's eggs, stated in a 1953 publication that the pitting on the eggshell is different from that on an ostrich egg. It also not an egg from the recently extinct giant elephant birds of Madagascar as it is much smaller than their eggs. It is likely to be a dodo egg but we will not know until it has been properly examined.

**Dodo skeleton**

There are no real stuffed dodos in existence anymore; the last one, in the Oxford University Museum in England, was destroyed in 1755 although hundreds of dodo bones exist in the collections of at least 30 museums in fifteen countries. In South Africa the Iziko South African Museum and the Durban Natural Science Museum have real dodo bones and skeletons in their collections and displays. All modern stuffed dodos are replicas made from the feathers, beaks and bones of other birds mostly by the London taxidermy company Rowland Ward of Piccadilly.

**What is a dodo?** The French naturalist Carolus Clusius, who interviewed sailors who had visited Mauritius, was the first scientist to mention the dodo, in a scientific paper published in 1605. In 1856 the famous British paleontologist Richard Owen, superintendent of the British Museum, arranged for about one hundred dodo bones to be sent to his museum from Mauritius. After carefully examining them he declared that the dodo was a large, short-winged, flightless, fruit-eating pigeon; other scientists later confirmed his finding. Research on the DNA of dodos at Oxford University has revealed that their closest living relative is the Nicobar pigeon from the Andaman Islands in the northern Indian Ocean east of Sri Lanka from which they evolved about 43 million years ago. Live Nicobar pigeons, which are attractive birds with blue heads and mauve, green and blue plumage, can be seen in the World of Birds in Hout Bay, Cape Town.

The dodo was a large bird standing about 75 cm tall and weighing 15-20 kg depending on the season as they stored fat in their bodies in anticipation of food shortages. (In their natural habitat they were not as fat as the birds depicted in Roelandt Savery's famous paintings as he used bloated specimens that had been overfed during long sea voyages as his models.) Their plumage was grey and they had small, vestigial wings and a large fluffy tail. According to sailors who witnessed their behaviour they flapped their wings during courtship and territorial defence but they could not fly. Dodos had an enormous, hooded beak which they used for feeding and defence, large yellow eyes and strong claws on their feet. Solitaires had an interesting defensive tool in the form of a heavy bony growth on the 'elbows' of their wings with which they could give attackers a hefty blow.

In 1634 Sir Thomas Herbert described the dodo as follows,

> "The dodo .... for shape and rareness may antagonize the phoenix of Arabia; her body is round and fat, few weight less than fifty pounds. It is reputed more for wonder than for food, greasie stomackes may seeke after them, but to the delicate they are offensive and of no nourishment. Her visage darts forth melancholy, as sensible of Nature's injurie in framing so great a body to be guided with complimentall wings, so small and impotent, that they serve only to prove her bird."

**Dodo habitat and biology**: Dodos lived along the seashore, in coastal lowlands and on low hills, occasionally sheltering in caves on higher ground during storms. They were therefore very accessible to sailors who landed on the shore. They ate fruit including plantains, wild figs and dates and almost certainly also insects foraged in the forest litter and crabs and other shellfish found along the seashore. They carried one large stone in their gizzards with which they ground up their food, like many modern birds.

Dodos laid one large oval egg about 10 cm long and guarded it in an open nest made of leaves on the ground (in contrast, solitaires made a nest on a raised mound). The dodo's incubation period is unknown but was probably similar to that of the solitaire (seven weeks). The parents, who apparently mated for life, fed the newly hatched young which took about nine months to mature. Dodos were sociable animals that gathered in groups. As they had not been exposed to predators during their recent evolution they were unafraid of humans and tended to approach sailors who walked towards them. They even tried to defend dodos that were attacked by sailors, to their peril. In contrast, solitaires were cautious and solitary and demonstrated a healthy fear of sailors although they also tried to defend captured mates.

Artwork of dodos mating

A sailor commented in 1631:

> "These burgermeesters are superb and proud. They displayed themselves to us with stiff and stern faces and wide-open mouths. Jaunty and audacious of gait, they would scarcely move a foot before us. Their war weapon was their mouth, with which they could bite fiercely ... many of them were brought on board to the delight of us all."

Although dodos were able to fight off attackers to a limited extent with their strong beaks and huge claws they were relatively easy to capture or kill as they could not fly or swim and ran relatively slowly. They were apparently very hardy and survived sea journeys of several months and were therefore captured as a source of fresh food. At least 14 live dodos reached ports in countries as far afield as England, India, Japan, Holland and Indonesia. In contrast, their temperamental cousins, the solitaires, were intolerant of sea voyages and none reached foreign shores.

The taste of dodo meat received different reviews. One sailor who caught dodos in 1602 lamented, "… even with long stewing they would hardly become tender, but stayed tough and hard with the exception of the breast and stomach, which were extremely good". Another sailor was more positive, "The sailors brought 50 birds back to the ship, of which 24 or 25 were dodos, so big and heavy that scarcely two were consumed at mealtime, and all that were remaining were flung into salt." One of the early names of the dodo was the Dutch *Walghvoghel* (tasteless bird). In contrast, solitaires were described as 'tasty'.

**Why did the dodo go extinct?** The original scientific name of the dodo was *Didus ineptus*, a name that suggests that they were so incompetent that they were destined to go extinct, but this is a harsh judgment as they were beautifully adapted for the environment in which they lived until humans arrived. In his satirical booklet *How to Become Extinct* (1941) Will Cuppy suggests that,

> "The dodo didn't have a chance. He seems to have been invented for the sole purpose of becoming extinct and that was all he was good for. I'm not blaming the dodo, but he was just a mess. For one thing, his appearance was against him. … He had an ugly face with a large, hooked beak, a tail in the wrong place, wings too small and weak for flight, and a very prominent stomach. You can't look like that and survive."

Early Portuguese mariners called them *doudous* (fools) because they would allow themselves to be approached closely and hit on the head with clubs.

The current scientific name of the dodo, *Raphus cucullatus* or 'extinct flightless bird with a cowl or hood', is kinder. It is true that dodos had a formidable combination of 'I'm likely to go extinct' characters such as living on only one island, being friendly yet relatively defenceless, having a slow breeding rate, nesting on the ground and an inability to fly, swim or run fast. Yet, looked at from the perspective of natural selection, these were all sensible traits for a bird that lived on an island richly endowed with food at ground level that lacked large predators.

**Why lose the power of flight?** Most birds are characterised by being able to fly but this ability comes at a huge physiological cost. To improve their power to weight ratio flying birds must have a low body weight, a sleek, streamlined body with a small head and retractable legs, powerful wings, light, hollow bones, minimal fat, a high metabolic rate and eat frequent small meals. Most other anatomical and physiological features become subservient to these demanding needs. As a wise old owl once said, "If you want to fly, give up everything that weighs you down".

When the dodo 'decided' to settle on a tropical island and not migrate anywhere else it had to 'consider' its options. Do I remain constrained by the severe requirements of flight or do I rather adapt to the circumstances in my safe, new, food-rich environment where I can become 'top dog' and become relatively fat and slovenly? They 'chose' the latter route and lost their ability to fly as a result of which their wings and tails were reduced, they lost their large, primary feathers, lowered their metabolic rate and, in doing so, saved a great deal of energy that could be directed into functions other than flight.

Having abandoned flight they could now do things that they could not do before such as develop a large, heavy beak and wider gape to eat bigger prey and larger meals, store food for lean seasons, become big and strong to defend their nests and young, lay larger eggs and produce larger young which are more independent when born and dig in the ground with their powerful claws. They transformed from being nimble and light to slow moving and ponderous. Of course, by losing their ability to fly they were no longer able to move as quickly, migrate across oceans, feed, nest or roost in trees or on cliffs or easily escape predators and storms.

The loss of the ability to fly is quite common in birds. There are over 40 species of living flightless birds including the ostrich, emu, cassowary, rhea, kakapo (a parrot), takahe, kiwi, penguin, a cormorant and a variety of ducks, teals, ibises, grebes, woodhens, rails, crakes and moorhens. There are also many species of extinct flightless birds including the diving puffin, moa, elephant birds and great auk. Oddly, immatures of the giant coot from South America can fly but the adults are flightless! Nor is the loss of a defining characteristic rare in Nature as various animals have lost seemingly vital organs or functions during their evolution such as sight, limbs, pigment and hair. Although scientists originally estimated that it would take about 10 000 years for flightlessness to be permanently acquired recent research on the lesser short-tailed bat in New Zealand has revealed that it evolved from flying to relative flightlessness in just a few centuries (Daniel, 1979)! Furthermore, bizarre fossils of a giant walking bat found near Central Otago on South Island in New Zealand in 2015 suggest that these winged mammals have been crawling around on forest floors for over 16 million years (Gough, 2015)!

The extinction of the dodo was not only caused by hungry sailors but also by the many animals that they introduced, deliberately or accidentally, onto Mauritius, which included crab-eating macaque monkeys, dogs, cats, goats, black rats and pigs. Dodo eggs and young and even adults would have been relatively easy pickings for these predators and omnivores which also competed with them for food and space.

**Could we bring the dodo back to life?** In 1901 Oliver Herford wrote a poem for his daughter which sums up the plight of the dodo:

> *"This pleasing bird, I grieve to own*
> *Is now extinct. His soul has flown.*
> *To parts unknown, beyond the Styx*
> *To join the Archaeopteryx.*
> *What strange, inexplicable whim*
> *Of Fate was it to banish him?*
> *When every day the numbers swell*
> *Of creatures we could spare so well:*
> *Insects that bite, and snakes that sting,*
> *And many other noxious thing.*
> *All these, my child, had I my say,*
> *Should be extinct this very day.*
> *Then would I send a special train*
> *To bring the dodo back again."*

This poem raises the question: could we use dodo DNA to bring the bird back to life, as Russian scientists plan to do with hairy mammoth DNA preserved in permafrost that they will implant in modern elephants? The answer is almost certainly 'no' with our present knowledge and techniques. We would need to find a living bird that is much more closely related to the dodo than the Nicobar pigeon, which is an efficient, light-weight flyer, to have any chance of bringing the dodo back to life. The same reason makes it impractical to consider rebreeding the dodo through artificial selection as Nicobar pigeons have no dodo-like characteristics (except in their DNA) that could be cross-bred as has been done with the quagga, a subspecies of Burchell's zebra in South Africa. The dodo and the Nicobar pigeon are too phylogenetically distant for this to work.

In retrospect it is a great pity that no-one chose to breed from the live dodos that were exported from Manutius in the 1600s. Dodos would probably have been easy to keep in captivity and could have rivalled chickens or even turkeys as a lucrative commercial product. While this suggestion might seem ludicrous it would at least have kept the dodo genome alive and kicking.

**Dodos and humans**: The dodo has had an interesting relationship with humans, a cultural history, since it became extinct. One event worth mentioning involves a mathematician, a little girl and the extinct bird. The year was 1862, the mathematician was Charles Dodgson who lectured for 47 years at Oxford University and the girl was Alice Liddell, daughter of another Oxford faculty member. The dodo was a specimen and painting on display in the Oxford University Museum, where Dodgson took Alice and her friends for walks.

Dodgson, whom we know better as Lewis Carroll, author of children's books, noticed the interest shown by Alice in the weird bird. Once, on a boat trip with Alice and her sisters, he assigned roles to each person so that they could act out a story that he was conjuring up in his mind. Dodgson, who apparently stammered ('My name is Do-Do-Dodgson') was the dodo and later even adopted the bird's name as his nickname. When he penned *Alice in Wonderland*, first published in 1865 at the height of the first dodo craze in England, he made Alice the heroine and the dodo was one of several characters. As Carroll's book has never been out of print since then the extinct bird has remained in the public consciousness for the past 150 plus years. In addition numerous other popular books have been published on the dodo including *The Annotated Alice – Alice's Adventures in Wonderland and Through the Looking Glass* edited by Martin Gardner (1965), *Dodo: from Extinction to Icon* by E Fuller (2002) and *Dodo. The Bird behind the Legend* by A Grihault (2005).

**Illustration of the dodo in *Alice in Wonderland***

Today the dodo is very much part of popular culture, epitomised by the expression 'As dead as a dodo'. It has featured in several movies including the 2001 animated feature *Ice Age* which won an Oscar for Best Animated Picture. Poems, songs and comics have been written about the bird and it has also featured in many works of art. The dodo is also widely used as a symbol of the plight of

Cartoon from the *Climate Change Action Blog* (2012)

endangered species as in the logo of the Jersey Zoo established by the late conservationist and writer Gerald Durrell and also appears on coins and the coat-of-arms of Mauritius.

**Lessons from the dodo**: The extinction of the dodo teaches us many lessons one of which is that, while natural selection is an efficient way of ensuring the survival of the fittest, its disadvantage is that it can only act on past and present selection pressures not on future ones, like hungry sailors and their pets. It only works in relation to ecological processes with which the animal concerned has co-evolved. This means that the millions of animals that did not co-evolve with humans are vulnerable to the multiple changes that we are making to natural environments and life-support processes on land, in the sea and in the air.

The fossil record reveals that there have been five major extinction events in the past caused by natural physical events such as meteorite strikes and volcanic eruptions. The animals affected were all innocent

Dodo stamp issued by Mauritius

Dodo on the National Arms of Mauritius

victims of bad luck. The sixth extinction event which is taking place now is being caused by one of the animals, *Homo sapiens*, you and me. The environmental changes that have been brought about by this human-made extinction can be handily summarised by the acronym HIPPO, i.e., Habitat destruction, Invasive plants and animals, Pollution, Population explosion (of humans) and Overhunting and overfishing. The imbalance that we have created has caused climate change, sea level rise, droughts, floods and spikes in catastrophic events.

Speciation (the formation of new species) and extinction (the permanent loss of existing species) are, of course, natural processes but the combination of our HIPPO impacts has increased extinction rates by thousands of times above normal. This is essentially a consequence of species such as the dodo having to share their living spaces with plants, animals and processes to which they are not adapted (or, more accurately, pre-adapted) by natural selection.

Our response as custodians of the biosphere should be to minimize our negative impacts on micro-organisms, plants, animals, ecosystems and their life-support systems but to do so we need to undergo a major mind-set change. We need to acknowledge that, although we arose from and survive through a biological process, most of us are no longer part of wild Nature. We have evolved from hunter-gatherers to shopper-industrialists and become servants of our machines, trapped in unsustainable urban environments. We are now the most abundant large animals that have ever existed using three-and-a-half times more resources than the planet can provide, and we are the first animal to domesticate itself and lose its ecological niche. Furthermore, we tend to take Nature's 'free' ecological services for granted and squander and abuse them for short-term gains.

Our recent evolution into urbanized information fetishists does not mean that we are outside of Nature and free from her checks and balances. On the contrary, through the disruptions that we have caused we are more vulnerable than ever to what Nature can throw at us. The Covid-19 pandemic has reminded us that we are still an integral part of Nature and very much subject to her checks and balances. Despite our superior intellect and advanced technologies an ancient, invisible but highly effective virus has brought our civilisation and its economic system to its knees and we are scrambling to pick up the pieces. The anthropocene extinction event also reminds us that *we* are the misfits in Nature, not the other species, and that the biosphere is a self-correcting system that favours the fittest and is

## 11. Lessons from the dodo

in the habit of rejecting misfits. Ask the dodo.

We need to recognise that we are a valuable species but not a superior one and that every living thing has the right to live or at least to *struggle to live*. Our role as *Homo sapiens*, 'wise man', if we are to live up to our lofty appellation, is to understand and work in harmony with Nature not to conquer it as the major religions and most socio-economic and political systems in the past have encouraged us to do.

The Covid-19 crisis has also taught us that humankind can work together and address even the most daunting challenges when it faces a single common enemy. In this Information Age our unprecedented connectivity combined with artificial intelligence has created an 'intellectual superorganism' comprising the community of human brains, computers and the Internet of Things that can potentially equip us to do something that we have never done before on this scale – call a truce on wars and other internecine conflicts and work together as a species in a fight to resolve the biggest crisis of our generation, climate change and biodiversity loss. Before we 'save the planet', as we so arrogantly claim to do, we need to learn how to fit into its ecological processes ourselves. We may not be able to bring the dodo back to life but we can at least learn some lessons from its tragic demise.

## 12.

# Which is South Africa's strangest animal? Animals without backbones

*'If we and the rest of the back-boned animals were to disappear overnight, the rest of the world would get on pretty well. But if the invertebrates were to disappear, the world's ecosystems would collapse.'*

David Attenborough, British broadcaster and conservationist

**INTRODUCTION:** Humans are fascinated by superlatives such as the largest or smallest, fastest or slowest or most or least long-lived. But the above are all measurable variables whereas 'strange' is a more qualitative assessment that combines many different traits and is often defined by the *lack* of a body organ or function that is characteristic of a particular animal group. 'Strangeness' implies that an animal is different from the norm, out of the ordinary and unusual, the opposite of conventional. Strange animals may be primitive or advanced, prominent or inconspicuous, aquatic or terrestrial and they may belong to any family, class or phylum in the animal kingdom. 'Strangeness' is an opinion applied by humans to other animals, most often to ones that live in different environments to us and therefore have dissimilar adaptations to maximise their survival. We need to bear in mind, though, that all living species, whether humdrum or extraordinary, cockroaches or blue whales, are the extraordinary endpoints of over 3.8 billion years of biological evolution as there is no more competitive process on the planet than natural selection.

Island countries tend to have a plethora of strange animals. Australia has its laughing kookaburra, cassowaries, kangaroos, wallabies, thylacine, numbat, wombat, koala bear, echidna and Tasmanian devil, New Zealand its walking bat, tuatara and kiwi and Madagascar its aye-aye, giraffe weevil, panther chameleon, indri lemur and satanic leaf-tailed gecko. Continental countries also have their share of strange animals. China has the giant panda, giant salamander, Siberian tiger and enormous sturgeons, India its freshwater dolphins, gharial and giant waterbug, and Canada its moose,

## 12. South Africa's strangest animal? Animals without backbones

**Axolotl**

bighorn sheep, beluga whale, wolverine, polar bear, beaver and live-bearing perch. The USA boasts the coypu, coatimundi, gila monster, alligator gar and hellbender salamander and Brazil its arapaima, rhea, tapir, capybara, jaguarundi, capuchin and tamarin. Not to be left out Mexico has its uber-weird axolotl, vaquita and agouti, Russia its freshwater seals and Amur lemming and the Congo its chimpanzee, mountain gorilla, okapi and bonobo.

But which is the strangest animal in South Africa? At least 100 000 animal species are known from the southern tip of the African continent (including its surrounding oceans) and each one represents a survivor that has found unique solutions to the challenges posed by evolution. But which one deviates most from the norm? In making this choice it is tempting to focus on the familiar animals on land and to ignore the zany denizens of the deep but, as an aquatic biologist, I have tried to avoid this trap.

**What is 'strange'?** 'Strange' is a multi-meaning word whose synonyms include curious, peculiar, unusual, odd, atypical, extraordinary, deviant, unorthodox, idiosyncratic, supernatural, unearthly, other-worldly, abnormal, outré, off-kilter and even eccentric, take your choice! I use it here in a positive way as 'different from the norm'. It differs from 'weird' in that it does not imply that the animal is eerie and creepy or makes one

**Platypus**

fearful. The degree of strangeness is difficult to compare between animal groups such as the major phyla or the classes among the vertebrates, i.e., fishes, amphibians, reptiles, birds and mammals. How, for instance, do you compare red bait with a blue whale, an earthworm with a chimpanzee, a coral with an elephant or a nudibranch with a flamingo?

Perhaps it is more useful, at least at the outset, to first compare animals within their taxonomic groups and determine their degree of strangeness by the extent to which they deviate from the norm within that group? Thereafter we can compare the most significant outliers in each group and come up with the winner of the 'strangest' title. During this playful romp through the animal kingdom we will also learn something about animal classification and biodiversity.

**The main *bauplans*:** The phylum (plural phyla) is the broadest level of classification below kingdom (such as Animal or Plant Kingdom) but above the class. It was first coined by the Prussian scientist Ernst Haeckel in 1866 based on general specializations of the body plan or *bauplan*. Each phylum or class of animals has a unique *bauplan* that distinguishes it from other phyla or classes. There are an astonishing 34 phyla in the Animal Kingdom and we cannot discuss all of them here although some of the strangest animals do occur in the so-called 'lower phyla' that comprise a wide variety of curious animals. The major phyla with which we are most familiar are the Annelida (segmented worms), Arthropoda (jointed leg animals, mainly insects and crustaceans), Echinodermata (starfish and their five-rayed relatives), Mollusca (snails, octopus, squid and their relatives) and the Chordata (animals with dorsal nerve chords, including all the backboned [vertebrate] animals. Within the most biodiverse phyla, such as the Arthropoda (1 250 000+ living species), Mollusca (85 000 species) and Chordata (55 000 species), it is most practical to compare animals in the next lowest classification level, the class.

Some animals such as velvet worms and the platypus of Australia appear to defy humanmade classifications as they possess traits from different phyla or classes. We need to remember that our classifications are artificial 'cubicles' into which we have shoe-horned animals that have different origins, are constantly evolving over time and sometimes exhibit convergent evolution (evolution of similar traits by unrelated animals as a consequence of living in similar environments). We also need to bear in mind that the majority of animals that have ever lived have gone extinct and that, in some animal groups, such as the lampshells and five-rayed animals, more extinct species are known from the fossil record than there are living species today. There is a good chance that the strangest animal from South Africa is an extinct species but they are not considered here.

**Zebdonk**

**Burchell's zebra**

**What about humans?** We could conclude, without going any further, that *Homo sapiens* is the strangest animal that has ever evolved as we have broken all the rules during our evolution and survive largely through artificial rather than natural selection. We also have the dubious distinction of being the first living organism to cause a major extinction event, the Anthropocene, which is happening now, on our watch. I have chosen to leave humans (and their pets and domesticated animals) out of the equation so that we can focus on wild animals. We will also not consider hybrids that have been created by humans such as zebdonks, zorses, zonkeys, donkras and hebras (by hybridising zebras with other horse-like animals) or ligers (lion x tiger). The extinct quagga is now regarded as a subspecies of Burchell's zebra. Through selective breeding a lineage has been developed that visually resembles the extinct animal but it is not sufficiently different from other zebras to be considered as strange here.

Let us now trawl through the different phyla and classes in search of the strangest animal.

**Sponges:** Thirty-three of the 34 animal phyla are creatures without backbones (invertebrates) and they comprise the great majority of animals with many of them taking strangeness to extremes. Sponges (Porifera; 'pore bearers') are among the most primitive animals as they lack internal organs and a nervous system although they do have spicules in their tissues that provide rigidity. Sponges consist of an intricate network of pores, channels and turrets through which water is drawn and from which oxygen and food are absorbed. Many sponges like our chilli pepper sponge contain toxic chemicals and are vividly coloured to advertise their toxic nature whereas others such as the vented sponge have tissues that show potential for cancer drug development.

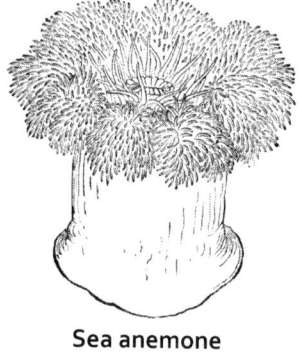

Sea anemone

**Cnidarians**: Cnidarians ('stinging nettles') are primarily marine animals that include anemones, sea fans, corals and jellyfish. They have a simple body plan with tentacles encircling a mouth that is the only opening to a sac-like gut cavity and they have no organs for respiration or excretion. Uniquely, cnidarians have stinging cells (nematocysts) which contain a coiled, thread-like sting that can explosively inject a toxin into an attacker or prey. How has such a complex organ developed in such a 'primitive' animal? Most sea anemones are attached although a few such as the hedgehog anemone float around unattached on vertical reefs where they feed on sea fans. Like some starfish our candy-striped anemone can evert its stomach to digest its food externally. Many anemones have commensal or symbiotic relationships with other marine animals. A good example is the symbiotic anemone which hitches a ride on the shells of hermit crabs which, themselves, have borrowed a 'home' from a mollusc, a nice inter-phyletic partnership. The giant anemone which at 500 mm across is much larger than other species plays host to anemone fish (nicknamed 'Nemo') which feed off food remnants and may repel predators.

**Zoanthids:** Zoanthids are anemone-like animals with polyps crowned by tentacles around the mouth. Unlike anemones they are all colonial and are joined by a living 'carpet'. They capture tiny prey floating in the water but like corals must also obtain food from symbiotic algae that live in their tissues. With their form and function they could not make a living in freshwaters or on land but like many other sedentary, reef-dwelling animals they are perfectly adapted for life in the sea. Our green zoanthid consists of a carpet of tightly packed squat polyps whereas at first sight sea fans look more like plants than animals. Their tall, fanlike branching colonies are covered with tiny polyps and, in the case of our palmate sea fan, form spectacular underwater forests that reach ages of 100 plus years. Soft corals have no hard internal skeleton and consist of soft polyps that catch tiny planktonic animals. Like many reef animals they depend on the buoyancy of sea water and their hydrostatic (water pressure) 'skeletons' to function efficiently although the leather-corals form hard crusts. Sea pens have a rootlike peduncle that anchors them in the sand or mud and an upright stem with feathery branches that carries rows of polyps. They are strange, otherworldly animals that resemble fantastical plants.

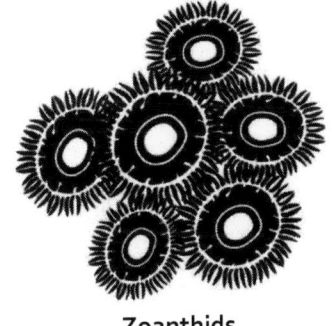

Zoanthids

It is interesting that on land almost all plants are attached and sessile (fixed in one place) whereas on the edges of the sea there are more species of animals (at least in the adult phase of their life cycle) that are attached and/or very slow moving. This is because a large proportion of marine invertebrates live in the wave-washed, high-energy coastal zone where they must be attached to avoid being dislodged and washed ashore. In contrast most marine plants are not attached but are free-floating in the phytoplankton. Humans, from the perspective of their stable terrestrial world, tend to be puzzled by the peculiar adaptations of life in the intertidal zone that is subject to the twice-daily ebb and flow of the tides. We also marvel at the ability of stationery predators to catch free-floating prey and of invertebrates to catch and eat vertebrates, both uncommon on land. The profusion of marine animals with one, three, five or multiple legs, anything but four, is also perplexing.

**Hard corals:** Hard corals come in a stunning variety of forms but most are colonial and comprise anemone-like polyps that produce hard limestone skeletons. They typically host single-celled algae (zooxanthellae) that absorb nitrogen from the coral and provide food and help to build the skeleton in return. This symbiosis with photosynthesising algae limits most reef corals to relatively shallow, sunlit warm waters which exposes them to the threats of global warming, increased UV radiation due to ozone depletion and bleaching. One of our most unusual hard corals is the mushroom coral which is oval, solitary and free-living and, unlike most corals, is able to move if they are competing for space. Some soft corals like the mushroom and cup corals are solitary whereas others (disc corals) form platelike colonies. One local species, the spiky coral, is aggressive and stings competitors with its long tentacles. Who would have thought that there is an aggressive coral?

Jellyfish

**Jellyfish:** Jellyfish are amongst the most surreal animals on the planet. They are bell-shaped gelatinous creatures with a circular body plan that swim by pulsing water from their bells. Despite their apparent simplicity they have light-sensitive organs and are carnivores that stun their prey with stinging cells on their tentacles or mouth-arms. Many jellyfish spend most of their lives as juveniles (medusae) that float in the plankton. Our most unusual jellyfish is probably the blue blubber whose bell margin has multiple scallops but no tentacles, like most other species. Hydroids are fernlike marine animals that form colonies of polyps that may be specialised for defence, feeding or reproduction.

**Bluebottles:** Bluebottles are also extraordinary animals, or rather groups of animals, with each one comprising a colony of individuals that are specialized to perform specific functions such as catching prey or defending against predators (dactylozooids, which form tentacles up to 10 m long), taking in, digesting and distributing food to the rest of the colony (gastrozooids), breeding (gonozooids) or forming a large blue-green float using a gas gland that inflates it with nitrogen and carbon monoxide. There is no animal on land or in freshwater that is fashioned in this way. As a result of differences in the shapes of their floats bluebottle colonies may be left- or right-handed and are blown by the wind to port or starboard respectively! Consequently, more left-handed bluebottles live off South Africa's west coast as they are less likely to be blown onshore by the prevailing south-easterly winds! I am relieved to know that.

If jellyfish lived on land we would gaze at it them in awe but such outré creatures are common in the sea. A close relation, the by-the-wind sailor which is also an anthozoan, is a solitary floating animal that has a small sail and is also blown around by the wind. Like jellyfish and bluebottles they have stinging cells that stun and capture tiny planktonic organisms. Bluebottles and by-the-wind sailors are often found in the company of the bubble-raft shell, a mollusc which hangs upside down from the sea surface suspended by a raft of mucous-coated bubbles and is one of their predators.

**Worms and leeches**: Worms both segmented and unsegmented are accommodated in several animal phyla. The annelids ('little ring') are worms with a *bauplan* consisting of multiple circular segments but without limbs; they include ragworms, earthworms and leeches. The most spectacular annelids in South Africa are probably the giant earthworms that live near Debe Nek in the Eastern Cape which far exceed the normal length of earthworms. I have dug them out myself and found specimens up to 3 m long although they are difficult to measure accurately as they are very stretchy. Leeches are odd in that each of their 32 segments has its own mini-brain! The freshwater leeches that live in some of our subtropical rivers have an anti-coagulant in their saliva that ensures that they enjoy a good blood meal whenever they attach themselves to an unsuspecting mammalian host.

*isiFonya* fishermen in the Phongolo floodplain in Maputaland equip themselves with 'boxes' made from gourds which they place over their sensitive anatomy to reduce the risk of leech attacks. I have had to use them myself while participating in *isiFonya* drives but they do not protect you from tigerfish or crocodiles! One of our segmented marine worms, the quirky orange thread-gilled worm, looks like a creature from science fiction. Its gills in the form of long threads hang outside the body and it typically lives tucked up between mussels on rocky reefs.

**Velvet worms**: The velvet worms are so unique that they have their own phylum, the Onychophora ('claw bearers'), the only animal phylum that is confined to land. They are a separate line of evolution that arose independently from a long forgotten marine ancestor and are regarded as 'living fossils' as they almost exactly resemble their 550-million-year-old extinct ancestors that we know from the fossil record. Velvet worms are

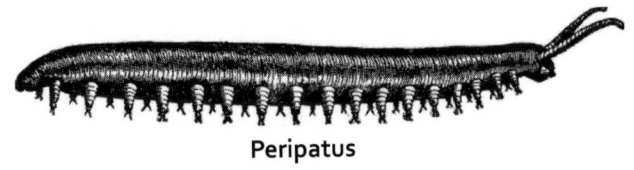

**Peripatus**

secretive, caterpillar-like animals with 13 to 25 pairs of stumpy jointed legs that live in forest leaf litter. Unusually among land animals (except some snakes like the python) but in common with many marine creatures (including sea cucumbers and hagfish) they have slime glands that secrete a milky-white goo that is used to ensnare prey and for defence. They appear to be intermediate between jointed-legged animals and segmented worms as they have jointed legs, compound eyes and breathing tubes (tracheae) like arthropods but a worm-like body, non-chitinous skin (cuticle) and 'kidneys' (nephridia) on each segment like a segmented worm (annelid)!

But the strangeness of velvet worms, which I have collected at Hogsback in the Eastern Cape, does not end there. They are of Gondwanic origin and are only found on the southernmost tips of South America, Africa and Australasia. Some lay eggs but many (including the ten South African species) are live bearers, which is unusual for a wormlike creature. They breathe through their skin and unlike arthropods do not have a rigid exoskeleton but a fluid-filled body cavity that acts as a hydrostatic skeleton, similar to those of many soft-bodied animals in the sea such as anemones. They are very strange creatures!

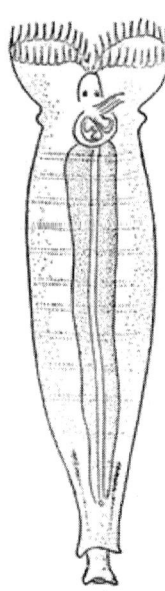

Rotifer

**Wheel animalcules:** Wheels have been useful transportation devices for humans on land and one wonders why no animals have evolved them. Perhaps there is no biological device available for connecting nerve endings to a spinning disk, as in the rotary electrical switches (commutators) in some of our electric motors or slip rings in the electrical motors of cars? The closest that animals have come to developing wheels is probably in the wheel animalcules (Rotifera, 'wheel bearing') which have a crown of cilia that moves in a circular sequence to propel them forwards; these cilia give the appearance of rotating without actually doing so. While some animals, such as woodlice, pill millipedes, armadillos and girdled lizards, may propel themselves by rolling none have functional wheels.

Rotifers are remarkable in other ways. They are common in freshwaters where they form an important part of the plankton but also occur in the sea. Like many other microscopic animals adult rotifers have the extraordinary quality of having a fixed number of cells within a species, usually about 1 000. Remarkably rotifer eggs and even the adult females of some species can survive complete desiccation for years or even decades. Other animals that can survive desiccation include brine shrimps, the eggs of some frogs, killifish eggs and African lungfishes in their cocoons.

**Arthropods:** The Arthropoda ('jointed foot') are animals with segmented bodies, jointed limbs and chitin exoskeletons (skeletons on the outside of the body) as opposed to our internal skeletons. They include millipedes, centipedes, insects, spiders and crustaceans.

**Millipedes:** The giant African millipede is the largest of its kind in the world, reaching 33.5 cm. It acquired its colloquial name *songololo* from the Nguni word *ukusonga* ('to roll up') from their habit of coiling up in a tight spiral when threatened, thus exposing only the hard exoskeleton. They also secrete an irritating liquid and are therefore one of the few invertebrates that driver ants cannot take as prey. Songololos have about 256 legs but this number changes with each moulting so it varies between individuals!

**Insects:** How does one choose South Africa's strangest insect? By how much it differs from the generalised insect *bauplan*, of course, while remembering that it is not only an insect's anatomy that makes it what it is but also its physiology, behaviour and especially its relationships with plants and other animals. Keep in mind also that new insect species (and even higher taxonomic categories) are still being discovered. In 2002 University of Cape Town entomologist Mike Picker stunned the world when he announced the discovery of a new order of insects, the 'heelwalkers', with their centre of distribution in South Africa. This would be the equivalent in the mammal world of finding the first primate or whale.

Most insects live on land, although a large number inhabit freshwaters during their juvenile and adult stages, but only two species commonly occur along our marine coast – the marine springtail and the kelp fly. As springtails are air breathers they shelter underwater in air pockets between rocks or in shells at high tide and scavenge on dead animals at low tide. Kelp flies are found exclusively along the marine coast where kelp and seaweeds have washed ashore; their maggots are important decomposers of rotting seaweed.

On land insects abound in every habitat, underground, above ground,

# Curious Notions

Praying mantis

in trees, lakes and rivers, in human abodes and as parasites of larger animals. Over 45 000 insect species are known from southern Africa but which is the strangest? Our longest insect, the giant stick insect, reaches 25 cm in length and like gargoyle mantids has mastered the art of camouflage by modifying its entire body to resemble a twig or leaf. Elegant grasshoppers go to the other extreme by using a kaleidoscope of bright colours to advertise their toxic nature. The defining character of praying mantises is their pair of raptorial forelegs, deadly weapons that are used to capture prey ranging from insects to frogs, lizards, fish and even rodents and also to defend themselves against predators. They are solitary ambush hunters and their huge eyes, set on the front corners of their triangular heads, combined with binocular vision enable them to judge the distance to prey and strike with great speed and accuracy. They also have unique hearing organs situated at the base of their legs which help them to detect bat calls at night and avoid being eaten. Their mating habits are truly bizarre as they practice sexual cannibalism. The females, which are larger than the males, feed off their mate's head while they are mating. This is probably not a good idea if you are a human but it seems to work among mantises!

Assassin bugs inject a paralysing saliva into their victims and suck out their body fluids and stink bugs live up to their name by ejecting foul-smelling formic acid at attackers. Ants' nest beetles secrete a chemical that causes ants to tolerate their company and giant water bugs are a good example of an invertebrate eating a vertebrate - they feed on tadpoles. Dung beetles collect dung for their newly hatched larvae but are carnivores as adults as are delicate, brightly coloured ladybirds. The horn of a rhinoceros beetle

Rhinoceros beetle

is strong enough for traditional fishermen to use them as hooks! A children's favourite is the tok-tokkie beetle which taps from four to 20 times on the ground to communicate with its kind. Many insects are parasitic on other species. Mosquitoes suck the blood of mammals, bee lice are external parasites of bees and snail-killing flies prey on aquatic snails including bilharzia vectors.

# 12. South Africa's strangest animal? Animals without backbones

During my childhood moth- and butterfly-collecting forays the stand-out lepidopterans were hawk moths (especially death's head and bee hawks), African monarchs, acraeas, swallowtails, painted ladies and charaxes. I kept ant colonies in carefully made Plaster-of-Paris nests covered with microscope slides and linked by transparent plastic tubes. The easiest species to domesticate were Argentine and black sugar ants but the most interesting were the red driver and harvester ants; I avoided keeping *bal-byter* ants because of their fearful reputation! The aquatic larvae of caddis flies make amazing sand grain-covered cases using their silk glands, mayflies have a typical aquatic nymph but a unique winged flying life stage that resembles an adult but is not sexually mature, and mud dauber wasps construct elaborate multi-celled mud nests.

**Death's head hawk moth**

Termites are social insects that live in colonies with the snouted harvester termite, which lives in savannah and grassland habitats, being one of our most common species. They forage on grass at night and store their food in earthen mounds. Harvester termites secrete a toxic chemical to reduce the risk of predation but some of their predators such as the aardwolf seem to be immune to it and eat up to 300 000 termites a night. The mound-building activities of termites have important consequences for the landscapes in which they live and may also increase the nutritional value of the soil as well as improve its drainage. In some cases these changes are strong enough to switch ecosystems from grasslands to shrub- or tree-dominated woodlands and, in wetlands such as the Okavango Delta, termite mounds may form the nuclei around which new, vegetated islands develop.

My vote for our strangest insect, although it is so familiar, is the honeybee. I have a beehive in my garden in Rondebosch, Cape Town, and find it endlessly fascinating to observe the extraordinary behaviour of these highly social insects. After I had set up my hive and smeared 'Swarm Lure' paste (comprising beeswax, propolis, lemongrass and wheat germ oil) on the entrance slits I waited patiently for five weeks for the bees to arrive. Then, exactly on schedule, one of Nature's miracles took place and on a bright Spring day a swarm of bees colonised the hive and they have been working industriously since then tending the queen and making honey.

They are aggressive though and I have been stung several times, mostly on the face, as they are savvy enough to home in on soft targets.

The honeybee is a very special insect. They have two sets of wings and use wing hooks to attach the wing sets together during flight for maximum efficiency. They have five eyes (two compound eyes and three single lens eyes), with a worker bee's main eyes having nearly 7 000 lenses! An electrostatic charge on the bee's hairs attracts pollen and it uses its leg brushes to scrape the pollen backwards where it collects in a pollen basket. They also have a long proboscis which sucks up nectar and feeds the larvae. The queen bee methodically lays an egg in each cell in the honeycomb and the eggs hatch into small larvae which are fed and cared for by nurse worker bees. After about a week the nurse bees close the cells with a wax lid and the larvae metamorphose into pupae which hatch after about another week into adult bees.

**Drone honeybee**

Honeybees communicate with one another by performing a series of elaborate dance moves with the number of turns, the duration of the dance and the moves themselves communicating the distance and direction (in relation to the sun) of food sources that they have located. Honeybees forage on the nectar (carbohydrates) and pollen (protein) of flowering plants and require a wide range of these nutrients to be healthy. The foraging bees suck nectar into a second, storage stomach from which worker bees in the hive remove it and then digest it in their own stomachs for about 30 minutes, which breaks down the complex sugars. Raw honey is deposited in empty comb cells where it is fanned by the bees to dry it after which the workers cap the cell with wax to store the honey for later use as food. Honeybees are less active in winter and remain in the hive where they feed on honey and generate heat by flapping their wings. The complexity and sophistication of their colonies is only exceeded by the societies of humans.

**Spiders and scorpions:** Did you know that some spiders live in the sea off South Africa? The scarlet sea spider (not a true spider) has four pairs of long spindly legs which are used to scurry around underwater and (by the males) to carry fertilised eggs. The formidable shore spider (a true spider) lives up to its name as it has huge poison fangs, about one-third of its body length. They trap bubbles of air in silk-lined crevices in which they shelter during high tide and then emerge at low tide to hunt for prey. A fish-eating spider which lives in freshwater lakes anchors its back legs on a waterside plant and rests its front legs lightly on the water surface. From this platform it pounces on prey including small fish and tadpoles that swim by and is also able to run rapidly across the water surface to catch insects. The most spectacular spider in our garden in Cape Town is the golden orb-web spider which makes a web strong enough to catch large insects and even small birds. Another unusual spider is the spindly daddy-long-legs, a common denizen of dusty homes.

**Crustaceans:** The Crustacea is a huge group of invertebrates that is common in the sea but also occurs in freshwaters and, to a limited extent, on land. They include barnacles, copepods, isopods, amphipods, crabs, shrimps, prawns, lobsters and crayfish which typically have five pairs of legs (decapods; 'ten legs') and a carapace or shell. The most familiar crustaceans on land are land crabs, ghost crabs, fiddler crabs and woodlice, with the latter living in compost heaps and forest leaf litter. Some

**Hermit crab**

terrestrial crabs venture far from water although they usually have burrows that reach into the water table. They return to water to lay their eggs which pass through several aquatic larval stages before returning to land. Crown crabs typically inhabit brackish estuaries and lagoons but also freshwater coastal lakes whereas paddler crabs live in freshwaters and estuaries and must return to the sea to complete their life cycle.

Hermit crabs are unusual in that they have opted to live in empty snail shells and need to find progressively larger shells as they grow. To fit into their adopted homes their abdomens are small and soft, which means that they are vulnerable when they change shells. Most hermit crabs spend their life at sea but the land hermit is an air-breather that lives above the high-tide mark and scavenges on dry land, only returning to the sea to wet its body and breed. The anemone hermit crab forms a mutually beneficial relationship with a sea anemone which it deliberately places on its shell to provide camouflage and protection as the anemone releases sticky threads when it is attacked. The anemones in turn feed on food scraps released by the crab. When we dived off the west coast of the Cape Peninsula National Park in the *Jago* submersible we saw many anemone hermits with their hitchhiking anemones. Other crabs that use 'living camouflage' are the cryptic sponge-crab which cleverly covers itself with unpalatable sponges and the decorator crab which nips off seaweed fragments, attaches them to its hairy back and then changes colour to match that of the algae!

Although giant coconut crabs do not reach as far south as South Africa they are worth mentioning. They are the largest terrestrial arthropods in the world, reaching 4 kg, which as adults hide in burrows or in coconut trees during the day and forage for ripe fruit (including coconuts) and carrion at night. They also eat turtle eggs and hatchlings. They lay their eggs in the sea and the juveniles use small snail shells as their first homes whereas the adults develop a tough exoskeleton on their abdomens and do not carry a shell. Coconut crabs use lungs rather than gills to breathe and are probably the most extraordinary terrestrial invertebrates in Africa. They are easy to catch and tasty to eat so they have sadly been exterminated along the African mainland coast and only survive on isolated islands such as Chumbe Island off Zanzibar where I saw them in December 2012.

The Columbus crab is strange in that it spends most of its life not on the ocean floor but in the 'blue community' near the ocean surface. They were first reported by Christopher Columbus on his voyage to the West Indies in 1492. Ghost crabs scurry around on sandy beaches at low tide and hide in deep burrows in the sand at high tide. Male fiddler crabs have one enormously enlarged nipper that is used for defence and attack and to make 'come hither' gestures to females during courtship (although this interpretation may hint at anthropomorphism!). Many crabs and shrimps live in cooperative relationships with other animals, such as the commensal shrimp which lives in sand prawn burrows, spotted porcelain crabs which hide among the tentacles of large anemones and cleaner shrimps that sidle up to large fishes and remove parasites and diseased tissues from them without being eaten. Some crabs and shrimps are strong enough to open mussel shells or rip starfish apart.

South Africa has several species of lobsters (incorrectly called crayfish) which were previously abundant in shallow water but are now mostly confined to protected areas and deeper offshore habitats. Jack Skead, a respected researcher on the historical usages of South African animals, noted that lobsters were so numerous and accessible during Jan van

## 12. South Africa's strangest animal? Animals without backbones

Riebeek's time that a law was passed that stated that they should not be served to slaves *more than twice a week* – the slaves preferred something tasty, like porridge! During dives in the *Jago* submersible in 1996 we encountered huge aggregations of east coast lobsters in water up to 400 m deep off Cape Point. On another dive at a depth of 30 m in Table Bay Hans Fricke and I watched west coast rock lobsters parade along the sandy bottom in single-file with their tails held high, like disciplined soldiers in orange-and-white tunics. When they are threatened by predators they form a defensive *kraal* with their tasty tails pointed inwards and their spiky antennae outwards.

The east coast rock lobster is unusual in that the male deposits a package of sperm on the female's chest before the breeding season and she scrapes the sperm onto the eggs when she lays them. The males attract the females by rubbing their antennae against their hard carapaces, a built-in violin! Pistol shrimps have powerful nippers that can be snapped together to make a loud sound, so sharp that it generates a 'heat flash' that stuns it prey. Another noisy crustacean on tropical reefs is the snapping shrimp, which often forms commensal relationships with gobies. Mantis shrimps have a pair of enormous sickle-like limbs resembling those of a praying mantis which they use to disable their prey including molluscs and crustaceans. Their strike, with the 'heel' of their raptorial limb, has a force approaching that of a low-calibre bullet and can crack the glass of an aquarium. Their eyes are extremely sophisticated with trinocular vision and better detection of colour than those of any other animal including humans.

One of the strangest isopods is the fish louse which attaches itself to the head of fish such as the hottentot and sucks out their blood. Another zany crustacean fish parasite is the extraordinary tongue-replacement isopod which lives in the mouth of its host where it eats up and gradually replaces the fish's tongue! I know of no other parasite that replaces an organ of its host.

Barnacles are highly specialised crustaceans which have minute planktonic larvae that float around in the plankton and adults that are permanently encased in a hard shell that is attached to rocks as well as to floating logs and living animals. Whale barnacles live partially embedded in the skin of their hosts, mainly humpbacks but

**Barnacle**

also fin, blue and sperm whales, whereas crab barnacles attach themselves to deep-water crabs and goose barnacles to logs or ships.

**Moss animals**: There's a joke among marine biologists that, if you can't identify a reef invertebrate, call it a bryozoan (lace or moss animal) as they come in so many different forms that they could be mistaken for anemones, sea fans, sponges, hydroids, corals, fan worms or even seaweeds. They build colonies of tiny individuals (zooids) each enclosed in a minute skeleton and crowned by filter-feeding tentacles, although some zooids are specialized to snap or thrash at predators that are much larger than themselves. One species, the nodular bryozoan, lives exclusively on the shells of a whelk forming an orange or purple cloak that contains highly toxic chemicals which protect the host from predators such as rock lobsters.

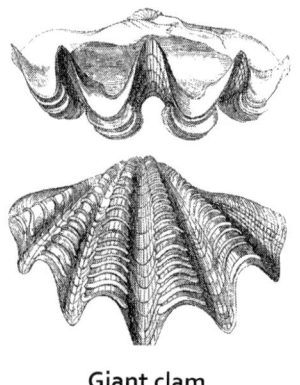

**Giant clam**

**Molluscs**: Molluscs, which include mussels, clams, oysters (all bivalves with two shells), chitons (with eight shell-plates), limpets, snails, winkles, whelks (with single shells) and sea slugs (no shells), as well as octopus, squid, cuttlefish and paper nautilus, are a huge phylum of unsegmented, soft-bodied animals with over 5 000 species in South Africa. They are characterised by having a unique rasping tongue (radula) and a large muscular foot although this appendage has been modified into eight or ten tentacles in the cephalopods. Most molluscs have shells but some have lost them and adopted other means of defence.

Our largest mollusc, the extraordinary giant clam, like many other marine reef species filter feeds and 'farms' microscopic algae on its brightly coloured mantle. As giant clams are very long-lived they have traces of radioactive elements in their tissues from past atomic and nuclear bomb explosions that help us to age them and, even further back, markers of past climates and even changes in the salinity and temperature of the sea that help us to reconstruct past events. The dwarf rusty clam starts life as a male and then changes to a female, so it has the best of both worlds. Like many bivalves Gilchrist's tellin lives communally with a goby. It is named after Dr JDF Gilchrist, a famous early Cape marine biologist who established the first public aquarium in sub-Saharan Africa at Kalk Bay in Cape Town in 1904 (it was sadly demolished in 1948). Another bivalve, the shipworm, is a filter-feeder but is also able to digest wood with the help of symbiotic bacteria in its gut and is notorious for destroying wooden wharfs and shipwrecks.

Many of the more advanced marine molluscs are carnivorous and are well-equipped to attack and subdue prey. Cone shells have hollow harpoon-like teeth through which they inject a potent neurotoxic poison into their prey; two species from South Africa, the textile cone and geographic cone, are potentially lethal to humans.

Chitons are flattened molluscs covered by eight overlapping shell plates like tiny armadillos, the end plate of which may resemble a set of false teeth! They rasp plants and animals off the rocks and some display homing behaviour, always returning to the same spot after feeding. Chitosan, a chemical extracted from one of our chitons,

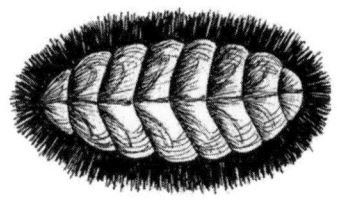

Chiton

shows promise as a treatment for liver cancer. Abalone (or perlemoen) have a large shell beautifully decorated with mother-of-pearl nacre on the inside. They are endemic to South Africa and are poached for their tasty flesh despite the fact that their commercial exploitation was banned in 2008 (personally I prefer the flesh of key-hole limpets boiled in sea water and garnished with vinegar!). Interestingly, prior to 1980 perlemoen were mainly exploited for their shells rather than their meat. Cowries are treasured for their beautiful egg-shaped shells but have also been used as an aid when darning socks and as money in West Africa!

Sea slugs, sea hares and nudibranchs differ significantly from other molluscs as their shells are either very reduced or absent. They have compensated for this loss of protection by secreting toxic chemicals and their flambouyant colours, among the most extravagant of any animal, warn predators that they are toxic. The sea swallow is a surrealistic beast that resembles a bluebottle more than a mollusc. It floats upside down on the sea surface, gulping air bubbles to keep it afloat and can re-use the stinging cells of its prey, including the Portuguese man-of-war jellyfish, to defend itself! Imagine a predator on land acquiring the defensive mechanisms of its prey after lunch! Sea swallows should be handled with care as they can give a nasty sting.

Nudibranchs ('external gills') are fantastic creatures that might have been invented by an abstract artist but their brightly coloured gill filaments and soft, wavy bodies belie the fact that most of them are ferocious predators on sponges, moss animals, hydroids and algae. Some nudibranchs are plant suckers that pierce the walls of algae and suck out their tissues including the chloroplasts (green, chlorophyl-containing organelles that carry out photosynthesis), which continue the process of converting light energy into food inside the mollusc and colour it green, yet another enviable adaptation that land animals have failed to emulate.

Octopus

Although there are many deviations from the general molluscan *bauplan* cephalopods ('merged head-foot'), which are exclusively marine, are probably the most radically different group. They have diverged so far from the typical molluscan form that it is difficult to appreciate that garden snails, slugs, Spanish dancers, oysters, octopuses and squids all belong to the same phylum. Cephalopods are more sophisticated than any other invertebrates and typically have a large head, two well-developed eyes, a muscular mantle that covers the body cavity, a siphon that propels them backwards using powerful jets of water, elongate, prehensile arms covered with two rows of suckers and seven hearts! Although molluscs are generally small squids reach lengths of over 20 m although extrapolated lengths up to 50 m based on metre-wide sucker marks on adult whales have been discounted as it was found that the sucker marks left by an attacking squid on a young whale's skin increase in size as the whale grows! Giant early relatives of octopuses and squids once dominated ancestral seas in the company of now-extinct marine reptiles that resembled modern-day dolphins and whales.

Two cephalopods that are poorly known except for their shells which are frequently washed on shore are the ram's horn and paper nautilus. The former is a small, deep-water squid which has light-emitting organs like many deep-sea creatures. The latter is a small midwater octopus in which the males are minute, shell-less and planktonic and the female produces a large delicate, translucent shell that serves as a brood chamber for the eggs. Our common cuttlefish, a close relative of squids and octopuses, produces a cuttlebone that is used to regulate its buoyancy by varying the proportions of its liquid and gas contents. Cuttlefish bones are mainly comprised of calcium carbonate and are a popular source of calcium for caged birds and have also been used as a grinding paste in jewellery manufacture and as an additive in toothpaste.

Our common octopus tends its eggs which hatch into miniature octopuses without a larval stage. They grow rapidly reaching their maximum mass of 6 kg within a year. The southern giant octopus, one of the largest in the world, has arms up to three metres long. During my directorship of the JLB Smith Institute of Ichthyology (now the South African Institute for Aquatic Biodiversity, SAIAB; 1982-1995) we were sent underwater footage taken from the remote-controlled submersible that searched for the black box of the South African Airways Boeing 727-244B Combi, the *Helderberg*, which crashed into the sea near Mauritius in 1987. The footage showed many bizarre long-legged deep-sea fishes and dumbo squid at depths of 4 900 m but most spectacularly a giant red octopus that attacked the submersible and tried to tear bits off it!

As Craig Forster's brilliant Oscar-winning documentary film *My Octopus Teacher* has shown octopuses are intelligent, can form friendly relationships with human divers, deliberately change the colour of their eyes and bodies and use an extraordinary range of techniques to catch their prey and avoid predators. When I worked at the Two Oceans Aquarium in Cape Town we were astounded by their ability to squeeze through cracks barely wider than the diameter of their eyes and to camouflage themselves so effectively that they were extremely difficult to find.

**Five-rayed animals**: The echinoderms ('spiny skin') are the largest phylum that occurs exclusively in the sea and they are probably the group that differs most from all other animal phyla. In the adults they have a unique five-rayed (pentaradial) body plan that is quite different from the bilaterally symmetrical *bauplan* of most other animals although their larvae are bilaterally symmetrical. Echinoderms usually have calcified spines and a calcareous endoskeleton consisting of ossicles but the exoskeleton is absent in the feather stars and sea cucumbers. Their central disks have a mouth on the underside and an anus facing upwards and the tube-feet running along the underside of each arm operate using water pressure.

In addition to being able to reproduce asexually echinoderms can regenerate tissues, organs or arms that are lost and in some cases can regenerate their whole body from a single arm, which no vertebrate and few other invertebrates can do. The otherworldly scaly-armed brittlestar sheds an arm if it is attacked and the writhing arm emits a flashing bioluminescent light to distract predators. Research has also revealed that brittlestars (and some starfish) are able to detect and react to light even though they have no eyes or brain. How weird is that! Most echinoderms produce minute planktonic larvae but some brood the young and give birth to miniature replicas of themselves.

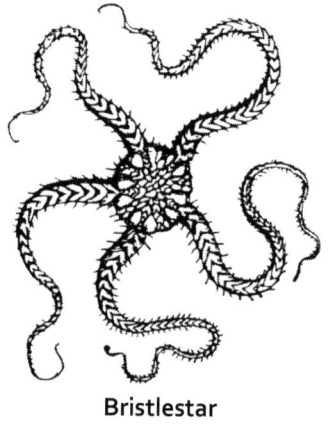

Bristlestar

The basket star has ten arms which branch into ever-finer tendrils that are spread out like a net to catch passing prey. Some brittlestars are commensal with other animals such as the hitchhiker brittlestar which lives with jellyfish and the pansy-shell brittlestar, a co-habiter with pansy shells. Our red starfish and striped brittlestar typically have five clearly defined arms whereas the sea urchins (including pansy shells) have wrapped their five arms together to form a spherical protective test. Some starfish like the speckled sandstar cannibalise burrowing starfish whereas the dwarf cushion-star and spiny starfish exude their stomachs through their mouths to digest their food externally! Like many marine reef animals the reticulated starfish often has tiny shrimps living with it. The short-spined urchin uses pieces of algae held over its body by its tube-feet for concealment whereas the Cape urchin uses dead snail shells as a sunshade.

The deadly crown-of-thorns starfish, which reaches 400 mm in diameter and has many arms, is a voracious predator of corals. Although their population explosions were originally thought to have been caused by human interventions such as pollution or the over-exploitation of their predators (triggerfish, wrasses and giant tritons) recent research has revealed that their periodic 'explosions' occurred long before humans evolved. Sand dollars have also wrapped their arms around themselves to form a shell whereas feather stars have tiny round bodies with hook-like limbs and a crown of feather-like growths which they use to catch food. Although they are usually attached to rock or coral they can crawl or even swim.

## 12. South Africa's strangest animal? Animals without backbones

The most atypical group of five-rayed animals is the sea cucumbers whose five-rayed pattern has been simplified into an elongate, sausage-like shape resembling a cucumber but they retain their radial symmetry by having five rows of tube feet along their bodies. Sea cucumbers have another peculiar trait in that some species like our golden sea cucumber play host to a small pearl fish that lives inside its intestines! Sea cucumbers can expel their internal organs (and regrow them) if they are attacked by a predator, to tempt it away from their main body. Few other invertebrates and no vertebrates exhibit this bizarre behaviour! Some species like our snake sea cucumbers can also eject long sticky threads from their anus to distract predators, reminiscent of velvet worms.

**Sea cucumber**

**The main invertebrate contenders are ...:** Taking the above into account, I select the following animals as the strangest invertebrate outliers in their respective groups: Sponges - chilli pepper sponge; Worms – velvet worm; Cnidarians – hedgehog anemone; Zoanthids – sea pen; Corals – spiky coral; Jellyfish and bluebottles - bluebottle; Spiders - fish-catching spider; Insects – praying mantis and honeybee; Crustaceans - mantis shrimp, tongue-replacement isopod and barnacle; Molluscs – Spanish dancer, sea swallow, cone shell and octopus; Echinoderms - sea cucumber.

Now let us consider the animals with backbones, the vertebrates.

## 13.

# Which is South Africa's strangest animal? Animals with backbones,

### and the winner ...

*'When we try to pick out anything by itself,
we find it hitched to everything else in the universe'*

John Muir, Scottish-American naturalist and author

**INTRODUCTION:** In this second essay on this topic we will discuss animals with backbones. The chordates ('with a cord') are characterised by having, during some stage of their life cycle, a notochord, dorsal nerve cord, slits in the throat (pharynx) for breathing and a tail behind the anus. They include simple sac-like animals such as redbait as well as the advanced vertebrates ('with a backbone') that are so familiar to us (fishes, amphibians, reptiles, birds and mammals). As adults these advanced vertebrates do not have pharyngeal slits or post-anal tails but these tell-tale features are found in their embryos. For example human embryos at the age of five weeks have gill slits and tails but these organs are lost before they are born whereas in fishes (and axolotls) they develop into gills and tails.

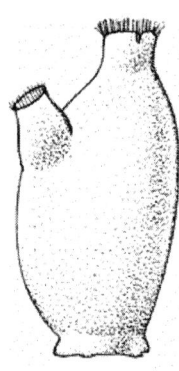

Sea squirt

**Sea squirts:** As adults sea squirts such as redbait look like many featureless sac-like reef animals and some like elephant's ears and blue lollipop ascidians look more like plants than animals. They are a paradox as the anatomy of their larvae, which are tadpole-like with a tail and dorsal nerve cord supported by a notochord, reveals that they are related to backboned animals. (Interestingly, sea squirts have purple blood whereas the blood colour in spiders and octopuses is blue and in various marine worms green!) Salps like the three-tailed salp and fire roller are small planktonic

## 13. South Africa's strangest animal? Animals with backbones

animals encased in jelly-like capsules that resemble alien spacecraft but their larvae have the give-away dorsal nerve cord. Tiny adult arrow worms and Cape lancelets (50 mm long) also have a dorsal nerve chord and notochord, gill slits and a tail behind the anus that place them firmly among the chordates. Humans are therefore more closely related to sea-squirts, although they have no musculature or nerve cells, than they are to jellyfish which have elaborate sensory organs and nerve and muscle cells.

**Vertebrates**: The vertebrates, the dominant group of chordates, are comprised of seven main groups: jawless fishes, cartilaginous fishes, bony fishes, amphibians (frogs and toads) reptiles (lizards, snakes, crocodiles, tortoises and turtles), birds and mammals (including humans). Each group of backboned animals has a unique combination of features that distinguishes it from the others: jawless, cartilaginous and bony fishes generally have scales, gills and fins and usually swim in water although the 28 000 plus living species have diversified to such an extent that many of them break these human-imposed rules. Amphibians are typically small, have moist skin, lack scales or fur, usually live between freshwater and land and have four legs as adults with strong back legs for jumping (although rubber frogs prefer walking, platannas favour swimming and tree frogs like climbing). Almost all amphibians have a full metamorphosis from egg to an aquatic tadpole to adult.

Reptiles have four legs (or did so ancestrally), do not have an aquatic larval stage, typically have scales, generally lay eggs and may or may not have shells or carapaces. Like fishes and amphibians, they are cold blooded which means that their body temperatures vary according to external temperatures. Birds have feathers, beaks, wings, can fly and breed exclusively by laying eggs, and mammals have fur, produce milk, suckle their young and lack scales (with one notable exception) or feathers. Birds and mammals (and some advanced fishes) are warm blooded in that they can maintain a steady, high body temperature independent of that in the outside environment.

Strangeness within a class of vertebrates is often characterised by the lack of the definitive traits of that group. They include scaleless fish, fish without gills or those that can breathe in water and air, fish that fly or crawl around on land, frogs that do not live near water, reptiles without scales or legs, reptiles that are live bearers, birds with reduced wings that cannot fly but can swim or run, and mammals without fur and/or whose legs have been transformed into wings or fins.

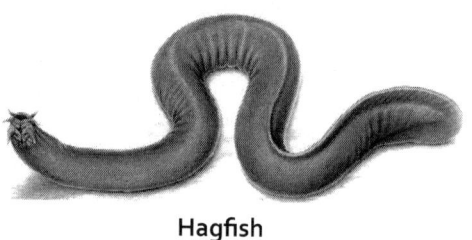

Hagfish

**Jawless fishes:** The sixgill hagfish is one of four species of Agnatha or jawless fishes in South Africa. They are eel-like parasitic fishes with a cartilaginous skeleton, no paired fins and barbels around the mouth. They lack jaws but have orange-yellow protrusible teeth. When disturbed or attacked they secrete a sticky slime from over 100 pairs of slime organs; fishing tackle that has been contaminated with this slime must be discarded as it repels fish. Together with lampreys they are the only living animals that branched off the vertebrate family tree before jaws had been developed and have managed to survive without them. The strangeness of hagfishes is derived from the fact that that they have remained virtually unaltered for over 300 million years rather than by the development of modern specialisations. I have observed Cape hagfish in the kelp forest off Smitswinkelbaai beyond Simon's Town. They have the same aura as coelacanths – unbelievably prehistoric creatures that are frozen in time yet remain competitive in today's cutthroat world.

**Cartilaginous fishes:** The cartilaginous fishes (Chondrichthyes; chimaeras, skates, rays, sawfishes and sharks) are aquatic backboned animals with jaws, paired fins, paired nares (nose openings), denticles instead of scales, a well-developed electroreceptive system, no swim bladder, a heart with its chambers in line and skeletons composed of cartilage. All cartilaginous fishes fertilise their eggs internally with the sperm transferred into the female using paired intromittent organs (claspers; absent in bony fishes). Unlike bony fishes they have high concentrations of urea in their blood and other tissues to maintain their ionic balance relative to sea water.

Unusual cartilaginous fishes in our waters include cow sharks with a single dorsal fin, filter-feeding whale sharks that reach lengths of 12 m and are the largest living fishes, remarkably well-camouflaged catsharks, hammerhead sharks with eyes on the ends of their laterally expanded heads, threshers with enormously extended tail fins, goblin sharks with a sharp rostrum on the snout, angelsharks, sawsharks, sawfishes, great

## 13. South Africa's strangest animal? Animals with backbones

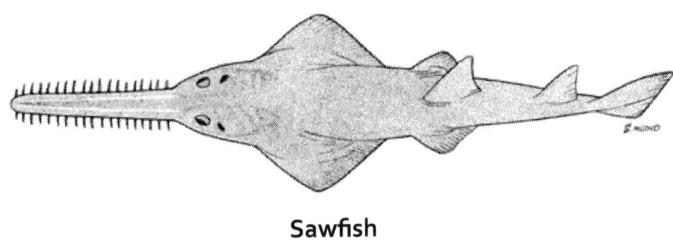

**Sawfish**

white sharks that reach lengths of 6 m, guitarfish, manta rays, eaglerays and chimaeras. Many of our sharks, including the great white, are live bearers with uterine cannibalism, i.e., developing pups eat fertilised eggs and younger embryos while inside the mother.

I have dived with whale sharks, which are filter feeders, off Ponto do Ouro in Mozambique and found them to be gentle creatures but experienced divers from Australia have warned me that it is dangerous to inadvertently swim into the slipstream that flows into their cavernous mouths! Whale sharks have a worldwide distribution and were first described in 1828 by Sir Andrew Smith based on a specimen caught in Table Bay, Cape Town.

Great white sharks reach 6.4 m in length and are extremely dangerous. Their diet seems to include anything that they can ingest as Professor JLB Smith, in the first edition of his *Sea Fishes of Southern Africa* (1949), reported on a specimen that had in its stomach a human foot, half a goat, two pumpkins, a wicker-covered scent bottle, two large bony fish and a small shark! When oarfish wash ashore people are astounded by their weirdness. They are elongate, ribbon-shaped, up to 8 m long and coloured a brilliant silver with blue and black streaks. They feed on small planktonic crustaceans while suspended vertically in the water column at depths from 20 to 200 m, maintaining their position with undulations of the long dorsal fin. You can't do that on land!

During my time at the Ichthyology Institute in Makhanda many unusual fish specimens were brought to us by anglers, aquarists and beachcombers but the strangest was probably the six-gill stingray. This flabby deep-water fish with a long, fleshy snout was found by journalist Dave Bickell on a beach in Port Elizabeth. Most remarkably it has six pairs of gill slits, different from the five in all other skates

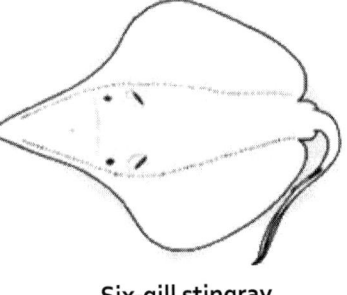

**Six-gill stingray**

and rays. After carefully examining the specimen and a few others found around the world taxonomist Phil Heemstra decided that it was not only a new species, which he and Margaret Smith named *Hexatrygon bickelli*, but also a new genus, family and suborder of fishes, equivalent to finding the first representative of all monkeys and apes!

**Bony fishes:** The bony fishes (Osteichthyes) typically have bony skeletons and fin supports, fins with spines and/or rays and a single pair of gill openings on either side of the head. They are by far the largest class of vertebrates, numbering over 28 000 species, and are divided into two groups, the lobe-fins (coelacanths and lungfishes) and ray-fins (all other bony fishes).

**Lobe-fins:** Over 90 species of extinct coelacanths have been found in their fossil record, which stretches back over 420 million years. The first living coelacanth known to scientists was caught off South Africa in December 1938 and they are now known from Mozambique, Tanzania, Comoros and Madagascar with another living species occurring in Indonesia. Coelacanths have an amazing potpourri of characters that make them difficult to place within our animal classifications. They have bony heads and fin supports as well as scales, lateral lines and fins with spines and rays like bony fish but a hollow cartilaginous notochord, heart arranged in a straight one, spiral valve in the intestine, a rectal gland for secreting salts and urea retention in the blood like sharks. Unlike any other fish they have three lobes in the tail fin and a joint that allows them to lift their upper jaw to increase their gape. They even have an inner ear that is shaped like that of four-legged animals and some modern classifications place them closer to the tetrapods than to fishes.

**Skeleton of the coelacanth**

Other remarkable features of the coelacanth include the lowest haemoglobin count and the smallest gill surface area of any fish (which means that they must inhabit well-oxygenated waters) and the slowest metabolic rate of any vertebrate. Despite their ancient origins they have an advanced breeding strategy, producing the largest eggs of any fish (the size of an orange) and give birth to large young (33 cm and 500 g) after by far the longest gestation period of any animal (36 months). They are social fishes that congregate in caves during the day and hunt at night, feeding on fish, squid and crabs. They grow to about 2 m and 100 kg (with females growing larger than males) and may live for more than 100 years, an age exceeded in fishes by only a few polar species. They swim with a unique sculling action of their eight fins and are capable of very rapid pouncing movements using their huge tail fin when they ambush prey. They also have a well-developed electro-sensory system like sharks that allows them to detect prey in the dark and under sand. It is truly a fish of superlatives! (More information on the coelacanth is given in the chapters on 'Finding old fourlegs'.)

Bony fishes typically have gills and absorb oxygen from the water but the otherworldly lungfishes have lungs; South American lungfish are compulsory air breathers whereas African and Australian species can breathe in air and water. Unlike most fishes their paired pectoral and pelvic fins are long and threadlike. Lungfishes, which first appeared in the fossil record about 416 million years ago, are the earliest known creatures to have single bones in their upper limbs (like our humerus), two bones in their lower limbs (like our ulna and radius) and the beginning of a multi-boned 'wrist' and 'ankle'. They are also unusual in that they excavate burrows during the dry season where they aestivate in mucous cocoons which allows them to survive desiccation, an unusual adaptation for a fish and otherwise known only in annual killifishes which have desiccation-resistant eggs and embryos. Lungfishes have an advanced breeding strategy in that they lay their eggs in a nest that is guarded by the male. They belong to an ancient lineage and are the closest living relatives of the amphibious creatures that first made that giant evolutionary leap from water onto land about 320 million years ago.

**Other unusual bony fishes:** Our other unusual bony fishes include shark remoras which have modified dorsal fin spines that form a powerful sucking disc on top of the head which they use to attach themselves to sharks; while

Mudhoppers

hitching a ride they benefit from scraps wasted by the host. The mimic blenny is a foxy creature that mimics the benign cleaner wrasse, which allows it to approach close to unsuspecting large fish and bite chunks from their fins! Bigfin mudhoppers regularly venture onto land, storing a supply of oxygenated water in their gill chambers. They live in mangrove swamps and estuaries and use their enlarged pectoral fins as limbs to shuffle around on the mudflats and climb up mangrove roots and even sloping tree trunks. Adult flatfish such as the Cape sole bizarrely have both eyes on one side of the head although their juveniles have the normal arrangement with an eye on each side.

Freshwater eels are unusual in that they breed at sea where the eggs hatch into transparent, leaf-like larvae called leptocephali which are carried towards shore on ocean currents. Near shore they metamorphose into elvers which enter estuaries and migrate up rivers where they live (in the case of the longfin eel) for up to 20 years. When they are ready to breed they migrate back down river into the ocean. Some primitive marine bony fishes such as the oxeye tarpon and bonefish also have transparent leptocephali that drift with the plankton in the open ocean. Remarkably, as these leptocephali transform into juveniles, they become smaller before growing larger again as they develop into adults.

I admit that I have a bias towards the sharptooth catfish as I studied this species in Zululand but one cannot deny that it is an astonishing fish. Besides having a handsome moustache of long barbels they are the largest freshwater fishes in South Africa, occasionally reaching 50 kg but more usually 20 to 30 kg. They have the widest natural distributional range of any freshwater fish in the world, from the Eastern Cape in South Africa through

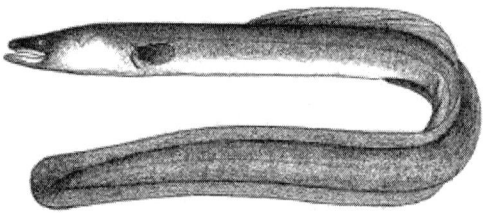

Longfin eel

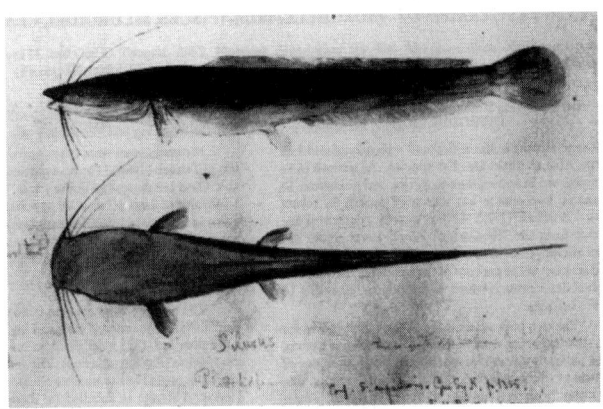

Painting by William Burchell of the type specimen of the sharptooth catfish

East and North Africa into Eastern Europe. Furthermore, they have been widely distributed beyond that range for fish culture and have become serious invasive pests in their introduced habitats due to their omnivorous habits and ability to survive in harsh conditions. Although they have gills they are air breathers that use a remarkable tree-like accessory breathing organ in a chamber behind the head to breathe air. They drown if they are confined underwater as this prevents them from rising to the water surface or swimming and passing water over their gills. Like the many-spined climbing perch sharptooth catfish can also migrate over land from one water body to another after rain (using their sharp pectoral spines as legs) and they are often the last species to survive in a drying-up lake or pan, as we observed in Lake Ngami in Botswana in December 1982.

They feed on an astonishingly wide range of prey from plankton to snails, crabs, small and large fishes, even snakes and water birds. In a slightly irregular series of experiments (using medical instruments intended for humans) I gazed into the stomachs of living catfish using a gastro-intestinal photoscope at the Mseleni Mission Hospital near Lake Sibaya to determine their digestion rates, which are fast. They gather in large groups to breed in the flooded grassy margins of lakes and rivers on dark nights after rain in early summer and also undertake mass upstream feeding migrations in rivers and swamps. As they are hardy, fast growing and long-lived, and produce a tasty white fillet, they are ideal aquaculture species although they do have an image problem as many people in Africa do not like to eat scaleless fishes with long barbels.

Snoutfish

Other unusual South African freshwater fishes include the snoutfish which have an elongate proboscis which they use to feed on bottom-living invertebrates. They communicate with one another using weak electric currents with each species having its own 'signature discharge'. Squeakers, a kind of catfish, emit a loud squeaking sound when they are removed from the water and are able to lock their sharp dorsal and pectoral fin spines into a triangle that makes it difficult for predators to swallow them. They are also known as upside-down catfish as they frequently swim belly-up near the water surface. The Cape galaxias, a small Gondwanaland relic whose relatives also live on the southern tips of South America and Australasia, inhabits coastal streams and rivers in the Cape where it tolerates a wide range of environmental conditions. The spotted killifish is an annual species that completes its lifecycle within a year. They live in temporary pools, often sharing their habitat with lungfishes, and lay desiccation-resistant eggs that hatch after the first rains.

Our many-spined climbing perch is able to shuffle overland in wet weather or at night using its serrated opercula as legs and a well-developed air-breathing organ above its gills to breathe. Neither of two related endemic species, the Cape kurper and Eastern Cape rocky, have air-breathing organs. Males of our freshwater pipefish, like the Knysna seahorse, brood their eggs and larvae in an abdominal pouch. Pipefish and seahorses uniquely have a body covering comprising hard plates arranged in a series of rings, more akin to the exoskeleton of an arthropod than the scaly skins of fish.

The 2 500 plus species of estuarine and marine fishes in South Africa have many interesting outliers. Parrotfish are unusual in that they have beaklike jaws with which they scrape algae off coral or rock and occasionally eat live coral. The coral, rock and sand fragments that they ingest are ground into a fine sediment in their pharyngeal mill, which is of considerable importance in sand production around reefs as parrotfishes are often abundant there.

Striped eel-catfish carry out elaborate, synchronised manoeuvres in dense shoals in shallow water to evade predation whereas devil firefish, as their name implies, have venomous spines that can cause painful wounds. Almost all fish are cold blooded but yellowfin tuna, opahs and some other fast-swimming gamefish are warm blooded in that they can maintain their body temperatures several degrees higher than that of the surrounding water using counter-current exchangers.

Although I have enjoyed memorable dives in False Bay, the iSimangaliso Wetland Park in Zululand and along the Mozambique coast at Inhaca Island and near Vilanculos my most spectacular underwater adventures have taken place near Sharm el Sheikh and Ras Mohammed in the Red Sea. Here the coral reef castles and vertical reef walls must be among Nature's most amazing creations, far surpassing any architecture by humans. Some of the most exciting endpoints of evolution both in terms of species and behaviours are on display there interacting in beneficial reciprocities and fierce rivalries. When we first glimpsed their secret world, it unfolded as a tranquil scene but closer examination revealed that its bewildering array of life forms is engaged in a ruthless life-and-death struggle, an arms race between predator and prey that involves all the guile, intrigue and subterfuge of a gangster movie.

About 80% of fishes produce eggs that are fertilised externally and released into the water. In our area many damselfishes and triggerfishes as well as several freshwater species guard their eggs until they hatch

**Shoal of striped eel-catfish**

whereas male sea catfishes and some freshwater cichlids brood their eggs in their mouths. Sea horses and pipefishes have adopted the marsupial habit of brooding their eggs in a pouch on the male's belly. About 5% of fishes including all cartilaginous fishes have internal fertilisation and are either bearers that give birth to young that resemble miniature adults or release the developing young in large egg cases (which measure 30 x 15 cm in whale sharks). The coelacanth is unique in that the female extrudes her oviduct to receive sperm from the male but is an internal brooder. Some internal bearers (spiny dogfish and smoothhounds) provide yolk via a yolksac for the developing young whereas others (most sharks and hammerheads) supply additional food through a placenta-like connection with the mother. In some sharks (including the ragged-tooth) the first-born young feed on unhatched eggs and juveniles as well as yolk secretions from the mother's uterus.

Other reproductive strategies that are common in fish but would be regarded as bizarre in land vertebrates are hermaphroditism (the occurrence of both sexes in the same individual) and sex change. Some fish (such as rockcods) can be male and female at the same time (but do not usually fertilise their own eggs) whereas others (many wrasses, parrotfishes, sea breams, emperors and threadfins) start their adult life as one sex and then change to the other. In the sea goldie, a common inhabitant of tropical coral reefs, a dominant male mates with a harem of females but if he is taken by a predator one of the females changes sex and colour and takes the alpha male's place within a fortnight.

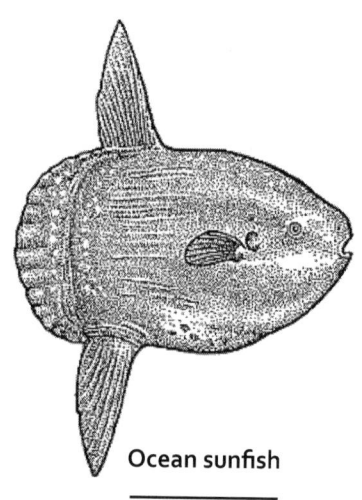

**Ocean sunfish**

Arguably the most bizarre looking fish in our waters is the ocean sunfish, a huge open-water species that reaches 3.3 m and 2 300 kg, one of the heaviest fish in the world. They resemble a giant fish head with two fins (its common name in Polish is *samogłów*; 'lonely head') with their large, circular shape reflected in their generic name, *Mola*, Latin for 'millstone'. In adults the tail fin is absent and is replaced by a rudderlike structure called a clavus. Their dorsal and anal fins, which are symmetrical and very tall, are sculled synchronously from side-to-side to propel the fish forward at a surprisingly fast speed. Their spinal column has fewer vertebrae than that of any other fish and, although they are descended from bony ancestors, their skeletons comprise mainly cartilage which is lighter than bone and allows them to grow to sizes that would be impractical for other bony fish. Ocean sunfish have small permanently open mouths with their teeth fused into a beak; they feed on jellyfish, salps, squid, fish larvae, small fish and crustaceans. They are often seen drifting on their sides on the water surface as if basking in the sun, hence their name, and may even allow seabirds to perch on them to feed on skin parasites.

Most remarkably, ocean sunfish produce the most eggs at a single spawning of any fish - over 300 000 000! Their juveniles resemble pufferfish (to which they are related) as they have large pectoral fins, prominent spines on the body and a normal tail fin; in fact, they are so different from the adults that they were initially classified as a different species. When we introduced an adult sunfish into the live displays at the Two Oceans Aquarium in Cape Town it caused a sensation as few landlubbers had seen these otherworldly fishes with their rolling eyes, stubby tails and sculling fin action. They seemed to come from another world.

The breeding strategy of the ocean sunfish is diametrically opposite to the 'low-number/high-investment-per-young' strategy of, for instance, sharks and coelacanths which produce a few dozen large eggs that hatch internally and are born as large juveniles. Surfperches, which I have collected on the Pacific coast of Canada near Vancouver, have taken this high parental investment strategy to the extreme as they produce only one egg at a time which hatches internally with the developing embryo being fed through a placenta-like organ as in mammals. The viviparous blenny is another far-out example - it suckles its young which attach themselves to a tube from the mother that supplies fats, proteins and oxygen-saturated fluids to the developing embryo!

**Amphibians**: The giant bullfrog, which, at 11 cm is twice the size of most frogs, is unique among South African frogs in that it shows parental care. Males guard their tadpoles, chase away predators and may even dig channels to deeper water to save the tadpoles from being trapped in drying-up pools. Most frogs lay their eggs in the ground or on vegetation but the foam nest frog lays its eggs in a foam nest in trees overhanging water and the leaf-folding frog deposits its eggs on submerged leaves which are folded over and sealed. Almost all frogs live in moist habitats and depend on water to complete their life cycle but some such as the hardy Namaqua caco and marbled rubber frog live in arid areas and are reliant on temporary, rain-filled pools.

The extraordinary platanna (or clawed toad) is a strictly aquatic swimming frog. I have been able to make prolonged observations from both above and under water on a colony of 18 adult platannas that live in the ecopool (artificial wetland) in my garden since the winter of 2019, which is unusual as most frogs are difficult to study over long periods as they are secretive and often nocturnal. Platannas have webbed toes with claws on the inner three toes and are excellent swimmers (but cannot hop). Mating takes place after the males attract the females with a buzzing call underwater and the eggs are laid in jelly capsules. The tadpoles are aggressive predators of aquatic insect larvae and form schools in midwater, hanging head down.

Platannas are facultative air breathers that take a gulp of air at the water surface when they are actively hunting but appear to absorb sufficient oxygen from the water using their gills and skin during hibernation, as I rarely saw them rising for air during winter. They are social feeders, scavenging in groups on the substrate, among the roots of wetland plants or at the water surface, usually from dawn to about 10h00, with a second, lower-level feeding peak at dusk. When they approach floating prey they rise slowly to near the water surface, hover there for a few seconds and then attack with a rapid lunge. The short front legs are used for holding or tearing the prey and placing it in the mouth. They also rip large prey (such as grasshoppers) with their powerful, clawed hind legs but these legs are mainly used for swimming. Platannas appear to be able to see both under and above water as they occasionally swim with their bulbous eyes protruding above the surface and sometimes catch prey on emergent plants. They have an acute sense of hearing and react quickly to noises. They hibernate over winter, typically after the first heavy rains of autumn, emerging again about five months later. In 2021 their pre-hibernation (with reduced feeding

**Cape platanna**

activity) started immediately after 17.8 mm of rain fell on 15$^{th}$ March and the water temperature dipped below 17°C, and they went into full hibernation by mid-April.

Platannas gained international fame when South African scientists found that they can be used for a reliable human pregnancy test! If a drop of urine from a pregnant woman is placed on the frog's back the gonadotrophin in the urine induces the frog to produce eggs. This finding was so significant that over 400 000 live platannas were exported from South Africa between 1940 and 1970. These exports and the secondary distribution of specimens on different continents resulted in institutions from 48 countries having live colonies and the platanna became the world's most widespread amphibian (as well as a serious alien invader)! Platannas were even launched into space in 1992 on the space shuttle *Endeavour* so that scientists could determine whether their breeding and development could take place in near-zero gravity. The horny frogs did not miss a beat and bred normally!

**Reptiles**: Many reptiles have deviated significantly from the basic reptilian *bauplan* and lifestyle of being four-legged animals with scales that live on land. Snakes, legless lizards and amphisbaenians have lost their legs, some snakes and lizards live in trees and crocodiles, tortoises, terrapins and turtles have either hard bony plates or carapaces in addition to scales, and crocodiles, terrapins, turtles, sea snakes and several water snakes live permanently or primarily in water. They are all cold blooded and typically heat-seeking animals and air breathers. Although most lizards live on dry land, Bouton's snake-eyed skink, which I found at Black Rock on the Maputaland coast, forages and hides from predators in marine intertidal pools. This is their only known locality in South Africa but they have a wide Indo-Pacific distribution due to their habit of floating across oceans on logs and coconuts! Another lizard with aquatic habits is the water leguaan which eats fishes and frogs.

My first scientific research project, back in 1969, was a study of the behaviour of rock agamas (*koggelmannetjies*) in Makhanda where I determined their temperature preferences in the laboratory and then confirmed them in the field. These lizards are common in the drier parts of South Africa and typically sit on exposed rocks nodding their heads and basking in the sun. The southern stiletto snake (or side-stabbing snake) also interests me as it was the only snake that bit and injected venom into me during my reptile-collecting adventures in Zululand in the 1970s even though we collected such deadly beasts as gaboon adders, boomslangs, puff-adders and cobras during those carefree days! These vipers can extend their fangs outside their mouth even when it is closed, which is handy when you are hunting moles in tight tunnels but also means that if a human handler holds it in the time-honoured pinch behind the head he can still get bitten! My hand swelled up and went blue, but I did not die. Another snake that brings back fond memories is the Mozambique shovelsnout, one of our rarest snakes but one that I got to know intimately as they regularly congregated at dusk on sandy roads near our research station at Lake Sibaya in Maputaland. When the famous herpetologist Dr Carl Gans from the University of Michigan visited us I told him that I could find this rarity at will but he thought I was joking! Late one afternoon we ventured out and found them with ease.

I took other visitors to witness one of the most spectacular sights in Nature the nesting of leatherback and loggerhead turtles on the Maputaland coast. Watching these primeval beasts haul out onto the beach, laboriously dig their nests with tears streaming down their cheeks (to remove sand from their eyes), lay over 100 ping-pong-ball-sized eggs and then return to the ocean is a timeless and life-changing experience. Leatherback turtles are the second largest reptiles in the world after various crocodiles, reaching over 3 m and 900 kg. They spend most of their life at sea but must return to land to nest, often returning to their natal hatching site. Leopard tortoises are the only tortoises in Africa that can swim - they have no scute above their neck like other tortoises which allows them to lift their heads (like terrapins) to breathe air while swimming.

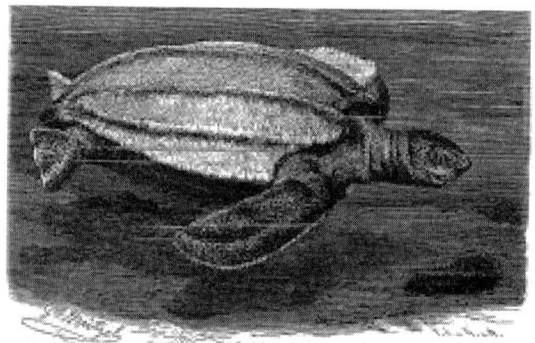

**Antique drawing of a leatherback turtle**

Legless lizards, like the extraordinary giant legless skink and African long-tailed seps as well as the otherworldly wedge-snouted amphisbaenian, have opted for the snakelike legless condition and mainly live underground in soft sand. Another remarkable fossorial reptile is Schlegel's blind snake which has lost its sense of sight as it spends its entire life underground. We often encountered pythons when I lived on the shores of Lake Sibaya in Zululand in the 1970s. They are constrictors that crush their prey before ingesting them. In our area where most game animals had been exterminated they were semi-aquatic, often catching small prey such as fish, frogs and waterbirds in the water or ambushing cane rats, goats and other small mammals on the water edge. We caught two pythons in our fish traps – they had gorged themselves with fish before they drowned.

Everyone adores chameleons, for good reasons. Their slow, deliberate movements, ability to freeze for long periods with one leg raised, extraordinary tongue-predation, rolling eyes and remarkable colour changes make them unique among animals. A dwarf chameleon that we had in captivity in Zululand gave birth to seven babies while we watched. Despite their camouflage they are easy pickings for predators such as vine snakes, boomslangs, hornbills and monkeys.

Pythons usually lay about 50 eggs (exceptionally up to 100) which they guard and incubate. They reach a large size – the largest we collected (as roadkill) measured 5.3 m – but are dwarfed by extinct Eocene pythons from Egypt that reached 15 m! At the other end of the scale flowerpot snakes only reach 170 mm and are 40 mm long when they hatch. Remarkably this tiny snake, which was introduced into South Africa from Australia with pot plants, only has females as they can produce viable eggs without mating.

The gaboon adder is a formidable foe. It has the largest volume of venom of any snake in the world, with both neurotoxic and haemotoxic properties, and is very cryptic, lying undetected on the leafy floor of the coastal dune forest in Zululand ready to strike at lizards, small mammals and ground-living birds. They are beautiful but deadly and a

**Gaboon adder**

Nile crocodile

herpetologist whom we knew Mtubatuba, who unwisely kept a colony of them in aquaria in his lounge, was killed by one of his pets. The more common puff adder which is responsible for about 60% of snake bites in South Africa is a surprisingly talented snake. Besides being able to slither along sinusoidally like any self-respecting serpent it can 'walk' forwards on the tips of its ribs without bending its body, jump at least 50 cm into the air and swim efficiently.

Crocodiles are directly descended from archosaur dinosaurs and retain their mean-spirited look, but their maternal behaviour belies their evil countenance. Once, while sitting in a fig tree overlooking Bande Bande Bay at Lake Sibaya (observing catfish predation), I witnessed a most extraordinary sight. A large crocodile, almost certainly a female, dug up a nest under the tree and helped several baby crocodiles to emerge from their eggs by gently biting their eggshells, with the youngsters assisting the process with their specialised egg teeth. The adult crocodile then waited at the water's edge for some of the hatchlings to climb onto her back and head. The next day I found the crocodile family in a shallow wetland nearby with the mother still guarding the young. At a crocodile farm in southern Mozambique I experienced first-hand the predatory instinct of a newly-hatched croc. The ranger Mario da Costa gave me an egg that was about to hatch and the first act by the little critter when it emerged was to bite me on the finger!

The most significant outlier among our reptiles is the yellow-bellied sea snake which has forsaken the land and lives permanently at sea, even giving birth to its young in the open ocean. They breathe air at the water surface and have a single, large elongate lung that enables them to remain underwater for several hours diving to depths of 50 m and more. They have a laterally flattened paddle-like tail, a nasal gland for purging excess salt and valved nostrils to prevent water from entering their lung. They feed on fishes and are the most widely distributed snakes in the world. Sea snakes are related to cobras and mambas but, along with the whales and dolphins (which are small whales), are the most completely aquatic of all air-breathing vertebrates. They are venomous and can survive for several hours if they are stranded on the beach, so they are best left alone if you find one.

**Sea snake**

**Birds**: South Africa has many strange birds which are out-of-kilter to various degrees with the norms for their groups. The ostrich is an unusual flightless bird which at 2.8 m and 156 kg is the largest living bird in Africa. They have muscular legs and can reach speeds of 70 km/hr and a stride length of 3 to 5 m, enough to outstrip any predators. They also use their powerful legs to kick, with their 10-cm long talons capable of doing serious harm to attackers. Their most vulnerable life stages are the eggs and newly hatched young but they have compensated for this threat by nesting communally and sharing the egg- and young-guarding duties.

The feathers of ostriches have retrogressed into decorative plumes and their wings are used for courtship displays and to distract predators. Although they are strange compared to other South African birds, they are not much different from other ratites such as the emu and cassowaries of Australia and rhea of South America although they are much larger than the kiwi and far smaller than the recently extinct elephant birds of Madagascar and moa of New Zealand. Their group name, ratites, comes from the Latin word *ratis* ('with no keel'), which refers to the absence of a bony keel on their breastbones (sternums) which is normally used to anchor the flight muscles.

**Ostrich**

Skimmers, which are occasionally seen in the Kruger National Park but are common further north along the Zambezi and Chobe Rivers, feed by skimming neuston (floating objects) off the water surface and have been observed bravely mobbing large herbivores such as elephants and buffaloes that approach too close to their ground nests. They also flop down on the sand and pretend to have a broken wing to distract predators like many other ground-nesting birds.

Pelicans have pushed the limits of their *bauplan* by developing a beak pouch with a cavernous capacity and by fishing in co-operative groups and even attacking ground-nesting birds on islands. The flamboyant design of flamingos equips them perfectly to tramp mud with their feet to suspend small animals which they sieve out of the water with a back-and-forth motion of their head, which is held upside-down. Black herons have the unique habit of shading the water with their outstretched wings so that they can detect prey more easily and I have seen a squacco heron 'flyfishing' by dropping a grasshopper into a stream and then catching the fish that rose to eat it! Male long-tailed widowbirds have tails 50 cm long which they display while executing slow wingbeats during flights over their grassland territories.

The beaks of kingfishers are so well-adapted for plunging at high speed into water that their shape, in a classic case of biomimicry, was used to design the nose cone of the super-fast Japanese *Shinkansen* bullet train! Penguins are classic examples of the ways in which an animal has been able to rejig its anatomy and behaviour to hunt underwater although they do roost, breed and care for their young on land. Their wings have become powerful flippers and their bodies are streamlined so that they can outswim and outmanoeuvre their prey, mainly fast-swimming, shoaling fishes, which they sometimes encircle to break up their defensive shoaling patterns. Furthermore, their feathers have been modified to form a scaly cover in which air bubbles are trapped which helps to incubate them in cold water. Each bird has a unique pattern of chest spots which allows researchers to identify them individually. Albatrosses and petrels have perfected the energy-efficient technique of gliding and spend long periods on the wing without alighting. Cape gannets plunge from a dizzy height to catch prey, mainly sardines, pilchards and midwater gobies, and when they return to their colonies have the incredible ability to recognise their mates among the squawking mass.

Sociable weavers have taken 'sociable' to the extreme by building huge, permanent community nests, probably the most spectacular structures built by any bird. The nests, which are constructed in trees and on telephone poles, may house over 100 pairs and can be more than 100 years old. Their intricate structure ensures that they maintain relatively stable internal air temperatures. Sociable weavers are known to share nest-guarding duties with other residents, another rarity among birds.

**Sociable weaver**

Owls such as our spotted eagle-owl are perfectly adapted for nocturnal hunting with their huge forward-facing eyes, binocular vision, the ability to swivel their heads through $270°$ and acute hearing, which enables them to pinpoint the faintest sounds. Their hollow faces reflect sounds towards their ears so that they can catch prey in almost total darkness using sound. If you see an owl bobbing its head up-and-down or from side-to-side it is trying to obtain a clearer picture of you. The surface area of their wings is large in comparison with their body size, which allows them to fly slowly if necessary, and their feathers have a soft, comb-like edging which gives them the ability to fly silently and ambush unsuspecting prey.

The secretary bird has an eagle-like body on crane-like legs. They perform elaborate courtship displays and build large nests up to 1.5 m across on top of thorn trees where they lay clutches of one to three eggs which are brooded by both parents. Unlike most birds of prey they hunt on the ground, stamping their prey to death. Adults typically hunt in pairs, stalking around with long strides hunting insects, lizards, snakes, tortoises, birds and small mammals. Although they have a wide range in Africa their numbers have declined sharply in recent years due to habitat loss and they are now classified as endangered by the IUCN. The secretary bird appears on the coats-of-arms of Sudan and South Africa.

Another idiosyncratic South African bird is the hamerkop, an endemic that is the only species in its genus and family and is most closely related to pelicans and shoebills. They are large, wading birds that feed along the shores of lakes, wetlands, estuaries and marine rocky shores. Their prey includes shrimps, insects (including flying termites and insects flushed by grazing cattle and buffalo), fish, frogs and even rodents, with platannas and their tadpoles being a favourite catch. They have also been observed fishing off the backs of hippopotamuses and hunting frogs with banded mongooses.

**Hamerkop and nest**

The strangest aspect of hamerkop behaviour is the huge nest that they build, over 1.5 m across, usually in the fork of a tree and often over water. The nest is made by both the male and female starting with a platform of sticks held together with mud followed by walls and a domed roof. A mud-plastered entrance, accessed from below, leads through a tunnel to a nesting chamber big enough for both the parents and young. The nests take from 10 to 14 weeks to build and continue to be refined even after the eggs have been laid. The pair engage in an elaborate courtship display, bowing to each other and flapping their wings while emitting a yapping, cackling call. Both parents share the duties of incubating the eggs and feeding the young.

Hamerkops are compulsive nest builders that make three to five nests per year whether they are breeding or not; this activity probably creates a stronger pair bond. As is the case with sociable weavers their nests may be used or even co-opted by other birds such as weavers, owls, starlings and pigeons and abandoned nests are inhabited by snakes and genets. Although they may make nests close to one another they are not colonial nesters.

**Mammals**: Mammals are characterized by having a neocortex for higher-order brain functions (such as sensory perception, cognition, spatial reasoning and language), three middle ear bones and mammary glands which (in females) produce milk for feeding the young. They usually walk on four legs although bats fly, whales, dolphins, seals, otters and dugongs swim, moles and mole-rats burrow, monkeys, squirrels and bushbabies climb and pangolins, some monkeys and spring hares walk or bound on two legs. Most of them have fur or hair but the pangolin, naked mole-rat, hippopotamus, dolphins and whales are virtually hairless.

**Dune mole-rat**

A strange small mammal that inhabits Rondebosch Common near my home is the Cape dune mole-rat which is neither a mole nor a rat and is more closely related to porcupines than to rodents. Their family name, Bathyergidae ('deep workers') hints at their subterranean lifestyle as they dig extensive interconnecting burrows and feed on roots and bulbs. A relative of theirs from East Africa, the naked mole-rat, is the only mammal that has a social structure like that of social insects (honeybees, termites and ants), i.e., they have one female (the 'queen') and one to three males that reproduce while the rest of the colony comprises sterile workers.

The African civet has a unique colour pattern with black-and-white blotches covering its coarse pelage, and rings on its tail and black rings around its eyes like a raccoon. They are one of the few carnivores capable of eating toxic invertebrates such as termites and millipedes. Bat-eared foxes have enormous antennae-like ears, of which the SKA project would be proud, which are used for detecting predators and prey and for thermoregulation. Elephant shrews hop along like miniature rabbits and can achieve speeds 28 km/hr. They are more closely related to elephants than to shrews and are endemic to Africa. Spring hares mimic the locomotory behaviour of kangaroos by bounding along on their strong back legs.

The riverine rabbit, the only indigenous burrowing rabbit in Africa, inhabits dense patches of bush along seasonal rivers in the semi-arid Karoo and is listed as critically endangered by the IUCN due to habitat destruction by farming. Another burrowing mammal that is indigenous to Africa is the ant bear or aardvark which resembles a pangolin but is covered with coarse hair rather than scales and has a long, pig-like snout. Bushbabies or galagos, which are only found in Africa where they are the smallest primates, are strictly nocturnal and feed on tree gum, insects and fruit.

Bats are the only mammals capable of true flight. The name of their order, Chiroptera, means 'hand-wing' as the 'fingers' on their forelegs have been greatly elongated and covered with a thin membrane. Their delicate wings are capable of more manoeuvrable flight than that of most birds and they can also achieve impressive speeds - the fastest bat, the Mexican

Egyptian free-tailed bat

free-tailed bat, can reach a ground speed of 160 km/h! Most African bats feed on insects although a few eat fruit and nectar. Many bats like our Egyptian free-tailed bat emit ultrasonic sounds to produce echoes that help them to create a detailed image of their surroundings and detect, localise and classify their prey in total darkness. Bats also have acute hearing and are able to use magneto-reception (detecting a magnetic field to determine direction, altitude or location) as they have a high sensitivity to the Earth's magnetic field, as in homing pigeons, eels, turtles and mole-rats.

The so-called 'charismatic megaherbivores' (giraffe, hippopotamus, black and white rhinos and elephants) are the darlings of wildlife photographers and have high symbolic appeal which is used by environmentalists to gain public support for their goals. Their most significant outliers are the giraffe and elephant. The specific name of the giraffe, *Giraffa camelopardalis*, refers to their size and marking being 'as big as a camel' but 'spotted like

Giraffe

a leopard'. The only other member of their family, Giraffidae, is the okapi, a long-necked inhabitant of lowland forests in Zaire. The giraffe is the tallest animal in the world exceptionally reaching heights of 5.88 m but usually between 4.9 and 5.2 m for males. They have the same number of neck bones as most mammals and humans (seven) but each bone is about 25 cm long. Giraffes have a long gestation period (about 457 days), the only ungulate in which this period is longer than a year. At birth the young already weigh 102 kg and stand 1.5 m tall at the shoulder!

Giraffes are extraordinary in many ways. They have a circulatory system that defies the effects of gravity by providing a regular flow of blood up the long neck to the brain. Their arteries above the heart are muscular and elastic to pump blood against gravity whereas the lower arteries are narrower with thick walls to bear higher pressures and prevent blood from pooling. Their hearts weigh 10.8 kg (compared to ours at 285 to 345 g) and have a muscular left ventricle that pumps blood to the head and around the body and a thinner right ventricle that pumps it the shorter distance to the lungs.

African elephants are the largest terrestrial animals in the world, reaching 4 m at the shoulder and a mass of over 6 000 kg in males. They have deviated significantly from the generalised mammalian *bauplan* by having trunks which are elongate extensions of the nose, huge ears which allow them to radiate excess heat, very thick skin, pillar-like legs and a pair of long, curved ivory tusks that develop from the upper incisor teeth. Their trunks are impressive multi-tools that are used for breathing, smelling, feeding, trumpeting, drinking and showering and even as a snorkel when crossing a deep river.

African elephants have a large brain (up to 5.4 kg) and a highly convoluted neocortex, a trait they share with humans, apes and dolphins. They are amongst the world's most intelligent animals and have been shown to exhibit a wide range of 'smart' behaviours including those associated with grief, learning, mimicry, art, play, a sense of humour, altruism, use of tools, compassion, cooperation, self-awareness, memory and possibly language. The gestation period of the African elephant (22 months) is the second longest of any living animal and is exceeded only by that of the coelacanth (36 months). The latest research reveals that dugongs are the most closely related living animals to elephants, not dassies.

**Mandala elephant face**

The hippopotamus ('river-horse'), like marine mammals, has taken advantage of the buoyancy of water to support its weight but feeds on grasses on land at night. Surprisingly this gentle-looking herbivore is responsible for killing more people in Africa than crocodiles. My experience with them in lakes and rivers in Zululand and in the Okavango Swamps is that they are dangerous and unpredictable and will attack unprovoked on land and in the water. Our 5 m aluminium boat was once lifted out of the water by a hippo at Lake Sibaya but fortunately did not tip over otherwise we would

**Hippopotamus**

have been goners. Carolynn and I were once chased, with evil intent, by a hippo when we tried to photograph it at night eating a farmer's maize crop. I have an indelible image in my mind of a desperate botanist, Pete Ashton, running for his life while being chased along the shore of Lake Sibaya by a surprisingly fast hippo!

Although hippos prefer to spend the day in freshwater lakes or rivers they occasionally venture into the sea. We once tracked an adult that walked and swam up the Maputaland coast from Lake St Lucia to Mbibi. They are also capable of long overland treks, the most famous of which was that by 'Huberta' which strolled 1 600 km over a period of two years from Zululand to the Eastern Cape near King William's Town where she was eventually shot by farmers in the Keiskamma River in April 1931.

**Dugong**

Dugongs, which I have observed near Vilanculos in Mozambique and off Bahrain, occasionally stray onto the Maputaland coast. They are large, reaching 3 m and 400 kg, and aquatic, feeding on sea grasses. Females give birth to single young (rarely twins) which they suckle while lying on their backs in the water, which may have given rise to the mermaid myth.

Elephant seals are huge sea-going seals that mainly live off Antarctica and the southern tips of South America and New Zealand. They occasionally visit South Africa – I photographed a stray on Jesser Point at Sodwana Bay in 1972. They reach 6 m and 4 000 kg with males having a prominent proboscis on their snout, like an elephant's trunk, hence their name.

All whales and dolphins (cetaceans) are remarkable exceptions to the basic mammalian *bauplan* and are extremely well-adapted for their aquatic existence. They have lost their hind limbs and their forelimbs are modified into flippers for stability and steering whereas the horizontal tail flukes provide propulsion. Whales communicate with one another using intricate songs as well as whistles and buzzes and dolphins emit clicks for echolocation and navigation. All cetaceans are air breathers but mate and give birth at sea; they all produce milk and suckle their young.

Blue whales, which rarely reach our seas and cannot therefore be regarded as a permanent part of our fauna, are the largest animals that have ever existed (as far as we know), reaching 33 m and 175 tonnes. Their numbers plummeted from over 125 000 in 1905 to less than 3 000 in 1998 due to commercial whaling but have increased again under protection to over 5 000. Their pregnancies (11 months) last longer than those of humans and by the end of the six-month weaning period (while the young are fed milk by the mother) they reach 16 m and 23 tonnes! Remarkably, scientists tracking their underwater songs have recently found a previously undiscovered population of blue whales off the east coast of Africa, extending from Mozambique northwards to the Gulf of Oman. Nowadays we hear blue whales more often than we see them.

Killer whales are the largest predators on the planet, reaching 8 m and 8 tonnes, and are further distinguished by being the most widespread mammals, occurring in all the oceans and to depths of 3 000 m. They are frequently seen off our shores where they feed on fish, squid, birds, seals, turtles and dolphins and may attack larger whales in organised packs. Other toothed whales in our seas include the enormous sperm whale which reaches 17 m and 46 tonnes in males and can dive to depths of 2 km for up to two hours, and the smaller false killer whale. Dolphins are small, toothed whales that are a prominent feature of our oceans, often forming huge schools. One of the best places in South Africa to view dolphins is Dolphin Point looking eastwards along the seashore towards Wilderness where they can often be seen surfing in the breakers.

**Killer whale**

The huge baleen whales, including humpbacks, southern rights and Bryde's whales, have baleen plates instead of teeth with which they sieve out marine organisms as food. Humpbacks circle underwater and release a curtain of bubbles that concentrates the prey and makes it easier to catch; in early April 2021 humpbacks feeding in Table Bay inadvertently herded an enormous shoal of fish into Cape Town harbour. The most commonly sighted great whale off our southern coast is the southern right which nurses its calves in Walker Bay at Hermanus and off Gansbaai and the De Hoop Nature Reserve. Our great whales were heavily exploited by commercial

**Southern right whale**

whalers in previous centuries. About 130 000 southern rights were killed between 1770 and 1850, reducing their population to about 300 individuals and threatening them with extinction, but strict protection since 1935 has allowed the species to recover at a rate of about 7% per year, with southern populations of the humpback recovering at an even higher 9%. We tend to take them for granted but they are truly extraordinary animals.

The pangolin or scaly anteater has also significantly broken the *bauplan* rules of its group. It has no fur (although fine hairs are found in the ears and on the eyelids and underparts of the body) and its body is covered by an overlapping armour of heavy scales, structurally different from those of reptiles, which also extend along the tail and head and down the legs to the feet. The scales are made of keratin and may constitute up to 20% of the animal's weight. Their English and Afrikaans names, pangolin and *ietermagog*, are derived from the Malay word *pengguling* ('one who rolls up') which refers to their habit of curling into a tight ball with the head inside when threatened. The sharp-edged scales, besides forming a protective layer, are also used for defence as the pangolin performs a cutting action if an attacker inserts its paw or snout between the scales. They also release a foul-smelling chemical from glands near the anus, similar to the smelly spray of a skunk.

**Pangolin**

In the pangolin the typical mammalian toothed mouth has been replaced by a tapering, toothless, two-jawed snout, with the slender lower jawbone playing little role in chewing which is carried out with the aid of ingested stones in a muscular, gizzard-like stomach lined with keratinous spikes. Pangolins have a sticky, 40 cm-long tongue which they use to collect ants (larvae, pupae and adults) and termites in their nests, which are excavated with their powerful front claws. They have poor vision but a well-developed sense of smell which they use to find ant nests.

South African pangolins give birth to one young at a time and protect the newly born by carrying them on their tails or backs or hiding them in burrows or within their coiled-up ball when danger threatens. Females have one pair of milk-secreting organs on their breasts and feed and care for their young for two years. They shelter in hollow trees and burrows, often those of other animals, and are solitary and typically nocturnal. When they walk they balance the body on their hind legs with the fore feet and tail held up, although the tail occasionally scrapes the ground to retain balance. When danger threatens they sit upright on their hind legs and tail and sniff the wind.

Pangolins are in high demand in China and Vietnam where their meat is considered a delicacy and their scales are used in traditional medicine; they are also hunted and eaten in Ghana and elsewhere in Africa as bushmeat. Despite their comparative rarity over one million pangolins have been traded over the past decade which makes them the most trafficked mammals in the world. Consequently, they have been declared as vulnerable to extinction by the IUCN due to both illicit hunting and habitat loss. In 2016 a treaty signed by 180 governments ended legal trade in pangolins and in 2020 China disallowed the use of pangolin scales in traditional medicine. This is a major triumph for conservation as an estimated 195 000 pangolins were trafficked in 2019 for their scales alone. The suggestion that pangolins may be an intermediate host for the SARS-CoV-2 virus variant that causes Covid-19 has further threatened their survival.

**Transition from water to land and back to water**: Some of the most extravagant adaptations of modern animals came about when they 'decided', through the process of natural selection, to leave the harsh environment of the land and return to the nurturing milieu of the sea. Sea snakes, marine turtles, seals, dugongs, whales and dolphins and, to a lesser extent, crocodiles and otters, converted their legs into fins or paddles and the penguins did likewise with their wings. Sea turtles, which evolved from land tortoises with heavy shells, have reduced the weight and reshaped their carapaces so that they act as streamlined hydrofoils underwater.

It is interesting that none of the vertebrates that have 'turned turtle' in this way have been able to 'retool' themselves to breath underwater using gills, like their distant fishy ancestors. As a result all marine vertebrates (with two minor exceptions) breathe air at the water surface which must be a real nuisance, especially for whales that feed at great depths. When we dived in the *Jago* submersible off the west coast of the Cape Peninsula National Park a Cape fur seal visited us at a depth of 170 m but had to return to the surface to breathe. This emphasizes what a difficult evolutionary experiment it must have been to evolve lungs from gills as the reverse process (gills from lungs) has not taken place over the 250 plus million years since tetrapods returned to the sea.

Two modest exceptions to air breathing include some sea snakes, which can absorb about 20% of their oxygen needs from the water through their skin, and a few Australian freshwater terrapins that have developed the partial ability to breathe through their anuses. Many marine vertebrates such as sea snakes, dugongs, whales and dolphins, have been able to adjust their breeding strategies so that they can give birth at sea and ensure that the new-born takes its first breath of air soon after birth, but the crocodiles, turtles, penguins and seals still have to go through the arduous process of hauling out onto land to lay their eggs or give birth to their young.

**And the finalists are ...:** Once again, taking the above into account, with a little personal bias, and realizing that I am opening myself up to severe criticism, I select the following vertebrates as the strangest outliers in their respective groups: Jawless fish – hagfish; Cartilaginous fish – six-gill stingray; Bony fish - Indian Ocean coelacanth, lungfish and ocean sunfish; Amphibians – bullfrog and common platanna; Reptiles – yellow-bellied sea snake and leatherback turtle; Birds – ostrich, hamerkop and African penguin; Mammals – free-tailed bat, giraffe, elephant, pangolin and southern right whale.

Taking the analysis further, I select the following eight quarterfinalists from both the invertebrates and vertebrates: sea cucumber, velvet worm, octopus, coelacanth, sea snake, pangolin, free-tailed bat and elephant, and the following four semi-finalists from both groups: velvet worm, coelacanth, pangolin and free-tailed bat.

**And the winner is ...:** The free-tailed bat and coelacanth are strong contenders but, **u**sing the main criterion that 'strangeness' is defined by the extent to which an animal has deviated most from the normal *bauplan* for its group, I vote for the pangolin and the velvet worm as the finalists in the 'Strangest Animal in South Africa' contest with the winner being the pangolin. Its unique combination of a scaly body cover, toothless mouth, 40-cm-long tongue and habit of walking on its hind legs and rolling up into a ball when threatened makes it one of the most peculiar animals on the planet. Sadly, our strangest animal is also one of the most traded and endangered species.

The velvet worm receives the silver medal due to its potpourri of arthropod and annelid traits and the free-tailed bat deserves recognition as the bronze medallist for the extraordinary adaptations it has made to achieve true flight. It is with a heavy heart that I assign fourth place, off the podium, to the coelacanth as it has, through time and space and at all levels of society, played a major role in enlightening us about the process of evolution. It is a bizarre animal by any measure if you care to look more than skin deep.

**Discussion:** This discussion on strange animals has highlighted the extraordinary endpoints of evolution with which we are fortunate to share the planet. Animals may be strange for many reasons, some because they 'froze' in time and retained their ancestral characters (bristletails, sharks, hagfish and coelacanths), others because they developed novel feeding methods (pangolins, southern right whales), adopted colonial habits (sponges, corals, jellyfish, bluebottles, termites, ants, bees and sociable weavers), acquired special powers (mantis shrimp, octopus and bats) or adopted unusual combinations of traits (velvet worms, barnacles, sea cucumbers, sea pens and coelacanths). Others are strange as they live in unusual habitats (cave catfish, sea swallow and ocean sunfish), have adapted to new biomes (turtles, sea snakes, penguins, dugongs, dolphins and whales) or lifestyles (platanna) or are highly specialised (stick insects and chameleons).

The five great extinction events that have occurred since the evolution of life began were all caused by cataclysmic physical events, such as meteorite strikes, earthquakes, volcanic eruptions, ice ages, tsunamis, reversals of the magnetic poles, floods, droughts, moving continents and sea level changes, or combinations of them, and they indiscriminately wiped out 70% or more of all species irrespective of their levels of specialisation. The human-induced Anthropocene extinction event that we are now experiencing is different and has caused humans to reflect on what we can do about it. The term 'Anthropocene extinction' suggests that humankind can be likened to a geological force that has had such an impact on reshaping the natural environment that it has 'moved' the planet from the Holocene to the Anthropocene era (Steffen *et al.*, 2011).

The impact of the Anthropocene is likely to be different from that of previous extinctions as it is being caused by a combination of ecological and physico-chemical events that are cascading through the biosphere in complex and poorly understood ways. These subtle but accumulating changes are likely to cause a higher proportion of specialised animals (and plants) to go extinct as these species tend to be partners in dependencies with other specialists, for instance, as host or parasite, symbiont or commensal, predator or prey, cuckoo or cuckold. As a result, when one species in the interdependent community becomes rare or disappears, the others are likely to follow suit. This could mean that many of the strange creatures that are discussed in this chapter may be the first to be exterminated whereas generalised, 'weedy' species, like cockroaches, Indian mynahs and rats (to take it to the extreme) are more likely to survive and colonise abandoned habitats and niches. This would be a disaster for biodiversity and for the essential ecological processes and life-support systems on the planet on which we, as well as all other living organisms, depend for our survival.

In recent centuries humans have caused the extinctions of at least 113 large animal species internationally, including the elephant birds of Madagascar, dodo of Mauritius and solitaire of Rodrigues, Steller's sea cow, auroch, blue antelope, Cape lion, California grizzly bear, great auk, Malagasy crowned eagle and Javan tiger, as well as many smaller species. Can you imagine a future in which we must tell our grandchildren about the extraordinary creatures that once lived with us in South Africa, such as velvet worms, hard corals, coelacanths, leatherback turtles, penguins, riverine rabbits, pangolins and southern black rhinos, which we put to the sword through our negligent behaviour?

Of course, extinction is not all bad. As I discuss in my autobiography, *When I was a Fish*, extinction has played just as important a role in evolution as speciation. Evolution is not only about making new species but also about creating a turnover of species, with many (over 99%) going extinct and others replacing them. Extinctions are therefore not unusual, recent or an exception to the rule, they are the rule. As the American science communicator, Carl Sagan, wrote in 1996, 'Extinction is the rule.

## 13. South Africa's strangest animal? Animals with backbones

Survival is the exception'. As far as evolution is concerned extinction is the beginning of new opportunities for other species as the demise of the old creates space for the novelty and artistry of the new. In fact, every major extinction event in the past has been followed by a rapid adaptive radiation of new species, with the direction of evolution often making major shifts during these transitions. For example, after the Cretaceous extinction about 65 million years ago the domination of the land and oceans by reptiles was effectively ended and birds followed by mammals then became the dominant large life forms.

But we should avoid oversimplifying trends in evolution. I remember listening to an inspiring talk by the American science writer Stephen Jay Gould in California many years ago. He said that the concept of the Age of Reptiles, Age of Birds and Age of Mammals is rubbish - we have always been, and will always be, in the Age of Microbes. His words ring true today as we continue to do battle with the SARS-CoV-2 virus. From a more recent perspective during human history we have had the Stone Age, Bronze Age and Iron Age and today, as suggested by Siegle (2018), we are living in the Plastic Age. But we cannot stand idly by as if the Plastic Age were just another natural phase of human evolution. Like all our other impacts on the planet it is a problem that we have caused and which we will have to solve. Equally, the fact that extinction is a natural process does not mean that we should sit back and allow our unsustainable lifestyles to escalate extinction rates above normal levels. Our role, as the most sentient custodians of the biosphere, is to allow natural processes including extinction and speciation to proceed at their *normal* rates.

There are many tragic stories around that document our devastating impact on animal diversity. A study recently published by Ross Crates from the Australian National University in Canberra has revealed that critically endangered regent honeyeaters have forgotten how to sing their own songs as there are not enough elder birds to pass them on. He found that about 27% of males sang songs that differed from their typical melodies and 12% had resorted to singing the songs of other birds! In Hawaii, a recording was made of the call of the last remaining male *kauïöö*, a forest bird that sings in vain as all the females of its species have gone extinct. We are unravelling millions of years of evolution! Future generations will not forgive us for our negligence.

This review of strangeness in animals started as a relatively flippant overview but has revealed some deeper insights. It has shown that many so-called 'primitive' creatures have powers and abilities that we as humans do not have except through the use of our tools. These abilities include sight without eyes, neural coordination without brains, production of ultrasonic sounds, echolocation, flight, electro-reception, magneto-reception, trinocular vision, use of electrostatic charges to gather food, use of blood counter-currents to maintain elevated body temperatures, voluntary expulsion and regrowth of internal organs, regrowth of limbs or whole bodies from fragments, deliberate buoyancy regulation, external digestion, capture and continued use of chloroplasts and nematocysts, light production, sound production without voice, and many others.

Perhaps we should acknowledge that, in addition to the Internet of Things (IoT) we also have an 'Internet of Living Things' (IoLT), a network of neural and genetic connections that exists between us and all other living endpoints of evolution. This interconnectedness with the rest of Nature should surely be the ultimate source of our communal spirit of self-preservation and biodiversity conservation? This approach would remove the human self-centredness that dominates our thinking and decision making and replace it with a humbler 'we are just part of the multispecies network' mindset that recognises that many other creatures have 'beyond human' capabilities that transcend our powers.

# 14.
# Africa's Nobel laureates

*'And this I believe: that the free, exploring mind of the individual human is the most valuable thing in the world.'*

John Steinbeck in *East of Eden* (1952)

**INTRODUCTION**: Ever since I attended the ceremony held in the Stockholm Konserthus where then-South African Minister of Water Affairs & Forestry Kader Asmal was awarded the Stockholm Water Prize in 2000 under the same circumstances as Nobel laureates I have been fascinated by who has won (and not won) the Nobel Prize, and why. I have also been surprised that so little has been written about our African Nobel laureates all of whom have inspiring stories to tell. As a result I have given several talks on the topic of Africa's top achievers; they are summarised here.

**Female Nobel laureates from Africa:** In the history of African innovation women have made significant contributions but their achievements have hardly been recognised. The Nobel Prize, one of the pinnacles of human achievement, has been awarded to only five African-born women, one in Chemistry, one in Literature and three in Peace, between 1964 and 2011. Liberian women have won two awards and one each has been won by women born in Egypt, Kenya and South Africa.

The Nobel Prize in Chemistry in 1964 was awarded to **Dorothy Hodgkin**, a Cairo-born English scientist who was recognised for her 'determinations by X-ray techniques of the structures of important biochemical substances' including penicillin and vitamin B12. Hodgkin's mother's four brothers were all killed in World War I and as a result she became an ardent supporter of the new League of Nations and was later awarded the Lenin Peace Prize by the Soviet government in recognition of her work on peace and disarmament.

At the age of 18 years Hodgkin joined her parents at an archaeological site in Jerash in present-day Jordan where she documented the patterns of mosaics in Byzantine-era churches. Her attention to detail in replicating these mosaics mirrored her subsequent meticulous work in recognising and documenting patterns in chemistry. When she was appointed to Oxford University in the 1940s one of her students was Margaret Thatcher (*nee* Roberts) who, while Prime Minister, hung a portrait of Hodgkin in her office at Downing Street out of respect for her former teacher. During the 350th anniversary of the founding of the Royal Society of London Hodgkin was the only woman to feature in a set of postage stamps celebrating ten of the Society's most illustrious members. She proudly took her place alongside Isaac Newton, Edward Jenner, Joseph Lister, Benjamin Franklin, Charles Babbage, Robert Boyle, Ernest Rutherford, Nicholas Shackleton and Alfred Russel Wallace. In 1934 at the age of 24 she was diagnosed with rheumatoid arthritis which became progressively worse, and eventually crippling, over time. In her last years she was largely confined to a wheelchair but remained scientifically active despite her disability. She died in Ilmington, England, in 1994.

South African author and political activist **Nadine Gordimer** received the 1991 Nobel Prize in Literature at the age of 68 years. She was largely home bound as a child because her mother, for 'strange reasons of her own', did not put her into school. Often isolated, she began writing at an early age and published her first short story for children, *The Quest for Seen Gold*, at the age of 15. As an adult she was active in the anti-apartheid movement, having joined the African National Congress while it was banned. She was never blindly loyal to any organisation but saw the ANC as the best hope for reversing South Africa's treatment of black citizens. Rather than criticising it from the outside she chose to join it and address its flaws from the inside.

Nadine Gordimer

She gave Nelson Mandela advice for his famous 1964 defence speech at the trial which led to his conviction for life and also helped him edit his famous speech, 'I am prepared to die', given from the defendant's dock at the trial. When Mandela was released from prison in 1990 she was one of the first people he chose to see. Gordimer was also active on HIV/Aids causes.

One of the most powerful voices against the injustice of apartheid in South Africa Gordimer published novels, short stories, essays and plays that are steeped in the drama of human life and emotion in a society warped by decades of white minority rule. Some of her most notable works include *The Conservationist* (1974, joint winner of the Booker Prize), *Burger's Daughter* (1979) and *July's People* (1981). The Swedish Academy referred to her as an author 'who through her magnificent epic writing has - in the words of Alfred Nobel - been of very great benefit to humanity'. Gordimer received 15 honorary degrees and many international honours. She was born near Springs in 1923 and died in Johannesburg in 2014.

In 2004 the Kenyan environmentalist **Wangari Muta Maathai** was awarded the Nobel Prize for Peace in recognition of her tireless contribution to sustainable development, democracy, peace, women's rights and international solidarity, the first African woman and the first environmentalist to be so honoured (As the Nobel committee does not award a prize for contributions to environmental conservation and management, the Peace Prize suffices.) Maathai launched a grassroots campaign in 1977 aimed at countering the deforestation that was destroying the lives of poor people especially women living in rural Kenya. Her Green Belt Movement (GBM), which encouraged women to plant trees and assert their rights, spread to other countries on the continent and led to the planting of tens of millions of trees. In 1984 she was awarded the Right Livelihood Award for 'converting the Kenyan ecological debate into mass action for reforestation.' In her 2010 book *Replenishing the Earth: Spiritual Values for Healing Ourselves and the World* she discussed the impact of the GBM and stressed the importance of communities taking responsibility for their actions and mobilizing their resources to address their local needs.

Wangari was born in the central highlands of Kenya in 1940 and attended rural schools but her education was disrupted during the Mau Mau uprising when her family was forced to move into an emergency village. In 1960 she was one of the first beneficiaries of the

**Wangari Maathai**

Kennedy Airlift project which enabled her to study and complete her MSc in the USA. After further study in Germany she returned to Kenya and in 1971 became the first East African woman to receive a PhD, with a doctorate in veterinary anatomy from the University of Nairobi. In 2002 she was elected a member of the Kenyan parliament with 98% of the vote in the Tetu constituency, serving as assistant minister in the Ministry for Environment & Natural Resources from 2003 to 2005. She died of complications from ovarian cancer in Nairobi in 2011.

Ellen Johnson Sirleaf

In 2011 **Ellen Johnson Sirleaf**, President of Liberia from 2006 to 2018, Yemeni journalist and activist Tawakkol Karman and Liberian peace activist **Leymah Roberta Gbowee** received the Nobel Peace Prize. The three women were honoured for their participation in peace-building processes in Liberia and Yemen and their non-violent efforts to improve women's safety. Sirleaf campaigned on a platform of fixing the mess that the men before her had created and was the first woman to be elected head of state in a modern African country. As president she embraced her dual roles as 'Ma Ellen', healing the damaged nation and 'Liberia's Iron Lady' for her authoritarian management style. She rebuilt broken infrastructure, kept the peace and ushered in an economic revival but she was also tarnished by accusations of corruption and cronyism. Sirleaf also did little to empower other female leaders; of the 19 candidates who ran to replace her only one was a woman. To her credit she did do something unprecedented in 70 years of leadership by Liberian men: she stepped aside for someone else when her time in power was up (Baker, 2020c).

Gbowee (born in 1972) is a Liberian activist who was responsible for leading a nonviolent peace movement, Women of Liberia Mass Action for Peace, that helped to bring an end to the Second Liberian Civil War in 2003. Her efforts along with those of her collaborator Sirleaf helped to usher in a period of peace and led to a free presidential election in 2005, which Sirleaf won. Gbowee received the Nobel Prize at the age of 39, the youngest Peace laureate at the time. In 2011 she published her memoir, *Mighty be our Powers: How Sisterhood, Prayer and Sex changed a Nation at War* and in 2012 she founded the Gbowee Peace Foundation Africa based in Monrovia which provides educational and leadership opportunities for girls and women. She is currently executive director of the Women of Peace and Security Program at the Earth Institute, Columbia University.

In March 2020 *Time* magazine produced a special issue in which it recognised 100 women who had made a significant impact on world affairs between 1920 and 2019, designating a Woman of the Year for each year. Four women born in Africa were selected: Zenzile Miriam Makeba ('Mama Africa'), the singer, songwriter, actress, activist and United Nations goodwill ambassador from South Africa (1967), Nawal El Saadawi, the feminist writer, activist, physician and psychiatrist from Egypt (1981), Wangari Maathai, the environmental activist from Kenya (2001) and Ellen Sirleaf from Liberia (2006). The portrait of Wangari Maathai which appeared on the cover of *Time* magazine on 16th March 2020 was created by American fibre artist Bisa Butler using quilted and appliquéd African Dutch-wax cottons, silk and velvet.

**Male Nobel laureates from Africa**: Nobel Prizes have been awarded to 22 Africa-born men, nine for Peace, five for Literature, four for Physiology or Medicine and two each for Chemistry and Physics. Of the 22 laureates ten were born or worked for most of their life in South Africa, five in Egypt, two in Algeria and one each in Ethiopia, Ghana, Madagascar, Morocco and Nigeria (Albert Luthuli was born in Zimbabwe but did most of his work in South Africa).

Max Theiler

**Max Theiler**, a South Africa-born American virologist and physician, won the 1951 Nobel Prize in Physiology or Medicine for developing a vaccine against yellow fever, becoming the first Africa-born Nobel laureate. Working at the Harvard Medical School in Boston, Massachusetts, he and his colleagues proved that the cause of yellow fever is not a bacterium but a virus. After joining the Rockefeller Foundation in 1930 Theiler worked on vaccines against the disease and eventually succeeded in passing the virus to mice, which allowed him to obtain a variant that could be used as a safe, standardised human vaccine. He was born in 1899 in Pretoria, one of four children of Sir Arnold Theiler, a well-known veterinary scientist, and his wife Emma (*née* Jegge). He died in New Haven, USA, in 1972.

The French philosopher, author and journalist **Albert Camus**, who was born in Algeria in 1913, won the Nobel Prize in Literature at the age of 44 in 1957, the second-youngest recipient after Rudyard Kipling, who was 42. Camus was active in the French Resistance during the German occupation of France in World War II and wrote for and edited the famous Resistance journal *Combat*. He divided his work into three phases, the cycle of the absurd ('that which is meaningless'), the cycle of revolt and the cycle of love. His breakthrough novel was *The Stranger* (1942) with other notable works including *The Plague* (1947) and *The Fall* (1956). The Nobel committee recognised him for his 'important literary production, which ... illuminates the problems of the human conscience in our times.' He died in 1960 in France at the age of 46 in a car accident while travelling with his publisher Michel Gallimard.

The teacher, lay preacher and politician **Inkosi Albert John Luthuli** was awarded the Nobel Prize for Peace in 1960, the first person of African heritage to be so honoured. Luthuli was born at a mission station near Bulawayo, Zimbabwe, in 1898 but emigrated to South Africa with his parents in 1908 at the age of ten. From 1911 he attended a mission school in KwaDukuza (Stanger) supported by his mother, who paid his school fees by selling vegetables and taking in laundry. After completing his schooling he trained as a teacher and was initially appointed as the principal and only teacher in a rural school in KwaZulu-Natal. He later became head of a teacher's training college where he taught Zulu history, culture and literature. Luthuli was passionate about music and with his student Reuben Caluza founded the Adams College School of Music in 1935 and lead choir tours

Albert Luthuli

throughout Natal. His Zulu name *Mvumbi* ('continuous rain') refers to his prodigious work ethic and the many benefits he brought to his community.

From 1936 to 1953 Luthuli acted as a Zulu chieftain and was a prominent practitioner of the philosophy of *ubuntu*, which recognises the humanity and interdependence of every person, and gained great respect for his wisdom and integrity. After an eye-opening tour of the American South in 1948/49 he returned to South Africa and became active in politics. He joined the African National Congress in 1944, serving as its president from 1952 to 1960, and emerged as a leading figure in the Defiance Campaign against apartheid.

Chief Luthuli was recognised by the Nobel committee for his 'role in the non-violent struggle against apartheid' by leading a campaign of civil disobedience against racial segregation and discrimination. He accepted the prize wearing traditional ceremonial attire from King Olaf of Norway and received a standing ovation for his acceptance speech in which he declared, "I regard this as a tribute to Mother Africa, to all peoples, whatever their race, colour or creed". At the conclusion of his address Chief Luthuli sang the liberation anthem *Nkosi Sikel'I iAfrika* with the august assembly joining him in singing or humming the anthem. He died in 1967 after being struck by a train near his home in KwaDukuza, KwaZulu-Natal.

Former Egyptian President **Anwar al-Sadat** won the Nobel Prize for Peace in 1978, the first Muslim Nobel laureate, together with the Israeli Prime Minister Menachem Begin. They were recognised for 'having taken the initiative in negotiating a peace treaty between the two countries'. As President, Sadat lead Egypt in the Yom Kippur War of 1973 to regain the Sinai Peninsula which Israel had occupied since the Six-Day War of 1967, making him a hero in Egypt and, for a time, in the wider Arab World. He subsequently engaged in negotiations with Israel that culminated in the Egypt-Israel Peace Treaty. Although reaction to the treaty, which resulted in the return of Sinai to Egypt, was generally favourable among Egyptians it was rejected by the country's Muslim Brotherhood and the left which felt that Sadat had scuppered efforts to develop a Palestinian state. Most of the Arab world and the Palestine Liberation Organisation strongly opposed Sadat›s efforts to make peace with Israel without prior consultation and his refusal to reconcile with other Arab states over the Palestinian issue resulted in Egypt being suspended from the Arab League from 1979 to 1989. The peace treaty was also one of the primary factors that led to Sadat's assassination by militants led by Khalid Islambouli, who opened fire on him during the 6 *October* parade. Al-Sadat was born in 1918 in Mit Abou El-Kom, Egypt, and died in Cairo on 6th October 1981.

**Allan MacLeod Cormack**, a South African-born physicist who mainly worked in the USA, won the Nobel Prize in Physiology or Medicine in 1979 with British electrical engineer Godfrey Hounsfield 'for the development of computer-assisted tomography'. Cormack, who was born in Johannesburg in 1924, attended Rondebosch Boys' High School in Cape Town and received his BSc in physics in 1944 and his MSc in crystallography in 1945 from the University of Cape Town before carrying out doctoral research at Cambridge University from 1947 to 1949. In 1955, while he was a lecturer at the University of Cape Town, he took up a part-time position at the Groote Schuur Hospital to supervise the use of radioactive isotopes. One of his tasks was to calculate accurate radiation dosages for cancer therapy as the methods in use at the time were inaccurate.

He realised that he needed to develop a method for assessing tissue density in the body which led him to construct accurate mathematical models of a cross-section of an irregularly shaped, non-homogenous object. This research eventually resulted in him developing the original algorithms for a three-dimensional x-ray machine but little interest was shown in his discoveries partly because computers at the time were incapable, in a reasonable amount of time, of carrying out the enormous calculations that were required. Hounsfield who worked for EMI in London brought Cormack's algorithms to fruition when he developed the first clinically usable computerised tomograph, the EMI Scanner.

Cormack and Hounsfield never collaborated directly and the only time they met was at the Nobel Prize ceremony where the former gave a lecture modestly titled, 'Early two-dimensional reconstruction and recent topics stemming from it'. They were an interesting choice for the Nobel Prize in Medicine as neither of them were medical doctors. As a result of their revolutionary invention scientists and medical practitioners were able to see cross-sections of the human body for the first time without invasive surgery. Then, like all great inventions, a variety of other uses was found for x-ray tomography including the three-dimensional imaging of a variety of materials. Cormack died of cancer at the age of 74 years in 1998 in Winchester, U.S.A.

**Sir Aaron Klug** was born in Lithuania in 1926, where his father was a cattle rancher, and emigrated to South Africa with his parents at the age of two years. He lived in South Africa until he was 25 years old, initially attending Durban High School and then studying at the universities of the Witwatersrand and Cape Town. Paul de Kruif's 1926 book *Microbe Hunters* aroused his interest in microbiology, which he initially studied, but he later moved into physics and mathematics. He received his PhD in physics from Trinity College, Cambridge, in 1953 and spent the rest of his career as a biophysicist in the United Kingdom. Africa can nevertheless claim him as one of its 'scientific sons'.

After completing his PhD Klug moved to Birkbeck College at the University of London where he worked with virologist Rosalind Franklin in the laboratory of crystallographer John Bernal. This experience aroused in him a lifelong interest in the study of viruses and he made important discoveries on the structure of the tobacco mosaic virus. In 1962 he moved to the new Medical Research Council laboratory in Cambridge. In 1969 Klug was elected a Fellow of the Royal Society and served as the society's President from 1995 to 2000. He was knighted by Queen Elizabeth II in 1988. Klug was awarded the Nobel Prize in Chemistry in 1982 for his development of crystallographic electron microscopy and his structural elucidation of biologically important nucleic acid-protein complexes. He used methods from x-ray diffraction, microscopy and structural modelling to develop sequences of two-dimensional images of crystals taken from different angles that were then combined to produce three-dimensional images of the target. He died in Cambridge, UK, in 2018.

Anglican cleric Archbishop **Desmond Mpilo Tutu** was born of mixed Xhosa and Motswana heritage to a poor family in Klerksdorp, South Africa, in 1931. He initially trained as a teacher but was ordained as an Anglican priest in 1960 and in 1962 moved to the United Kingdom to study theology at King's College, London. In 1966 Tutu returned to southern Africa where he worked in several seminaries and universities before serving as general-secretary of the South African Council of Churches from 1978 to 1985, emerging as one of South Africa's most prominent anti-apartheid activists. He was the Bishop of Johannesburg from 1985 to 1986 and then the Archbishop

**Archbishop Desmond Tutu**

of Cape Town from 1986 to 1996, the first black African to hold these positions. Tutu polarised opinion when he rose to fame in the 1970s. White conservatives despised him while many white liberals regarded him as too radical. Black radicals accused him of being too moderate, focused as he was on cultivating white goodwill, but he was widely popular among South Africa's black majority and was internationally praised for his anti-apartheid activism.

Archbishop Tutu won the Nobel Prize for Peace in 1984 for his opposition to apartheid in South Africa. His objective was to see South Africa 'as [a] democratic and just society without racial divisions'. Despite violent attacks committed against the black population he adhered to his non-violent approach and encouraged the application of economic pressure by countries dealing with the apartheid authorities. After receiving the Prize Tutu used his international stature to accelerate the campaign against apartheid, leading calls for punitive sanctions against South Africa as one of the few strong voices inside the country while others were imprisoned or forced into exile. In 1995 Nelson Mandela appointed Tutu head of the country's Truth and Reconciliation Commission, a body set up to investigate apartheid-era crimes. Since his retirement in October 2010 on his 79th birthday he has continued to be an outspoken critic of any form of racial discrimination.

This author has had the good fortune to meet the 'Arch' on several occasions in Cape Town. Once during a visit by Queen Elizabeth II and the Duke of Edinburgh the Queen attended a church service given by the Archbishop in St George's Cathedral during which a child choir sang beautifully for the royal guest. After the performance there was a stony silence. In a trice the Archbishop sidled up to the Queen and said in a loud stage whisper, "You're allowed to clap in my church!"

Claude Simon

**Claude Simon**, who was born in Antanarivo, Madagascar, in 1913 of French parents, was awarded the Nobel Prize in Literature in 1985. His father, a career army officer, was killed during World

War I before Simon was ten years old. He grew up with his mother and her family in Perpignan in France and attended secondary school in Paris. After brief sojourns at Oxford and Cambridge he took courses in painting at the André Lhote Academy and then travelled widely in Europe and the Soviet Union. Early in World War II he took part in the Battle of the Meuse (1940) and was taken prisoner but managed to escape and joined the French resistance movement.

Much of Simon's writing is autobiographical dealing with personal experiences during World War II and the Spanish Civil War. His early novels were largely traditional in form but, with *Le Vent* (1957) and *L'Herbe* (1958), he developed a style associated with the *nouveau roman* movement, with fragmented storylines that nevertheless retain a strong sense of narrative and character. In *Triptyque* (1973) three different stories are mixed without paragraph breaks. Conflict is a constant theme throughout his work and his novels also dwell on images of old age, often seen through the uncomprehending eyes of children. Simon was an accomplished equestrian and fought in a mounted regiment during World War II - a major theme in *La Route des Flandres* (1960) and *Les Géorgiques* (1981) is the ridiculousness of mounted soldiers fighting in a mechanised war! He was also strongly influenced by Marcel Proust and William Faulkner but despite these influences his work is thematically and stylistically highly original. Simon was recognised by the Nobel committee for his particular writing style 'which combines the poet and the painter's creativeness ... in the depiction of the human condition.' He died in Paris in 2005.

Nigerian playwright, poet and essayist **Akinwande Oluwole Babatunde Soyinka**, widely known as Wole Soyinka, won the Nobel Prize in Literature in 1986, the first black African to win this prize. In its citation the Swedish Academy referred to him as a scholar 'who, in a wide cultural perspective and with poetic overtones, fashions the drama of existence.' Soyinka was born in Abeokuta, Nigeria, in 1934 and spent long periods of voluntary exile in Europe and the USA. With his shock of

**Akinwande Soyinka**

white hair and Afro hairstyle he is still a prominent member of Africa's literature community in his mid-80s despite having prostate cancer. He has a long history of campaigning for social justice and human rights in Nigeria and has been an outspoken critic of apartheid and the politics of racial segregation in South Africa. His Nobel acceptance speech, 'This past must address its present', was devoted to South African freedom-fighter Nelson Mandela.

Soyinka's major works include *The Invention* (1957), *The Lion and the Jewel* (1959), *A Dance of the Forests* (1960), *The Strong Breed* (1964), *Kongi's Harvest* (1964), *Poems from Prison* (1969), *Madmen and Specialists* (1970) and *Death and the King's Horseman* (1975). In November 1994 he fled Nigeria to the USA to escape persecution. In 1996 his book *The Open Sore of a Continent: A Personal Narrative of the Nigerian Crisis* was published and in 1997 he was charged with treason by the government of General Sani Abacha in Nigeria. He later became the second president of the International Parliament of Writers, established in 1993, to provide support for writers victimised by persecution. Soyinka has received many honours and awards including honorary doctorates from Harvard and Princeton universities. Today the biennial Wole Soyinka Prize for Literature in Africa is the most prestigious literature award on the continent.

Egyptian author **Naguib Mahfouz**, who was born in Old Cairo in 1911, won the Nobel Prize in Literature in 1988, the first writer in Arabic to receive the award. Although his mother Fatimah was illiterate she took her son on numerous excursions to cultural locations such as the Egyptian Museum and the Pyramids. The Mahfouz family were devout Muslims and he had a strict Islamic upbringing, remarking in an interview later in life, "You would never have thought that an artist would emerge from that family". He married late, at the age of 43 years, as he was afraid that the social pressures of married life would interfere with his writing and initially lived with his wife on a houseboat on the west bank of the Nile River.

**Naguib Mahfouz**

Mahfouz is regarded as one of the first

contemporary writers of Arabic literature together with Taha Hussein to explore the nature of existence by emphasising the experience of the living human subject. He published 34 novels, over 350 short stories, dozens of movie scripts, hundreds of op-ed columns for Egyptian newspapers and five plays over a 70-year career, with many of his works being made into Egyptian or foreign language films. He was recognised by the Nobel committee for work 'rich in nuance' that 'formed an Arabian narrative art that applies to all mankind'. His first three novels, published in Arabic in 1939, 1943 and 1944 were set in ancient Egypt but his best-known work *The Cairo Trilogy* (1956-1957) is a saga about a modern Egyptian family living under British colonial rule from World War I until after the 1952 military coup that overthrew King Farouk.

In his writing Mahfouz expressed sympathy for socialism and democracy but strong antipathy against Islamic extremism. He did not shrink from controversy outside of his work and as a result many of his books were banned and, like many Egyptian writers and intellectuals, he was on an Islamic fundamentalist death list. He defended the British-Indian writer Salman Rushdie after the Ayatollah Khomeini condemned him to death in a 1989 fatwa but also criticized Rushdie's novel *The Satanic Verses* as 'insulting' to Islam. He nevertheless joined 80 other intellectuals in declaring that "no blasphemy harms Islam and Muslims so much as the call for murdering a writer". Despite being given police protection he was stabbed outside his Cairo home in 1994 by an extremist. Although he suffered nerve damage he survived the attack and died in Cairo in 2006.

South African anti-apartheid revolutionary, political leader and philanthropist **Nelson Rolihlahla Mandela** was jointly awarded the Nobel Peace Prize in 1993 for his work in ending the country's apartheid system of racial segregation and discrimination and for 'laying the foundations for a new democratic South Africa'. He shared the prize with **Frederik Willem (FW) de Klerk**, South Africa's last white president, whose negotiations with Mandela in the early 1990s helped end apartheid. After being expelled for political activism from the University of Fort Hare Mandela studied law at the University of the Witwatersrand in Johannesburg, one of very few black students at the historically white institution at the time. Fort Hare later made amends by naming the Nelson R Mandela School of Law at the university after him.

One of the world's most recognisable activists Mandela spent 27 years in prison for his active opposition to the racist regime in South Africa. A year after winning the Nobel Prize he was appointed as South Africa's first democratically elected president (from 1994 to 1999) and the country's first black head of state, a position from which he voluntarily retired after one term. His most enduring legacies include the Mandela Rhodes Scholarship, a flagship programme of the Nelson Mandela Foundation that has sent more than 530 outstanding young African students on postgraduate studies abroad.

One of his most famous quotes, from his book *Long Walk to Freedom* (1994), sums up his life philosophy, "No one is born hating another person because of the colour of his skin, or his background, or his religion. People must learn to hate, and if they can learn to hate, they can be taught to love, for love comes more naturally to the human heart than its opposite". The iconic leader was born in 1918 in Mvezo in South Africa and died in 2013 in Johannesburg.

This author has fond memories of meeting Madiba ('father' in isiXhosa), encounters that revealed his generosity of spirit and humility. As head of the science and technology portfolio of the MTN Foundation, I attended AGMs of the Nelson Mandela Children's Fund to which MTN was a generous donor. After the meeting and despite his busy schedule Mandela shook hands with every attendee and thanked them for their contributions.

Nelson Mandela

Once, at the opening of Parliament, an 80-strong children's choir sang for Madiba. After the performance, and to the horror of his schedule organisers, he took the trouble to greet every member of the choir. On another occasion, at the official opening of the Robben Island Museum in September 1997, he was offended when various members of his cabinet arrived late for the ceremony. He gently admonished each one as they belatedly took their seats, emphasising that Africa would never reach its full potential if we are not on time. At this event he thanked every attendee in person and said to me, after I had thanked him for his immense contributions to South Africa, "No, it is for me to thank you for all that you have done". His example was a major factor in my decision to switch careers from research to science education.

The late Pat Tebbutt, a retired supreme court judge, once told me the delightful tale of Mandela's arrival at an anniversary event at the Law School at the University of the Witwatersrand where they were both alumni. Even though he was at the height of his fame and had won the Nobel Prize and innumerable other awards Mandela's greeted his welcoming party with the words, "Do you remember me?" Oxford University was the first of eight British universities to confer an honorary doctorate on him. At the Oxford graduation ceremony in 1996 he broke all tradition at the 900-year-old institution by giving a short, unannounced speech and by encouraging the august assembly to join him in the traditional protest dance, the *toyi-toyi*!

FW de Klerk was born in Johannesburg in 1936 to an influential Afrikaner family many of whom were active in politics. He had a secure and comfortable upbringing and held relatively conservative opinions when he debated political issues with his family. He studied at Potchefstroom University before pursuing a career in law. De Klerk served as State President of South Africa from 1989 to 1994. As South Africa›s last head of state from the era of white-minority rule he and his government began to dismantle the apartheid system and introduced universal suffrage. He released Nelson Mandela and other key political prisoners in 1990 and was a co-recipient of the Nobel Prize for overseeing South Africa's peaceful transition from apartheid rule. His political choices were underpinned by his commitment to the common good and his demeanour was marked by humility and calm with a natural cordiality and sense of courtesy. He is still active on the international diplomatic circuit.

The Palestinian leader **Yasser Arafat** shared the 1994 Nobel Peace Prize with the Israeli Prime Minister Yitzhak Rabin and Foreign Minister Shimon Peres for reaching the 1993 Oslo interim peace agreement and for 'their efforts to create peace in the

**FW de Klerk**

Middle East'. He was born to Palestinian parents in 1929 in Cairo where he studied at the University of King Fuad I and embraced Arab nationalist and anti-Zionist ideas. He opposed the 1948 creation of the State of Israel and fought alongside the Muslim Brotherhood during the 1948 Arab-Israeli War. Arafat became a Palestinian political leader and served as Chairman of the Palestine Liberation Organization from 1969 to 2004 and President of the Palestinian National Authority from 1994 to 2004. Ideologically an Arab nationalist he was a founding member of the Fatah political party which he lead from 1959 until 2004 when he died in Paris.

**Claude Cohen-Tannoudji**, the Algerian-born French scientist, was awarded the Nobel Prize in Physics in 1997 for the 'development of methods to cool and trap atoms with laser light', the first physics Nobel laureate born in an Arab country. He shared the award with US physicists Steven Chu and William Phillips. The three laureates were recognised for developing innovative techniques that use laser light to cool atoms to extremely low temperatures at which they move slowly enough to be studied in detail. Cohen-Tannoudji was born in 1933 in Constantine, Algeria, and taught quantum mechanics at the University of Paris where he co-authored the popular textbook *Mécanique Quantique* which was based on his lecture notes. He is still an active researcher working at the École Normale Supérieure in Paris.

Egyptian **Ahmed Zewail** was the recipient of the Nobel Prize in Chemistry in 1999 for developing a rapid laser technique that helped scientists to study the action of atoms during chemical reactions. His work led to the development of a new field of physical chemistry known as femtochemistry, the study of chemical reactions on extremely short timescales, and he is widely regarded as the father of this field. Zewail was born in Damanhur, Egypt, in 1946 and received an MSc in chemistry from Alexandria University before moving to the US to complete his PhD at the University of Pennsylvania. After carrying out

Ahmed Zewail

postdoctoral research at the University of California Berkeley he joined the California Institute of Technology in 1976 where he became the Linus Pauling Professor of Chemistry and the director of the Physical Biology Center for Ultrafast Science and Technology. He received many awards including Egypt's highest state honour the Grand Collar of the Nile (1999) as well as the King Faisal International Prize (1989), Albert Einstein World Award of Science (2006), Priestley Medal (2011) and honorary doctorates from eleven universities. He died in Pasadena, California, in 2016.

Ghanaian diplomat **Kofi Atta Annan** was the secretary-general of the United Nations from 1997 to 2006 and was, with the UN, a co-recipient of the 2001 Nobel Peace Prize in the centennial year of the Nobel committee. The winners were recognised 'for their work for a better organised and more peaceful world'. Annan was born in Kumasi, Ghana, in 1938 and joined the UN in 1962 initially working for the World Health Organization in Geneva before serving in several capacities at the UN headquarters. He was the first secretary-general to be elected from the UN staff itself. In this capacity he reformed the UN bureaucracy, worked to combat HIV/Aids and launched the UN Global Compact.

Kofi Annan

In 2007, after leaving the UN, he founded the Kofi Annan Foundation to promote international development. He died in Bern, Switzerland, in 2018.

Biologist **Sydney Brenner** was born in Germiston, South Africa, in 1927 and shared the Nobel Prize in Physiology or Medicine in 2002 with Howard Robert Horvitz of the USA and Sir John E Sulston of the United Kingdom. They were recognised for their discoveries on how genes regulate tissue and organ development via a mechanism called programmed cell death or apoptosis. His work made it possible to link genetic analysis to cell division and organ formation. Brenner made significant contributions to work on the genetic code and other areas of molecular biology while working in the Medical Research Council laboratory in Cambridge, England. He established the roundworm *Caenorhabditis elegans* as

Sydney Brenner

a model organism for the investigation of developmental biology and founded the Molecular Sciences Institute in Berkeley, California.

Brenner's father, a cobbler, emigrated to South Africa from Lithuania in 1910. Sydney Brenner was educated at Germiston High School and the University of the Witwatersrand. Having started university at the age of 15 he would have been too young to qualify for the practice of medicine at the end of his six-year medical course so he was allowed to complete a BSc in anatomy and physiology. He subsequently stayed on for two more years to complete an Honours degree and then a MSc in cytogenetics, supporting himself by working part-time as a laboratory technician. During this time he was taught by Joel Mandelstam, Raymond Dart and Robert Broom. In 1951 he received his medical degree and subsequently completed his DPhil at the University of Oxford.

Known for his penetrating scientific insights and acerbic wit Brenner wrote a regular column (*Loose Ends*) in the journal *Current Biology* for many years. This column became so popular that a compilation, *Loose Ends from Current Biology*, was published in 1994 and became a collector's item. Brenner also wrote a popular paperback, *A Life in Science*, and was noted for his generosity of spirit and the large number of students and colleagues who were stimulated by his ideas. He died in 2019 in Singapore at the age of 92.

**John Maxwell ('JM') Coetzee**, the South African novelist, essayist, linguist and literature critic, won the Nobel Prize for Literature in 2003 for developing a style that 'in innumerable guises portrays the surprising involvement of the outsider'. He was the fourth Africa-born writer to be so honoured and the second South African after Nadine Gordimer. Coetzee is descended from 17th century Dutch immigrants on his father's side and from Dutch, German and Polish immigrants through his mother. He was born in 1940 and spent his early boyhood in Cape Town and at the age of eight moved with his family to the picturesque town of Worcester which features in his fictionalised memoir, *Boyhood* (1997). He attended a Catholic school, St Joseph's College in Rondebosch, Cape Town, and later studied mathematics and English at the University of Cape Town.

Coetzee moved to the United Kingdom in 1962 where he worked as a computer programmer for IBM in London. In 1963 the University of Cape Town awarded him an MA degree for a thesis on Ford Madox Ford and his novels. His experiences in England were later recounted in *Youth* (2002), his second volume of fictionalised memoirs. In 1965 Coetzee went to the University of Texas at Austin on a Fulbright Scholarship and was awarded his doctorate there in 1969 for a computer-aided stylistic analysis of Samuel Beckett's English prose. In 1968 he began teaching English literature at the State University of New York at Buffalo where he wrote his first novel *Dusklands*. In 1972 Coetzee returned to South Africa to take up a lectureship at the University of Cape Town, eventually becoming a Distinguished Professor in the Faculty of Humanities. After retirement in 2002 he emigrated to Adelaide in Australia where he is an honorary research fellow in the English Department at the University of Adelaide.

**JM Coetzee**

The Nobel committee hailed Coetzee as a 'scrupulous doubter, ruthless in his criticism of the cruel rationalism and cosmetic morality of western civilisation' and he is now one of the most critically acclaimed and decorated authors in the English language. He won the Booker Prize in 1983 for his *Life & Times of Michael K* and again in 1999 with *Disgrace*, the first author to win the prestigious British literary award twice. A recluse, he shunned both Booker Prize ceremonies but did accept the Nobel Prize in Stockholm.

The Egyptian law scholar and diplomat **Mohamed Mustafa ElBaradei**, a former director-general of the International Atomic Energy Agency (IAEA), was awarded the Nobel Peace Prize in 2005 together with the IAEA, the international nuclear watchdog. They were recognised for their 'efforts to prevent nuclear energy from being used for military purposes

Mohamed Mustafa ElBaradei

... and for ensuring that nuclear energy is used in the safest possible way'. ElBaradei was born in Cairo in 1942 and briefly served as the Egyptian vice-president in 2013.

After a stellar career in Egyptian and international politics and diplomacy he joined the IAEA in 1984 and became an outspoken critic of nuclear proliferation. In an October 2003 interview published in the *Cairo Times* he wrote, "The ultimate sense of security will be when we come to recognise that we are all part of one human race. Our primary allegiance is to the human race and not to one particular color or border". In a 12th February 2004 op-ed in the *New York Times* on the dangers of nuclear proliferation, he stated that we "must abandon the unworkable notion that it is morally reprehensible for some countries to pursue weapons of mass destruction, yet morally acceptable for others to rely on them for security. ... If the world does not change course, we risk self-destruction". In June 2009 *Egypt Post* issued a set of 16 postage stamps commemorating African winners of Nobel Prizes including ElBaradei..

The Nobel Prize in Physics in 2012 was awarded jointly to the Moroccan-born French scientist **Serge Haroche**, a pioneer of laser spectroscopy, and the US physicist David J Wineland 'for ground-breaking experimental methods that enable the measuring and manipulation of individual quantum systems'. The two scientists who worked separately paved the way for new experiments in the field of quantum physics after showing how individual quantum particles may be observed without being destroyed. Haroche was born in Casablanca, Morocco, in 1944 and was descended from a well-educated family of teachers, physicians and lawyers. He emigrated to France in 1956 and from 1967 to 1975 worked at the Centre national de la recherche scientifique while also defending his doctoral thesis in physics at the University of Paris VI in 1971. After spending a post-doctoral year at Stanford University he took up a professorship at Paris VI University. In 2001 he was appointed as Chair of Quantum Physics at the Collège de France and since 2012 he has been administrator of the Collège. In 2006 he published his authoritative book *Exploring the Quantum. Atoms, Cavities and Photons* with Jean-Michel Raimond.

Serge Haroche

The Nobel Prize in Chemistry in 2013 was awarded to Martin Karplus (Austria/USA), **Michael Levitt** (South Africa/Israel) and **Arieh Warshel** (Israel/USA) who were recognised for their ground-breaking work on computer programmes that simulate complex chemical processes which revolutionised research in areas as diverse as drugs and solar energy. Previously chemists made models of molecules using plastic balls and sticks but today modelling is carried out on computers. Working in the 1970s the trio laid the foundation for the powerful computer programmes that are used to understand, predict and simulate chemical processes today. Their work is ground-breaking in that they managed to make Newton's classical physics work side-by-side with fundamentally different quantum physics, devising methods that use both classical and quantum physics. Today the computer is just as important a tool for chemists as the test tube as simulations are so realistic that they predict the outcome of traditional experiments.

Michael Levitt was born in 1947 in Pretoria to a Jewish family from Lithuania. He attended Pretoria Boys High School and moved with his family to England when he was 15 years old. He returned to South Africa to study applied mathematics at Pretoria University and then attended King's College London where he obtained his honours degree in Physics in 1967. He completed his PhD at the University of Cambridge in 1971 on the development of a computer programme for studying the conformations of molecules which has underpinned much of his later work. From 1979 he conducted research at the Weizmann Institute of Science in Rehovot, Israel, becoming an Israeli citizen in 1980 and serving in the Israel Defence Forces for six weeks during 1985. In 1986 he began teaching at Stanford University and since then has split his time between Israel and California.

The Nobel Peace Prize in 2018 was awarded to **Denis Mukwege** of the Democratic Republic of the Congo (DRC) and Iraqi human rights

Denis Mukwege

activist and author Nadia Murad for their efforts to end the use of sexual violence as a weapon of war and armed conflict. Mukwege is a Congolese gynaecologist and Pentecostal pastor who founded and works in Panzi Hospital in Bukavu, DRC, where he specialises in the treatment of women who have been raped by armed rebels. Murad was herself a victim of rape but refused to accept the social code that requires women to remain silent and ashamed of the abuses to which they had been subjected. They have both campaigned to provide greater visibility of war-time sexual violence so that the perpetrators can be held accountable for their actions.

Mukwege was one of nine children born to a Pentecostal minister and his wife in Bukavu, DRC. He studied medicine after witnessing the complications of childbirth experienced by women who had been sexually abused and had no access to specialist healthcare. After graduating in medicine from the University of Burundi in 1983 he worked as a paediatrician in a rural hospital near Bukavu. Thereafter he studied gynaecology and obstetrics at the University of Angers in France, obtaining his MSc and completing his medical residency there in 1989, and then earned his PhD in 2015 from Université libre de Bruxelles. Since he founded the Panzi Hospital he and his colleagues have treated over 82 000 patients with about 60% of their injuries caused by sexual violence mostly in conflict zones.

Mukwege has become the foremost international spokesperson on the struggle to end sexual violence in armed conflicts with his basic principle, 'Justice is everyone's business', putting the blame for ongoing abuses on people at all levels of society. After receiving death threats he was placed under the guard of UN security forces in the DRC but in October 2012 four armed men attempted to assassinate him, killing his bodyguard and holding his daughters hostage. He and his family were forced into exile in Europe but he returned to Bukavu using a return ticket bought by his ex-patients who had sold vegetables by the roadside to raise the funds. In addition to the Nobel Prize Mukwege has received many other accolades including the Nigerian *Daily Trusts'* African of the Year award (2009), the Sakharov

Prize for Freedom of Thought from the European Parliament (2014) and being listed as one of *Time* magazine's 100 Most Influential People in 2016.

In 2019 the Nobel Peace Prize was awarded to **Abiy Ahmed Ali**, prime minister of Ethiopia since April 2018, for his decisive initiative to resolve (at least, temporarily) the 20-year border conflict with neighbouring Eritrea. A former army intelligence officer he launched a wide programme of political and economic reforms and worked to broker peace deals throughout the Horn of Africa. He spent his first 100 days in office lifting the country's state of emergency, granting amnesty to thousands of political prisoners, discontinuing media censorship, legalising outlawed opposition groups, dismissing military and civilian leaders who were suspected of corruption and significantly increasing the influence of women in Ethiopian political and community life.

Ali was born in Beshasha, Ethiopia, in 1976 and was the 13$^{th}$ child of his father, a farmer, and the sixth child of his mother, who was the fourth of his father's four wives. He has a BSc in computer engineering (2009), an MA in transformational leadership (2011), a MBA (2013) and a PhD in conflict resolution (2019). He and his family are deeply religious and he is also a fitness fanatic who frequents gyms in Addis Ababa and believes that physical fitness goes hand in hand with mental health. His many accolades include *The African* leadership magazine African of the Year in 2018, one of the 100 Most Influential People 2018 listed by *Time* magazine and one of 100 Global Thinkers of 2019 by *Foreign Policy* magazine.

Unfortunately, within a year of receiving the Nobel Peace Prize, Ali's reforms began to unravel when fighting broke out between government forces and a powerful regional government in northern Ethiopia, with hundreds reported to have been killed. Ali ordered the government offensive after accusing the rival Tigray People's Liberation Front of launching an attack against Ethiopia's military. The conflict later escalated into a civil war that destabilized an already fragile region.

**Abiy Ahmed Ali**

**Nobel Prize in Economics**: No African has as yet won the Nobel Prize in Economics which was first awarded in 1969, 68 years after the first Nobel Prizes in 1901, or the Abel Prize in Mathematics (modelled after the Nobel Prize), which has been awarded annually by the King of Norway since 2003. It is a pity that the identity of the inventor of the revolutionary mobile money transfer system *M-Pesa,* which was introduced by Safaricom in Kenya in 2007, is not known as this individual revolutionised commerce among people who do not own a bank account in Africa and could have been a candidate for the economics prize. Nyagaka Anyona Ouko, a student at the Jomo Kenyatta Institute of Agriculture & Technology in Nairobi, claims to have pitched the concept for *M-Pesa* to Safaricom but his certificate of copyright is dated 2012 by which time *M-Pesa* was well established. Another candidate is former Safaricom CEO Michael Joseph who sparked the thought process that led to its invention as the company at the time was focussing on building innovative new products. He asked his creative team to come up with ideas that had not been implemented before in Kenya (Bruton, 2021).

**Organisational awards involving Africans:** The **Tunisian National Dialogue Quartet**, a group of four organisations that was pivotal in attempts to build a democracy in the wake of the Jasmine Revolution in Tunisia in 2011, was awarded the Nobel Peace Prize in 2015. The quartet was made up of the Tunisian General Labour Union, Tunisian Confederation of Industry, Trade and Handicrafts, Tunisian Human Rights League and the Tunisian Order of Lawyers and comprised Abdessattar Ben Moussa, Noureddhine Allege, Houcine Abbassi and Wided Bouchamaoui, three men and one woman, all Tunisian. Through their mediation the quartet succeeded in creating a peaceful dialogue that bridged political and religious divides and established a pluralistic democracy. The 2007 Nobel Peace Prize was shared by the **Intergovernmental Panel on Climate Change**, the UN body that assesses science related to climate change, and US environmentalist and then-vice president, Al Gore, 'for their efforts to build up and disseminate greater knowledge about man-made climate change, and to lay the foundations for the measures that are needed to counteract such change'. Several Africa-born scientists were members of the IPCC, including Philip John Donne 'Taffy' Lloyd, Guy Midgley, Bruce Hewitson and Richard Washington from South Africa. Philip Lloyd received many other honours including African Intellectual of the Year from the Conrad Gerber Foundation in 2012.

**Stockholm Water Prize**: It is surprisingly that the Nobel Foundation has not introduced a Nobel Prize in Environmental Sciences (as they did belatedly for economics) to recognise accomplishments in this important field of endeavour. Some Peace laureates such as Wangari Maathai (and Al Gore) would surely have won this award? The Stockholm Water Prize is widely regarded as the nearest equivalent to a Nobel Prize in Environmental Sciences and has been awarded annually since 1991 by the Stockholm Water Foundation. It is presented by His Majesty Carl XVI Gustaf, King of Sweden, under the same circumstances as the Nobel Prizes. In 2000 the then-South African Minister of Water Affairs & Forestry Abdul Kader Asmal became the first African to receive this award in recognition of his outstanding contributions to water conservation and management.

This author and his wife Carolynn had the good fortune to be invited to attend the ceremony in the Stockholm Konserthus at which Asmal received his prize. It was a festive occasion with Zulu dancers entertaining the crowd before the presentation and a grand dinner served by a bevy of fashionably clad waiters, and was a fitting tribute to one of South Africa's most capable politicians. As a professor of human rights rather than a water scientist Asmal regarded the provision of adequate potable water to all communities as a basic right. His initiatives included the highly successful Working for Water (WfW) campaign that increased run-off in river catchments by over 28% through the felling of alien invasive trees, with the timber being used to make rustic furniture. Asmal also initiated the 2020 Vision for Water project, the educational arm of the WfW campaign, which encouraged young people to reduce their water consumption.

When he was Head of Education at the Two Oceans Aquarium in Cape Town this author was appointed as project manager of the 2020 Vision project through Guy Preston, then head of the WfW campaign. The aim of the project was to enable schools throughout South Africa to measure their water consumption and implement water-saving measures. We developed a water audit resource kit contained in a 20-litre bucket that included all the apparatus necessary to measure the amount of water that was being used or wasted from taps, pipes, toilets, urinals, showers, wash basins, irrigation systems and reservoirs. The kits also contained an instruction manual on how to determine how much water (and money) had been saved by the water conservation methods implemented. We assembled over 1 500 water audit kits and launched a national road show to introduce the kits to schools. The project was an immediate success and the kits were eventually used by over 60 000 learners countrywide. Soon we were hearing inspiring stories from schools that had reduced their water consumption by 30% or more by fixing leaking taps and pipes, installing push-button urinals, low-flush toilets and aerator shower rises, planting xerophytic indigenous shrubs and changing their patterns of water use. The project made it possible for thousands of learners to become directly involved in a hands-on, quantitative scientific project that had many beneficial outcomes.

Other Africa-born scientists who have won the Stockholm Water Prize include Peter Morgan (2013, the sanitation engineer and inventor from Zimbabwe), John Briscoe (2014, the water management expert from South Africa) and Jackie King (2019, a leading South African expert on river management).

**Templeton Prize**: The Templeton Prize is another prestigious international award that has been won by Africans. The Prize was initially established in 1972 by the late American-born investor and philanthropist John Templeton, the financier who pioneered global investment strategies. It was originally awarded to people working in the field of religion (Mother Teresa was the first recipient, six years before she received her Nobel Peace Prize) but in the 1980s the scope was broadened to include people working at the interface between science and religion. It was traditionally presented by Prince Philip at a ceremony in Buckingham Palace. The prize is awarded annually to a living person 'who has made an exceptional contribution to affirming life's spiritual dimension' and 'whose exemplary achievements advance Sir John Templeton's philanthropic vision: harnessing the power of the sciences to explore the deepest questions of the universe and humankind's place and purpose within it'.

The 2004 Templeton Prize was awarded to Professor George Francis Rayner Ellis, then distinguished professor of complex systems in the Department of Mathematics & Applied Mathematics at the University of Cape Town. Ellis is a leading theoretical cosmologist who specialises in general relativity theory, the area first investigated by Albert Einstein. He is also renowned for his bold and innovative contributions to the dialogue between science and religion and has strongly advocated balancing the rationality of evidence-based science with faith and hope. His view has been partially shaped by his first-hand experiences in South Africa as it peacefully transformed from apartheid to a multi-racial democracy without succumbing to racial civil war.

When defending his notion that rationality and reason must be balanced with faith and hope to accurately understand the universe Ellis cites his own country's history. "There were very many times in the past when it was rational to give up all hope for the future - to assume that the nation would decay into a racial holocaust that never happened", Ellis wrote in a statement released to the press on 17th March 2004. "It did not occur because of the transformatory actions of those marvellous leaders, Desmond Tutu and Nelson Mandela, confounding the calculus of rationality". Ellis has also directly challenged the notion that the powers of science are limitless and has noted the inability of even the most advanced physics to fully explain factors that shape the physical world including human thoughts, emotions and even social constructions such as the laws of chess.

Ellis' book, *The Large Scale Structure of Space-Time* (1973), written with Stephen Hawking, has become the standard reference on the subject. His research on the origin of the universe, evolution of complexity, the functioning of the human mind and how and where they intersect with areas beyond the boundaries of science is discussed in his ground-breaking book, *On the Moral Nature of the Universe* (1996), written with Nancey Murphy. In 2002 he edited *The Far-Future Universe* which developed from a symposium held at the Pontifical Academy of Sciences in The Vatican that examined cosmological, biological, human and theological aspects of the future.

Since the rise of democracy in South Africa Ellis has devoted much of his energy to helping to develop the nation's social, political, cultural and educational future particularly in making mathematics and science education more broadly available to his fellow citizens. This author has had the pleasure of knowing Professor Ellis, accompanying him on hikes on Table Mountain and listening to his inspiring views. He is South Africa's top scientist and one of our leading thinkers.

South African anti-apartheid campaigner and Nobel laureate Desmond Tutu won the Templeton Prize in 2013 for helping to inspire people by promoting forgiveness and justice. The Templeton Foundation praised Tutu as a moral voice for people around the world and pointed out that he not only preaches but also lives the idea that every human being is unique and can be an agent for spiritual progress and positive change.

**Discussion**: When one combines the Nobel Prizes for African men and women they total 28: thirteen for Peace, six for Literature, four for

Physiology or Medicine, three for Chemistry and two for Physics. All the male and female laureates in the science categories (Chemistry, Physics, and Physiology or Medicine) listed above were born and/or educated in Africa but did their seminal work in sophisticated laboratories in Europe and/or the USA.

Do 28 Nobel Prizes, and a sprinkling of other elite prizes, adequately reflect the novel contributions of Africans to world affairs? No, as many of their contributions were probably 'under the radar' of the Nobel and other committees as they were made in the context of African conditions and may not have been of broader, global significance. The fact that nearly half (46%) of the Nobel Prizes awarded to Africans were peace prizes does reflect the strong contributions that they have made to conflict resolution on the continent. The six literature prizes also attest to the significant contributions by African writers and poets to world literature, stimulated no doubt by their diverse experiences on the vast, spirited continent.

## 15.

# Discussion: What is science?

*"We are just an advanced breed of monkeys on a minor planet of a very average star. But we can understand the Universe. That makes us something very special."*

Stephen Hawking, theoretical physicist, *Der Spiegel*, October 1988

INTRODUCTION: Science means different things to different people. To me it is the study of the natural world based on facts learned through experimentation and observation. It is our attempt to understand Nature's mechanisms and to live and work within them. It also includes the study of the built environment and the impact of humankind's interventions on the functioning of the natural world and how we can mitigate them. The important point is that science generates knowledge and understanding as opposed to ignorance, misunderstanding and fake news.

Science has two components, a method for finding things out and a body of knowledge. Everyone needs to understand the scientific method if they are to appreciate the value of the body of knowledge as many misunderstandings and controversies about major science-related issues such as climate change, sea level rise, food and water security, GM foods, vaccinations and even the spread of the SARS-CoV-2 virus have arisen as a result of misinterpretations of the scientific method.

What many people do not understand is that we do not only advance science by 'standing on the shoulders of giants', as suggested by Isaac Newton. Rather, science often advances by kicking the feet out from under those 'giants' and re-examining their ideas and explanations. It is therefore not always a steady accumulation of knowledge and may progress through a series of pulses that respond to new opportunities and understandings. Furthermore, it is often an on-going reaction *against* prior knowledge and understanding and is and always will be 'work in progress'.

Many significant advances in science are made when the current wisdom is discarded and replaced by a new (but still imperfect) understanding, often at the interface between disciplines. Science is therefore not about certainty, stability and top-down authority but about doubt, skepticism, questioning the *status quo* and constantly testing alternate ideas to find better explanations for the facts at hand. It is therefore healthy for scientists to disagree with one another and to propose alternative ideas for the explanation of the same set of facts. Scepticism about science is nothing new as open and objective debate on the purpose, method and value of science has been carried out ever since the Greek elitist hegemony was broken by mediaeval scientists. Advocates of pseudoscience often point to the disagreements among scientists as evidence that science does not work. This is because they fail to understand that these disagreements are healthy as they often lead to more insightful thinking, better experimentation and observation and further breakthroughs.

During my work in Africa and the Middle East I am constantly asked to explain how science works and why it is so important - answering these questions is one of the biggest challenges facing science educators.

**Why is science important?** Science is important for many reasons, as it,

- Promotes curiosity and creativity and encourages us to question everything.
- Is the only rational way of finding things out and figuring out how things work.
- Facilitates significant discoveries including discoveries on things that we did not know we did not know.
- Hates top-down authority, guards against authoritarianism and encourages the free exchange of ideas, which kills ignorance.
- Is a reliable source of information because it uses consistent methods.
- Determines cause and effect and therefore helps us to make accurate predictions, assess risk and make good evidence-based decisions not only in science but also in governance and everyday life.
- Provides novel solutions to intractable problems and helps to address the big problems like poverty, the digital divide, human migrations, climate change, biodiversity loss, infectious diseases, habitat and biodiversity loss and the threats posed by invasive plants and animals and infectious diseases.
- Accelerates technology take-up and increases our computing power

while at the same time helping us to soften our environmental impact and live more sustainably.

Together with technology, science is the engine of prosperity that is responsible for over 67% of economic development. Furthermore, science speaks a universal language and therefore promotes international collaboration. Today we are not just connected to but entangled with other people through the internet as well as with networked objects through the Internet of Things. This means that science is no longer carried out by isolated individuals but by a multi-brained, multi-generational super-organism that is capable of addressing the most challenging problems of the day. As a result, the practice of science is more multi-generational than ever before and it is constantly attracting new globally connected players, which keeps it young-at-heart. This 'group intelligence' is epitomised by a paper that was co-authored by 5 154 CERN scientists in the May 2015 edition of *Physics Review Letters* in which a more precise estimation of the size of the Higgs boson (an elementary particle in the Standard Model of particle physics) was proposed.

As a field of human endeavour science has made the world a better place. Over the past century scientific advances and technological innovations have doubled our life span, reduced the cost of food by thirteen-fold, dropped the cost of energy by twenty-fold, of transport by one-hundred-fold, and of communications by over one-thousand-fold. Most importantly, science has also taught us to be humble, to know our roles and responsibilities, recognise our shortcomings and acknowledge our ignorance.

**Environmental crisis:** The environmental crisis is a classic example of a challenge that can be addressed by rational scientific thinking and problem solving. Humanity is at the crossroads as we are now the most numerous large animals that have ever existed and our ability to manipulate and alter the natural environment is unprecedented. Due to our unsustainable lifestyles and destructive resource-use practices we have exceeded the planet's ability to meet our needs and compensate for our actions. We also need to realize that we cannot solve the problem by emigrating to another planet, as suggested by Stephen Hawking and Elon Musk, we must solve it at home.

We must acknowledge that the environmental crisis is not due to the spontaneous breakdown of natural ecological processes but to a series of bad decisions that we have made, decisions that we must reverse. We need to make a suite of good decisions so that the widest diversity of use of the natural environment can be enjoyed by present and future generations and options can be kept open for the survival of all living organisms on the planet. Humans may have the tools to stave off the impacts of climate change for a while but other animals (and plants) do not. As Greta Thunberg has recently written, "We are all in the same storm, but we are not in the same boat" (Thunberg, 2021).

The custodianship ethic that we must develop should emphasize that we are an integral part of Nature, a valuable species but not a superior one, and that every living thing has the right to live, or at least *to struggle to live*.

Furthermore, this right is not dependent on its actual or potential use to humans. Our role is to understand and work in harmony with Nature, not to conquer it, as many past religious and politico-economic systems dictated that we should do. This modern custodianship ethic is close to the traditional Nguni concept of *ubuntu*, 'A person can be a person only through his/her experience with other persons and with the broader environment'. This valuable concept encompasses respect, concern, compassion, empathy, understanding, cordiality, generosity and sharing.

**Anti-science campaigns**: Considering the extreme importance of science in modern society it is worrying that its value is being questioned by anti-science sentiments and pseudoscience initiatives. A cover story in the March 2015 edition of *National Geographic* magazine entitled 'The War on Science' pointed out that science is being undervalued by modern society and deliberately undermined by radical groups. The concept of 'woke', which has come to the fore during the recent US 'culture war', rejects the idea that there is such a thing as objective knowledge and truth. Wokes argue that dominant cultures construct truth and knowledge in such a way as to defend their interests and privileged positions. As Zille (2021) has pointed out, wokes reject the scientific method as the only effective way of testing the validity of an hypothesis and seeking the truth as this approach lends authority to dominant cultures. Rather, they elevate 'lived experience' to a higher standing in research and analysis than scientifically verifiable facts.

The #ScienceMustFall campaign at South African universities in 2016 was launched by a University of Cape Town student who stated that she feels alienated from science and that "Science as a whole is a product of western modernity and the whole thing should be scratched off. Especially in Africa". While I agree that the way in which science is taught in South Africa needs to be addressed, with far more discoveries, inventions and insights by African scientists and technologists being part of the curriculum, I do not agree with her that science is a 'Western' project as many cultures have contributed to its development. There is no 'Western science' or 'African science', there is only 'science'. There are African, Western, Islamic and many other contributions to science, but science is a universal endeavour to which all humans have the potential to contribute.

**Indigenous knowledge**: Science and indigenous knowledge are different but complimentary knowledge systems. Indigenous knowledge systems (IKS) are sets of understandings embedded in cultural practices, language, naming, customs and world views and comprise knowledge about plants, animals and the environment and how to use them. They include scientific, cultural, spiritual and social components that are passed on through practice, example and oral tradition and are typically useful in everyday life. I believe that science and indigenous knowledge are compatible and that each can learn from the wisdom of the other. Furthermore, IKS has helped traditional cultures to survive for centuries under harsh conditions and should be preserved as part of our culture's identity. In my books on South African and African inventions (Bruton, 2010, 2016, 2017, 2021) I give many examples of inventions and innovations (both products and services) that are derived from indigenous knowledge that have been developed and taken to market by science. Science and IKS are among the most precious of all human endeavours and we cannot afford to allow them to be undermined and undervalued.

**Conclusion**: Science is more relevant than ever before and our ability to communicate it is better than ever, yet the value of science continues to be questioned. During my work in interactive science centres I proposed that the dissemination of scientific knowledge should follow an Information Value Chain from Information → Knowledge → Wisdom → Changed mindset → Changed behaviour → Influencing others to change their mindset and behaviour. We will only be able to address the major problems facing us today if information is communicated and transformed in this way.

This point is particularly important as we face the threats and opportunities provided by the Fourth Industrial Revolution, which I prefer to call the First Post-Industrial Revolution (PIR). The PIR requires a major change of mind-set as the new tools and unprecedented connectivity that is now at our disposal provide us with the opportunity to solve previously intractable local, regional and global problems and also to undo the many wrongs created by previous socio-political regimes and the first three industrial revolutions. The cross-generational and interdisciplinary collaborations, rapid flow of ideas, gig economies, contributions of disruptive millennials and totally new ways of thinking that characterise the PIR make me optimistic about the future, and especially about Africa's role in it.

# Acknowledgements

I am grateful to David Hilton-Barber of Footprint Press for allowing me to indulge so freely in my fantasies and passions and for agreeing to take these personal reminiscences and reflections to market. Anthony Cuerden of Flying Ant Designs is thanked for his innovative cover and text designs. Ndabuko Ntuli, Lucky Netshidzati, Sarao and Ellen Marcus are thanked for permission to use images of their work on the covers, and the Albany Museum, Biowise, Birmingham Central Library, Cybertracker, Mulalo Doyoyo, Dirk Durnez, Robert Gess, David Griessel, Peter Hyslop, John Keogh, Mauritius Natural History Museum, MOCAA, Uwe Pfaff, Pratley Putty, Rhodes University, Mark Shuttleworth, Soho House Museum, Sue Swain, Tellumat, Kosie Thiart, Triggerfish Animations, Terence Visagie, Mary Wadley von Hirschberg, Louise Williamson and Siya Xusa for permission to use their images in the text. Many people willingly acted as 'bouncing boards' for my ideas and I am grateful to them for their support. They include Nicholas Ellenbogen, who also kindly wrote the Foreword, Nancy Tietz, Ticky Forbes, Peter Steyn, Tracey Bruton, Ryan Bruton, Craig Bruton, Tony Bruton, Rik Nulens, Dennis Nick, Hugh Amoore and many others. I am grateful to the venerable Owl Club in Cape Town for permitting me to cite information and reproduce images from its archives and to the Owl Club President Richard Morris and his wife, Beverley, for making useful editorial comments on the text. I also thank the South African Institute for Aquatic Biodiversity in Makhanda and the East London Museum for permission to use images in their care. George and Margo Branch, through their magnificent books and talks, have taught me a great deal about marine life. As always, I thank my wife, Carolynn, for keeping the home fires burning and for carrying out numerous administrative tasks while I indulged my passion for science communication.

# References and further reading

## The funny side of science: 1. The ancients and 2. The moderns

Abrahams, M 2002. *Ignobel prizes.* Orion, London, 320 pages.

Bryson, B 2006. *Seeing Further: The story of science, discovery, and the genius of the Royal Society.* William Morrow, New York, 506 pages.

Challoner, J 2019. *Genius inventions.* Andre Deutsch, London, 160 pages.

Darwin, C 1859. *On the origin of species by means of natural selection, or the preservation of favoured races in the struggle for life.* John Murray, London, 502 pages.

Dyson, F 2006. *The scientist as rebel.* New York Review Books, New York, 361 pages.

Gribbin, J 2003. *Science – a history.* Penguin, London, 645 pages.

Gribbin, J (ed) 2008. *The Britannica guide to the 100 most influential scientists.* Constable & Robinson, London, 333 pages.

James, FAJL 2010. *Michael Faraday. A very short introduction.* Oxford University Press, 162 pages.

Kespert, D 2015. *Genius! The most astonishing inventions of all time.* Thames & Hudson, London, 96 pages.

Mahajan, S 2008. *The story of inventions from antiquity to present.* HF Ullman, Berlin, 120 pages.

Mason, P 2001. *Thomas Edison.* White-Thomson, Lewes, 48 pages.

Mosley, M & J Lynch 2010. *The story of science. Power, proof and passion.* Mitchell Beazley, London, 288 pages.

Rezende, L 2007. *Chronology of science.* Checkmark Books, New York, 502 pages.

Sandler, C 2007. *The illustrated timeline of inventions.* Sterling, New York, 120 pages.

Steer, M, H Birch & A Impney 2008. *Defining moments in science.* Cassell, London, 798 pages.

Strickland, S & E Strickland 2006. *The illustrated timeline of science.* Sterling, New York, 120 pages.

Suplee, C 2000. *Milestones of science.* National Geographic, Washington DC, 287 pages.

Winston, R 2010. *Bad ideas? An arresting history of our inventions.* Transworld Publishers, London, 417 pages.

www.britannica.com/science/history-of-science

www.en.wikipedia.org/wiki/Science

www.sciencedirect.com/topics/social-sciences/history-of-science

## 3. Creativity in the arts and sciences

Ansell, D 2019. Dorothy Masuku: Africa has lost a singer, composer and a hero of the struggle. *The Conversation,* 25 February 2019: 1-5.

Baker, A 2019. Art, restored. Museum of Black Civilizations. *Time,* 2-9 September 2019: 76.

Baker, A 2020. Zenzile Miriam Makeba. Sound of South Africa. Time, 16-23 March 2020: 64.

Balogun, S 2020. Victor Olaiyo: Nigeria's master trumpeter, gifted composer and genius showman. *The Conversation,* 3 March 2020: 1-4.

# References and further reading

Bouwers, T 2020. Heritage dance to shine at carnival. *News24 online*, 10 March 2020: 1-4.

Chikafa-Chipiro, R 2020. Tsitsi Dangarembga and writing about pain and loss in Zimbabwe. *The Conversation*, 20 August 2020: 1-3.

Chinyamurindi, WT 2019. Tribute to Oliver Mtukudzi – Zimbabwe's 'man with the talking guitar'. *The Conversation*, 24 January 2019: 1-5.

Durán, L 2020. Mory Kanté, Guinea's hero, found new ways of playing the old music. *The Conversation*, 1 June 2020: 1-4.

El Saadawi, N 1969. Women and Sex. *Women's Press, London.*

El Saadawi, N 1983. *Memoirs from the Women's Prison*. Women's Press, London.

Gedye, L 2020. Matorokisi ignites a global groove. *Mail & Guardian* 36(9): 44.

Goodman, JE 2020. Idir: how a song from the village took Algerian music to the world. *The Conversation*, 6 July 2020: 1-4.

Hampton, D 2019. Groundbreaking filmmaker. *Time*, 25 November 2019: 54.

Haynes, B 2020. Art@Africa. So much talent in our country. Art@Africa, Cape Town, 84 pages.

Himmelman, N, N Sarmiento & T Thipe 2020. A tribute to poet and professor Harry Garuba: we continue to learn with you. *The Conversation*, 31 March 2020: 1-7.

Hincks, J 2019a. Dina el Wedidi. Voice of hope. *Time*, 27 May 2019: 42.

Hincks, J 2019b. Zainab Fasiki. Comics crusader. *Time*, 21-28 October 2019: 82.

Hlalethwa, Z 2020. The rebirth of Lazi Mathebula. *Mail & Guardian* 36(3): 42.

Jamodien, R 2021. Dr Nawal El Saadawi: A daughter of Egypt remembered. *Mail & Guardian online*, 27 March 2021: 1-5.

Javan, M 2017. Competition keeps traditional riel dancing alive. *Brand South Africa online*, 5 May 2017: 1-4.

Jedlowski, A 2020. What Netflix's involvement in Nigeria's massive film industry really means. *The Conversation*, 31 December 2018: 1-3.

Johnson-Jones, E 2011. The Nama Stap Dance; an analysis of continuity and change among Nama Women. *South African Dance Journal* 1(1): 1-19

Johnstone, L 2020. Wanuri Kahiu, the Kenyan film director taking on the world. *Mail & Guardian online*, 17 December 2020: 1-3.

Krige, T 2020. So, kwaito, where to from here? *Mail & Guardian* 36(3): 32.

Makwa, DDB 2020. Uganda's musicians are fighting COVID-19 – why government should work with them. *The Conversation*, 7 May 2020: 1-4.

Malsin, J 2017. Lina Attalah. Muckraker of the Arab world. *Time*, 23 October 2017: 64.

Mgujulwa, N 2020. Splash of colour for Salt River's streets. *Tatler*, 20 February 2020: 14.

Morrison, D 2006. Nadine Gordimer. *Time*, 13 November 2006: 75.

Mtuta, L 2020. Street murals with a mission. *Weekend Argus*, 15 February 2020: 11.

Ncube, G 2020. How artists have preserved the memory of Zimbabwe's 1980s massacres. *The Conversation*, 4 August 2020: 1-3.

Omajola, OF 2020. Victor Olaiya: Stadium Hotel, Highlife, and nostalgia. *The Conversation*, 22 April 2020: 1-3.

Omoyele, I 2020. Garuba finds his 'port of death'. *Mail & Guardian* 36(12): 42.

Ramirez, E 2018. Southern Africa's biggest comics and graphic novels so far... *YouNeek Studios online*, 19 January 2019: 1-4.

Robinson, S 2006. Fela Kuti. *Time*, 13 November 2006: 73.

Rockwood, K 2018. A new home of African art. *Time*, 3-10 September 2018: 79.

Sassen, R 2020. RIP 'Fefe' – singing was your life. *Mail & Guardian*, 6-12 March 2020: 42.

Shelemay, KK 2016. Travelling music: Mulatu Astake and the genesis of Ethiopian jazz. pp. 239-257. In: PV Bohlman & G Plastino (eds) *Jazz World/World Jazz*. University of Chicago Press, Chicago.

Sosibo, K 2020a. Shabalala's coat of many colours. *Mail & Guardian*, 14-20 February 2020: 46.

Sosibo, K 2020b. Remapping African musics. *Mail & Guardian*, 14-20 February 2020: 43.

Sosibo, K & S Allison. 2020. Your soundtrack to the pandemic. *Mail & Guardian*, 3-9 March 2020: 13.

Soyinka, W 2006. Chinua Achebe. *Time*, 13 November 2006: 76-77.

Sullivan, H 2020. 'Why should you cry?' Ghana's dancing pallbearers find new fame during Covid-19. *The Guardian online*, 14 May 2020: 1-5.

Zelalem, Z 2020. Provocative Ethiopian singer slain. *Mail & Guardian*, 3-9 July 2020: 11.

## 4. 'Lunarticks', and Owls

Darwin, E 1794-1796. *Zoonomia; or the laws of organic life*. J Johnson, London.

Henry, JA (ed) 1996. *The Unjealous Years. An Owl Club anthology*. The Owl Club, Cape Town, 386 pages.

Malcolm, J 1959. *The early history of the Owl Club, Cape Town 1894-1900*. Africana Notes and News, Cape Town.

Murray, T 2019. *The glow of brotherhood. The Owl Club 1994-2018*. The Owl Club, Cape Town, 263 pages.

Ranby, WE 1952. *The Owl Club 1894-1950*. The Owl Club, Cape Town.

Rosenthal, E 1982. *The third Tuesday. A history of the Owl Club 1951-1981*. The Owl Club, Cape Town, 95 pages.

Schofield, RE 1963. *The Lunar Society of Birmingham: a social history of provincial science and industry in eighteenth-century England*. Clarendon Press, Oxford, 504 pages.

Shell, SR 2017. *Protean paradox. George Edward Cory (1862-1935), negotiating life and South African history*. Rhodes University, Makhanda, 286 pages.

Tebbutt, P 2015. *A life spiced with variety. My memoirs*. Footprint Press, Cape Town, 272 pages.

Uglow, J 2002. *The Lunar Men. The friends who made the future 1730-1810*. Faber & Faber, London, 588 pages.

www.birminghammuseums.org.uk/soho/highlights/the-lunar-society

www.lunarsociety.org.uk

www.en.wikipedia.org/wiki/Lunar_Society_of_Birmingham

www.en.wikipedia.org/wiki/Owl_Club

www.owls.org.za

## References and further reading

### 5. Boneshakers to bloomers: Evolution of the bicycle

Bruton, MN 2016. *What a great idea! Awesome South African inventions.* Jacana Media, Cape Town, 253 pages.

Challoner, J (ed) 2013. *1001 inventions that changed the world.* Cassell, London, 960 pages.

Ellyard, D 2006. *Who invented what when.* New Holland Publishers, Sydney, 320 pages.

Embacher, M 2011. *Cyclepedia. A tour of iconic bicycle designs.* Thames & Hudson, London, 224 pages.

Galbiati, F & N Ciravegna 1989. *Bicycles. Le Biciclette.* Chronicle Books, London, 142 pages.

Humphrey, CC 1972. *Back to the bike.* 101 Productions, San Francisco, 96 pages.

Philbin, T 2003. *The 100 greatest inventions of all time. A ranking past to present.* Citadel Press, New York, 294 pages.

Sandler, C 2007. *The illustrated timeline of inventions.* Sterling, New York, 120 pages.

Uhlig, R (ed) 2001. *James Dyson's history of great inventions.* Constable, London, 188 pages.

www.en.wikipedia.org/wiki/History_of_the_bicycle

### 6. Who is South Africa's greatest inventor?

Addison, G 2002. *The hidden edge. Quest for progress. Innovation in South Africa 1900-2000.* The Engineering Association, Meyersdal, 244 pages.

Bozzoli, GR 1997. *Forging ahead - South Africa's pioneering engineers.* Witwatersrand University Press, Johannesburg, 120 pages.

Bruton, MN 2008. Tellurometer. Inventor: Trevor Wadley. *The South African*, January 2008: 40.

Bruton, MN 2008. The Catscan. *The South African*, March 2008: 40.

Bruton, MN 2008. Cybertracker. *The South African*, April 2008: 48.

Bruton, MN 2010. Great South African Inventions. Cambridge University Press, Cape Town, 96 pages.

Bruton, MN 2017. *Traditional fishing methods of Africa.* Cambridge University Press, Cape Town, 96 pages.

Bruton, MN 2017. *What a great idea! Awesome South African inventions.* Jacana Media, Cape Town, 253 pages.

Bruton, MN 2021. *Harambee: The spirit of innovation in Africa.* Human Sciences Research Council, Cape Town, 285 pages.

Buckley, D, M Lombard, M Lomberg, K Meiring & R Theron. 2005. *Africa's giant eye. Building the Southern African Large Telescope.* SALT Foundation, Cape Town, 192 pages.

Cartwright, AP 1900. *South Africa's Hall of Fame.* Central News Agency, Cape Town, 248 pages.

Castleden, R 2011. *Events that changed the world.* Canary Press, London, 448 pages.

Smith, JR, B Sturman & AF Wright. 2008. *The tellurometer. From Dr Wadley to the MRA7.* Tellumat, Cape Town, 243 pages.

Vance, A 2015. *Elon Musk. How the billionaire CEO of SpaceX and Tesla is shaping our future.* Virgin Books, London, 392 pages.

Vaughan, CL 2008. *Imagining the elephant. A biography of Allan Macleod Cormack.* Imperial College Press, London, 304 pages.

Vick, K 2020. Wangari Maathai. Seeding a movement. *Time*, 16-23 March 2020: 84.

Wadley von Hirschberg, M 2009. *Trevor Lloyd Wadley. Genius of the tellurometer.* Fish Hoek Printing, Cape Town, 112 pages.

Wild, S 2013. *Searching African skies. The Square Kilometre Array and South Africa's quest to hear the songs of the stars.* Jacana Media, Auckland Park, 181 pages.

Wild, S 2015. *Innovation. Shaping South Africa through science.* Pan Macmillan, Johannesburg, 194 pages.

www.en.wikipedia.org/wiki/List_of_South_African_inventions_and_discoveries

www.sagoodnews.co.za/top-10-south-african-inventions/

www.thesouthafrican.com/lifestyle/12-south-african-science-inventions/

## 7. The comical art of naming new species

Bruton, MN 2015. *When I was a fish. Tales of an ichthyologist.* Jacana Media, Cape Town, 310 pages.

Bruton, MN 2019. Plants and animals named after Marjorie Courtenay-Latimer. *Inyathi*, August 2019: 10-12.

Jubb, RA 1967. *Freshwater fishes of southern Africa.* A.A. Balkema, Cape Town, 248 pages.

Skelton, PH 1993. *A complete guide to the freshwater fishes of southern Africa.* Southern Book Publishers, Johannesburg, 388 pages.

## 8. How well do you know your ologys?

Quinion, M 2002. *Ologies and isms. A dictionary of word beginnings and endings.* Oxford University Press, Oxford, 280 pages.

## 9. and 10. Finding Old Fourlegs: Act 1 and Act 2

Atz, JW 1976. *Latimeria babies* are born, not hatched. *Underwater Naturalist* 9(4): 4–7.

Balon, EK, MN Bruton & H Fricke 1988. A fiftieth anniversary reflection on the living coelacanth, *Latimeria chalumnae*: some new interpretations of its natural history and conservation status. *Environmental Biology of Fishes* 23: 241–280.

Barnett, P 1953. *Sea safari with Professor Smith.* Business Services, Durban, 158 pages.

Bell, S 1969. *Old man coelacanth.* Voortrekkerpers, Johannesburg, 141 pages.

Bruton, MN 1982. *The life and work of Margaret M Smith.* JLB Smith Institute of Ichthyology, Grahamstown, 12 pages.

Bruton, MN 1989. The living coelacanth fifty years later. *Transactions of the Royal Society of South Africa* 47: 19-28.

Bruton, MN 1994. Lungfishes and coelacanths. pp. 70–74. In: JR Paxton & WN Eschmeyer (eds) *Encyclopaedia of animals: Fishes.* University of New South Wales Press, Sydney.

Bruton, MN 2015. *When I was a fish. Tales of an ichthyologist.* Jacana Media, Auckland Park, 310 pages.

Bruton, MN 2016. *Traditional fishing methods of Africa.* Cambridge University Press, Cape Town, 96 pages.

Bruton, MN 2018. *The amazing coelacanth.* Struik Nature, Cape Town, 96 pages.

Bruton, MN & SE Coutouvidis 1991. An inventory of all known specimens of the coelacanth

*Latimeria chalumnae*, with comments on trends in the catches. *Environmental Biology of Fishes* 32: 371–390.

Bruton, MN & RE Stobbs 1991. The ecology and conservation of the coelacanth *Latimeria chalumnae*. *Environmental Biology of Fishes* 32: 313–339.

Bruton, MN, AJP Cabral & H Fricke 1992. First capture of a coelacanth, *Latimeria chalumnae* (Pisces, Latimeriidae) off Mozambique. *South African Journal of Science* 88: 225–227.

Courtenay-Latimer, ME 1979. My story of the first coelacanth. *Occasional Papers of the California Academy of Science* 134: 6–10.

De Vos, L & D Oyugi 2002. First capture of a coelacanth *Latimeria chalumnae* Smith, 1939 (Pisces: Latimeriidae), off Kenya. *South African Journal of Science* 98: 345–347.

Erdmann, MV 1999. An account of the first living coelacanth known to scientists from Indonesian waters. *Environmental Biology of Fishes* 54: 439–443.

Fricke, H 1997. Living coelacanths: values, eco-ethics and human responsibility. *Marine Ecology Progress Series* 161: 1–15.

Fricke, H 2007. *Die Jagd nach dem Quastenflosser, Der Fisch, der aus der Urzeit Kam*. Verlag CH Beck, München, 302 pages

Fricke, H & R Plante 1988. Habitat requirements of the living coelacanth *Latimeria chalumnae* at Grande Comore, Indian Ocean. *Die Naturwissenschaften* 75: 149–151.

Heemstra, PC, H Fricke, K Hissmann, J Schauer, M Smale & K Sink 2006. Interactions of fishes with particular reference to coelacanths in the canyons at Sodwana Bay and the St Lucia Marine Protected Area of South Africa. *South African Journal of Science* 102: 461–465.

Hissmann, K, H Fricke, J Schauer, AJ Ribbink, MJ Roberts, K Sink & PC Heemstra 2006. The South African coelacanths – an account of what is known after three submersible expeditions. *South African Journal of Science* 102: 491–500.

Hubbs, CL 1968. James Leonard Brierley Smith, 1897-1968. *Copeia* 1968(3): 659-660.

Musick, JA, MN Bruton & EK Balon (eds) 1991. *The biology of* Latimeria *and evolution of coelacanths*. Kluwer Academic Publishers, Dordrecht, 446 pages.

Nulens, R, L Scott & M Herbin 2011. An updated inventory of all known specimens of the coelacanth, *Latimeria spp*. *Smithiana Special Publication* 3: 1–52.

Smith, JLB 1939. A living fish of Mesozoic type. *Nature, London* 143 (3620): 455–456.

Smith, JLB 1953. The second coelacanth. *Nature, London* 171: 99–101.

Smith, JLB 1956. *Old fourlegs. The story of the coelacanth*. Longmans Green & Co, London, 260 pages.

Smith, MM 1969. JLB Smith, his life, work, bibliography and list of new species. *Rhodes University Department of Ichthyology Occasional Paper* 16: 185–215.

Stobbs, RE 1996. Eric Ernest Hunt – the aquarist. *Ichthos* 48:3-4.

Stobbs, RE 1997. Explosives and other destructive fishing practices. *Ichthos* 55: 4-8.

Thomson, KS 1991. *Living fossil. The story of the coelacanth*. Norton, New York, 252 pages.

Weinberg, S 1999. *A fish caught in time*. Fourth Estate Limited, London, 239 pages.

www.dinofish.com

www.en.wikipedia.org/wiki/Coelacanth

www.nationalgeographic.com/animals/fish/facts/coelacanths

www.ocean.si.edu/ocean-life/fish/coelacanth

## 11. Lessons from the dodo

Bruton, MN 2019. Lessons from the dodo. *Very Interesting* 49: 70-71.

Carroll, L 1965. *The Annotated Alice – Alice's Adventures in Wonderland and Through the Looking Glass*. Edited by Martin Gardner. Penguin, London, 352 pages.

Clarke, G 1865. Account of the late discovery of dodo remains in the island of Mauritius. Ibis 2: 141-146.

Courtenay-Latimer, ME 1953. A dodo egg. *South African Journal of Science* 49(6): 208-210.

Cuppy, W 1941. *How to become extinct*. University of Chicago Press, Chicago, 114 pages.

Farrar, S 1999. DNA science could rebuild dead dodo. *Sunday Times*, 21 March 1999.

Fuller, E 2002. *Dodo: from extinction to icon*. Harper Collins, London, 180 pages.

Grihault, A 2005. *Dodo. The bird behind the legend*. Imprimerie & Papeterie Commerciale, Mauritius, 171 pages.

Grihault, A 2007. *Solitaire. The dodo of Rodrigues Island*. Précigraph, Mauritius, 117 pages.

Hume, JP & AS Cheke. 2004. The white dodo of Réunion Island, unraveling a scientific and historical myth. *Archives of Natural History* 31(1): 57-79.

Newton, A 1874. On a living dodo shipped for England in 1628. *Proceedings of the Zoological Society of London* 307: 447-449.

Owen, R 1866. *Memoir on the dodo*. Taylor & Francis, London, 36 pages.

Prosper, L 2000. A walk in the south-east - Mare aux Songes. The silence of the dodos. *Le Mauricien*, 27 June 2000.

Quammen, D 1996. *The song of the dodo*. Pimlico, London, 704 pages.

Roberts, DL & AR Solav 2003. When did the dodo become extinct? *Nature* 426: 245.

Staub, F 1996. Dodos and solitaires, myths and reality. *Proceedings of the Royal Society of Arts and Sciences, Mauritius* 6: 89-122.

www.news.uct.ac.za/article/-2017-08-31-secret-life-of-the-dodo-revealed

www.thedodo.com

## 12. Which is South Africa's strangest animal? Animals without backbones, and 13. Animals with backbones, and the winner…

Appleton, CC 1996. *Freshwater molluscs of southern Africa*. University of Natal Press, Pietermaritzburg, 64 pages.

Branch, GM, CL Griffiths, ML Branch & LE Beckley. 1994. *Two oceans. A guide to the marine life of southern Africa*. Struik Nature, Cape Town, 456 pages.

Bruton, MN 2015. *When I was a fish. Tales of an ichthyologist*. Jacana Media, Auckland Park, 310 pages.

Bruton, MN 2017. *The annotated old fourlegs. The updated story of the coelacanth*. Struik Nature, Cape Town, 336 pages.

Bruton, MN 2018. *The amazing coelacanth*. Struik Nature, Cape Town, 64 pages.

Bruton, MN & KH Cooper (eds) 1980. *Studies on the ecology of Maputaland*. Rhodes University, Grahamstown, and the Wildlife Society of Southern Africa, Durban, 560 pages.

Cole, M 2019. Yellow-bellied sea snake stranded on East London beach. *Inyathi*, February 2019: 2-3.

Fitzsimons, VFM 1962. *Snakes of southern Africa*. Purnell, Johannesburg, 423 pages.

Gess, RW 2021. Fossil lamprey larvae from South Africa overturn assumptions about vertebrate origins. *The Conversation*, 10 March 2021: 1-2.

Griffiths, CL, J Day & M Picker 2015. *Freshwater life. A field guide to the plants and animals of southern Africa*. Struik Nature, Cape Town, 368 pages.

Jubb, RA 1967. *Freshwater fishes of southern Africa*. A.A. Balkema, Cape Town, 248 pages.

Lubke, RA, FW Gess & MN Bruton (eds) 1988. *A field guide to the Eastern Cape coast*. Grahamstown Centre of the Wildlife Society of Southern Africa, Grahamstown, 520 pages.

McLachlan, GR & R Liversidge (eds) 1971. *Roberts Birds of South Africa*. John Voelcker Bird Book Fund, Johannesburg, 643 pages.

Mills, MGL & L Hes (eds) 1997. *The complete book of southern African mammals*. Struik Nature, Cape Town, 356 pages.

Mittermeier, RA & CG Mittermeier 1997. *Megadiversity. Earth's biologically wealthiest nations*. Cemex, Prado Norte, 501 pages.

Picker, M & CL Griffiths 2019. *Field guide to insects of South Africa - The most complete guide to South African insects*. Struik Nature, Cape Town, 526 pages.

Pienaar, U de V, WD Haacke & NHG Jacobsen 1978. *The reptiles of the Kruger National Park*. National Parks Board, Pretoria, 236 pages.

Richmond, MD (ed) 2011. *A field guide to the seashores of eastern Africa and the Western Indian Ocean islands*. Sida/WIOMSA, 464 pages.

Scholtz, CH & E Holm (eds) 1985. *Insects of southern Africa*. Butterworths, Durban, 502 pages.

Siegle, L 2018. *Turning the tide on plastic pollution: How humanity (and you) can make our globe clean again*. Trapeze, London, 273 pages.

Skelton, PH 1993. *A complete guide to the freshwater fishes of southern Africa*. Southern Book Publishers, Johannesburg, 388 pages.

Smith, MM & PC Heemstra (eds) 1986. *Smiths' Sea Fishes*. Macmillan, Johannesburg, 1047 pages.

Smithers, RHN 1983. *The mammals of the southern African subregion*. University of Pretoria Press, Pretoria, 736 pages.

Steffen, W, J Grinevald, PJ Crutzen & J McNeill 2011. The Anthropocene: conceptual and historical perspectives. *Philosophical Transactions of The Royal Society A. Mathematical Physical and Engineering Sciences* 369(1938): 842-867.

Van der Elst, R 2019. *Beachcombing in South Africa*. Struik Nature, Cape Town, 144 pages.

## 14. Africa's Nobel laureates

Asmal, K, D Chidester & W James (eds) 2004. *South Africa's Nobel laureates. Peace, literature and science*. Jonathan Ball Publishers, Johannesburg, 289 pages.

Baker, A 2019. The Great Green Wall of Africa. *Time*, 23 September 2019: 40-42.

Baker, A 2020a. Zenzile Miriam Makeba. Sound of South Africa. *Time*, 16-23 March 2020: 64.

Baker, A 2020b. Nawal El Saadawi. For a more equal Egypt. *Time*, 16-23 March 2020: 72.

Baker, A 2020c. Ellen Johnson Sirleaf. A first for Africa. *Time*, 16-23 March 2020: 87.

Maathai, W. 2010. *Replenishing the Earth. Spiritual values for healing ourselves and the world.* Doubleday Religion, London, 208 pages.

Musila, GA (ed) 2020. *Wangari Maathai's Registers of Freedom.* HSRC Press, Cape Town, 334 pages.

Zewail, A 2002. *Voyage through time. Walks of life to the Nobel Prize.* The American University of Cairo Press, Cairo, 287 pages.

www.africa.com/africas-nobel-peace-prize-laureates/

www.aljazeera.com/news/2019/12/9/africas-nobel-prize-winners-a-list

www.news24.com/news24/africa/news/africas-nobel-prize-winners-a-list-20191209

www.thoughtco.com/africas-nobel-prize-winners-43298

## 15. Discussion: What is science?

Zille, H 2021. *#Stay woke go broke. Why South Africa won't survive America's culture wars.* Obsidian Worlds Publishing, Cape Town, 200 pages.

# Index

**A**CEP (African Coelacanth Ecosystem Programme) 205, 208
Aceso breast imaging system 131, 143
Achebe, C 69
Adiche, CN 63
Afro-beat 55, 68
Afroes 133
Agassiz, L 181, 182
Aidoo, B 62
Aldini, G 23
Ali, AA 315
Amoils, SP 144, 145
Amoore, H 98, 105
Amoriguard 148
Amphibians 183, 261, 271-3
Anemones 242, 259
Anglo-Boer War 122, 123
Annan, KA 309
Annelids 240, 245, 289
Anti-science campaigns 324, 325
Arafat, Y 307, 308
Arago, DF 41
Archimedes 14, 15
Aristotle 13, 14, 81
Art and science 9, 10, 48-51, 59, 73-79, 83
Art@Africa gallery 74, 77-79
Arthropods 240, 246-254, 289
Asmal, K 317
Astatke, M 55, 56
Attalah, L 70, 71
Avery, G 112

**B**acon, F 19, 20
    R 19
Baghdad battery 22
Baigrie, J 80, 111
Bailly, JS 25

Banks, J 30, 89, 90
Bardeen, J 42, 43
Barlow-Wadley radios 141, 156, 157
Barnacles 253, 254, 259, 289
Barnard, KH 167, 186
Bats 232, 238, 280, 282, 288, 289
Beaumont, W 25
Beethoven, L van 24, 49, 104
Bell, AG 35, 37, 93
    I 140, 151
Bezos, J 136, 137
Bicycle 10, 112-129, 131, 179
    Economic benefits 127
    Efficiency 117, 118, 121-126
    Environmental benefits 113, 127, 128
    Future 125-129
    History 109, 114-119
    In war 122, 123, 137, 138
    Popularity 117-127
    Production 127
    Racing 124-126
    Social benefits 113, 123, 124, 127
    Women's emancipation 123-125
Biodiversity 16, 50, 73, 78, 109, 110, 238-292, 322
Biomimicry 133
Biot, J-B 41
Biowise 133
Birds 183, 261, 277-280
Bluebottle 244, 245, 259, 289
Boeksma, AH 101
Bohr, N 13, 32, 39, 40-42, 93
Bonaparte, N 22-24
Boneshaker 116, 120
Bony fish 264-271
Booker Prize 67, 295, 311
Boucum, H 6
Boulton, M 82-88, 91-94
Bouwens, T 53, 54

Boyle, R 20, 21, 294
Brahe, T 17
Brenner, S 309, 319
Brittlestar 258
Buckland, W 31
Bulala, P 74

Camera obscura 16, 17
Camp, SD 73
Camus, A 298
Cartilaginous fish 262-264
Catfish 167, 266, 267, 270, 289
*isiCathamiya* 56, 57
CATscanner 34, 141-143, 300
*Celérifère* 114, 115
Cenocell 148
Chadwick, J 32, 33
Chameleons 275, 289
Chiton 254, 255
Chu, PCW 21
Clemens, S 36, 99, 100, 106, 118
Climate change 51, 107, 109, 321, 322, 324,
Cluver, M 80, 105, 109
Cnidarians 242, 259
Coelacanth, Indian Ocean 10, 106, 109, 171, 179, 180, 222, 262, 271, 283, 288, 289
    Anatomy 181, 187-192, 212-214, 264, 265
    Biology 190, 192, 208, 213-217, 264, 265
    Discovery 181, 187, 194, 203, 205, 209
    Conservation 194, 204, 205, 209, 210, 214-218, 221, 222, 240
    Cultural history 218-222
    Détente 201, 202
    Evolution 181, 213, 214, 221, 264, 265
Coelacanth, Indonesian 204, 208, 210, 211
Coetzee, JM 310, 311
Cohen-Tannoudji, C 308

Colindictor 44, 45, 130
Computicket 143, 144
Conservation 50, 51, 58, 59, 109, 110, 128, 221-226, 229, 233-237, 279, 285-287, 290-292, 295, 317, 322
Cope, ED 31, 180
Corals 242-244, 259, 269, 289, 290
Cormack, AM 34, 141-143, 151, 300
Cory, G 108, 183, 184
Courtenay-Latimer, MED 109, 164, 180, 185-193, 198
Cousteau, J-Y 162, 190
Covid-19 pandemic 51, 61, 62, 105, 127, 131, 236, 237, 287, 291, 321
Crabs 250-252
Creativity 10, 18, 48-79, 322
Crocodile 273-276, 283, 287, 288
Crustaceans 247, 250-253, 259
Cryoprobe 144, 145
Cryptozoology 106
CSIR 44, 152, 157, 160, 165, 186
Curie, M 32
    P 32
Custodianship ethic 78, 128, 133, 222, 236, 237, 290-292, 323-326
CyberTracker 146, 147

Daimler, G 120
Dance 49, 53, 63, 64
Dangarembga, T 67
Darwin, C 16, 28-30, 82, 87, 88, 103, 170, 176, 180, 181, 215,
    E 30, 82-86, 91-94
Da Vinci, L 18, 19, 21, 114
Davy, H 27, 28
De Klerk, FW 199, 305, 307
Democritus 13
Denard, B 206, 207
Dickman, L 44, 45
Dinosaurs 31, 181, 225

Dlamini, S 57
Dodgson, C 234
Dodo 10, 109, 179, 223-237
Dolos 130, 131, 147, 148, 150
Doyoyo, M 147, 148
Drais, Baron von 115
*Draissiene* 115
Dreaming 26, 27, 47
Dugong 280, 283, 284, 287-289
Durnez, D 77, 78
Dyson, F 27
      J 137

East London Museum 185, 186, 188, 190, 228
Echinoderms 240, 241, 257-259
Edgeworth, RL 87, 91
Edinburgh, Duke of 44, 45
Edison, TA 35-38, 93, 163
Eel-catfish 289
Eels 266
Einstein, A 13, 14, 37, 38, 93, 101, 109, 129, 318
ElBaradei, MM 311, 312
Elephant 239, 282, 283, 288
Ellis, GFR 98, 318, 319
Enlightenment 81, 85, 90, 93
Environmental crisis 50, 73, 148, 317, 323-325
Erdmann, M 210, 211
Erdős, P 38
Eskom 138
Ethio-jazz 55, 56
Evolution 28-31, 82, 107, 180-182, 187, 226, 229, 236, 238, 240, 265, 287, 290, 291
Extinction 10, 31, 106, 214, 221, 224-226, 231, 233, 235, 236, 240, 241, 289-291

Faraday, M 28, 39
Fasiki, Z 67, 68

Fejer, J 158
Feminism 63, 66
Feynman, R 27
Films 56, 67, 68, 71, 305
Fine art 18, 49, 57, 59, 65, 68, 73, 76, 77, 83, 99, 218
Five-rayed animals 240, 241, 257-259
Fleming, A 33
Flightlessness 232, 233
Flying fishcart 195, 197-199
Folk music 56, 59
Ford, H 38, 120
Fossils 112, 181, 187, 190, 211-215, 223, 224, 233, 235, 241
Franklin, B 25, 28, 33, 81, 84, 87, 294
      R 13, 180, 301
French Revolution 24, 25, 31, 81, 86, 91
Fricke, H 185, 204-210, 216, 219
Frogs 261

Gaboon adder 274-276
Galileo 108, 180
Galois, E 18
Galton, F 30, 31
Garuba, HO 66, 68
Gates, B 136, 170
Gay rights 66-68
Gbowee, LR 296, 297
Generation Xers 72
Geraud, JL 210
Gess, R 211
Gill, D 97, 108
     EL 108, 167
Giraffe 282, 283
Goosen, H 185
Gordimer, N 69, 70, 294, 295, 310
Graphic novels 72
Gray, S 22
Greathead, JH 93, 135, 136, 151
Green Belt Movement 295

Griffiths, M 43, 44
Gueller, B 48, 49, 108
*Gukurahundi* 65

**H**agfish 262, 288, 289
Hamerkop 279, 280, 288
Haroche, S 312, 313
Haytham, Ibn al- 16, 17, 19, 20
Heisenberg, W 42, 48
Herschel, J 188
Hertz, H 43
Hewitt, F 154
*Hexatrygon bickelli* 166, 211, 263, 288
High-wheeler 117, 118, 125
HIPPO impacts 236
Hippopotamus 282-284
Hip2B$^2$ 149
Hissmann, K 207-209
HIV/Aids 60-62, 69, 70, 106, 294, 309
Hodgkin, D 293, 294
Hölscher, HD 163
Honeybee 249, 250, 259, 289
Hooke, R 19-21
Hounsfield, GN 34, 142, 300
House music 60, 64
Hundessa, H 70
Hunt, E 194, 195, 197
Hunting 285-287
Hutton, J 88
Huxley, T 29, 180, 181

**I**dir (Hamid Cheriet) 52
Iger, B 136
Ig Nobel Prize 45-47
Indigenous knowledge 130, 324, 325
Industrial revolutions 81, 86, 88, 91, 93, 109, 110, 326
Information age 237
Information technology 131, 237

Information Value Chain 325
Insects 247-250, 259, 289
Intergovernmental Panel on Climate Change 316
Internet of Living Things 237, 292
Internet of Things 237, 292
Invasive organisms 233, 236, 267, 273, 317, 322
Inventions 10, 12-47, 82-86, 109, 128, 130-163, 325
iSimangaliso Wetland Park 205, 208, 217, 269
Iziko South African Museum 80, 105, 108, 123, 186, 228

**J**ago submersible 169, 208, 209, 252, 253, 288
Jāḥiẓ, al- 15, 16
Jawless fish 262, 288, 289
JB1 Radar Transmitter 153, 154
Jellyfish 242, 244, 255, 259, 289
Jobs, S 136
Johnson, H 135, 136
Joliot-Cutie, I 32
Juritz, J 98, 101

**K**ahiu, W 68
Kanté, M 54
Kasparov, G 79
Kekule, A 26
Kentridge, W 76, 77
Kenyatta, U 62
Keogh, J 46, 47
Kgatle, R 133
Khoisan 53, 146
Kibel, M 95, 106
Kidjo, A 56, 57
Kipling, R 100, 298
Klug, A 143, 300, 301
Kovalevsky, S 37-39

# Index

Kuti, FA 57, 68, 69
*Kwaito* 64

**L**adysmith Black Mambazo 56, 57
Laser 44, 147, 308
   Digital 44, 147
*Latimeria chalumnae* 164, 182, 189, 204-206, 212-218, 220
*Latimeria menadoensis* 204, 206, 210, 211
Laudi, M 58, 59
Lavoisier, P 24, 25, 180
Lawson, S 79
Leeches 245
Levitt, M 313
Liebenberg, L 146, 147, 151
Li, N 61, 62
Lloyd, PJG 106, 316
Lobe-fins 264, 265
Lobsters 252, 253
Loewi, O 26, 27
Lunar Society of Birmingham 10, 25, 80-95, 109, 110
Lungfish 166, 167, 183, 203, 215, 265, 288
Luthuli, IAJ 298, 299

**M**aathai, WM 295-297, 317
Mahfouz, N 304, 305
Makeba, ZM 56, 64, 65, 297
Makwa, D 61, 62
Malan, DF 165, 195, 198, 199
Mammals 183, 187, 261, 280-289
Mancoba, E 59
Mandela, NR 57, 69, 74, 77, 144, 151, 169, 170, 294, 302, 305-307, 319
Marconi, G 35, 37, 155, 163
Marsh, O 31, 180
Martin, A St 25
Maseko, O 63
Mashesha stove 132
Masuka, D 59,

Matlawa, K 63
Maxwell, JC 23, 43
Mazibuko, K-N 74, 78
Mbele, J 75
McMillan, K 115, 116
Mehta, A 210, 211
Mendeleev, D 25, 26
Michaux, P 116
Michell, J 84, 94
Military 18, 19, 21, 23, 66, 69, 73, 80, 122-125, 159, 161, 303, 311-315
Millipedes 247
Millot, J 190, 200, 201
Mole-rat 281
Molluscs 240, 254-259
Moss animals 254
Motorbike 120
Motorcar 120
Mott, N 42
*M-Pesa* 316
Mtshabe, N 54, 55
Mtukudzi, O 60,
Mudhopper 266
Mukwege, D 313-315
Mulato, C 74
Mullis, K 27
Murray, T 111
Museums 32, 51, 56, 76-79, 83, 89, 93, 105, 108, 113, 185-188, 201, 212, 224, 226, 228, 306
Music 18, 24, 48, 49, 52, 54-64, 68, 69, 79, 99, 101, 102, 298
Musk, E 36, 135-137, 150, 323

**N**atural selection 16, 28, 29, 30, 82, 83, 181
Newton, I 13, 17, 20, 21, 294, 313, 321
Ngcobo, SS 44, 147
Ngwenya, M 64
Nicobar pigeon 229, 233
Nirghin, K 133

Nobel Prize 11, 27, 32, 33, 34, 39, 40, 42, 45, 69, 106, 142, 143, 150, 151, 293-316
Ntuli, N 74, 78
Nudibranchs 255, 256
Nulens, R 205, 206
Nyokong, T 134

Octopuses 254-259, 288, 289
Okediran, W 63
Okigbo, C 66
Okupe, R 72
Olaiyo, V 57, 58
*Old Fourlegs* book 180, 193, 199, 202, 203, 206, 210
Ology words 10, 109, 171-179
Ørsted, H 28
Ostrich 228, 252, 277, 288
Owen, R 181, 229
Owl Club of Cape Town 10, 48, 49, 80, 95-111
Owls 279

Palais de Lomé 78
Pangolin 280, 281, 285-290
*Pantsula* 64
Pasteur, L 46
Pauli, W 39-42
Pelican 278
Penguin 232, 278, 288-290
Penny-farthing 117
Performing arts 38, 49, 56, 59, 79, 98
Periodic Table of Chemical Elements 25, 26
*Peripatus* 216
Phlogiston Theory 24, 25, 90, 180,
Pienaar, H 184
Plastic pollution 107, 291
Platanna 272, 273
Platypus 216, 239, 240
Pliny the Elder 15

Poetry 52, 59, 66, 68, 74, 82, 83, 96, 103, 218
Pollution 73, 109, 127, 236
Poverty 71, 72, 148, 322
Pratley, K 145, 146, 151
Pratley Putty 145, 146
Praying mantis 248, 259
Preston, G 317
Priestley, J 84, 88, 91
Prosperi, F 200
Pseudoscience 176, 181, 324, 325
Python 275

Quagga 241
Qubeka, JXT 71

Ralivhona, N 60
Reel gardening 132
Reid, C 132
Relativity 38, 42
Religion and science 318
Renaissance 81
Reptiles 183, 261, 273-277
Rhinoceros 282, 290
*Riel dance* 53
Röntgen, WC 34
Rose, JG 122, 123, 137, 138
Rosenthal, E 95, 98, 102, 106, 111
Rotifers 246, 247
Royal Society of London 19, 21, 30, 38, 81, 83, 89, 97, 294, 301
Royal Society of South Africa 80, 182
Rutherford, E 13, 39, 40, 93, 294

Saadawi, N El 66, 68, 297
Sadat, A al- 299
Safety bicycle 118, 119, 120, 125
SAIAB 189, 192, 193, 199, 202, 210, 257, 263,
SASSI 133
Schauer, J 207-209

# Index

Schonland, BFJ 103, 139, 140, 151, 154, 157, 165
Schwab, K 110
Science centres 17, 22, 325
Science communication 50, 51, 61, 322
Science education 9, 50
Science, history 10, 12-47, 109, 322
    role in society 321, 322, 325
    value 8, 321, 322, 324, 325
Scientific method 14, 17, 19-21, 26, 33, 38, 49, 164, 321, 322, 326
Scott, D 62
    RF 101
Sea cucumbers 259, 288, 289
Sea snake 273, 276, 277, 288, 289
Sea squirts 260
Sea urchins 258
Secretary bird 219
*Serenichthys kowiensis* 211,
Shabalala, MMB 56, 57
Sharks 262-265, 270, 271, 289
Shuttleworth Foundation 149
Shuttleworth, MR 148-151
Simon, C 302, 303
Sink, K 133, 151, 208
Sirleaf, EJ 296, 297
Small, W 87, 88, 91, 92
Smit, S 74, 75
Smith, FS 34
    JLB 41, 108, 164-171, 180-194, 202-204
    MM 108, 165, 166, 184, 189, 191-194, 197-199, 203, 264
    W 165, 189
Smuts, J 99, 138, 154, 199
Soddy, F 39, 40
Solander, DC 90
Solar energy 150
Solitaire 226, 229, 230, 231
Song 49, 51, 52, 55, 56, 60-62, 70
South African Institute for Aquatic Biodiversity (see SAIAB)
South African inventors 44, 45, 122-123, 130-163, 300
Soyinka, AOB 63, 66, 303, 304
Spargo, P 103
Spiders 247, 250, 259
Sponges 241, 242, 259, 289
Squids 256, 257
Stahl, G 24
Starfish 258
Starley, J 117, 119, 120
Steller's sea cow 225, 226, 290
Stingray, six-gill 166, 211, 263, 288
Stockholm Water Prize 317
*Stofskoppers* 53, 54
Strange animals 10, 238-292
Street art 74, 75
Struchkov, Y 43
Sunfish, ocean 270, 271, 288, 289
Sustainable living 50, 51, 110, 128, 148, 150, 236, 295, 323, 324
Swain, S 133

**T**ebbutt, P 8, 98, 307
Telephone 37, 44, 138
Tellurometer, infrared 162, 163
    radio 157-163
Templeton Prize 318, 319
Tesla, N 35, 36, 93, 163
Thawte Internet Security System 149
Theiler, A 134, 298
    M 134, 297, 298
Thiart, K 78
Thunberg, G 324
Timm, P 208
*Tour d'Afrique* 127
Tour de France 124-126
Triggerfish Foundation 75, 76
Tucker, P 143, 144

Tunisian National Dialogue Quartet 316
Turtles 273, 274, 287-290
Tutu, DM 77, 301, 302, 319
Two Oceans Aquarium 257, 271, 317
Tyrrell, J 100, 108

**U**buntu 324
Ubuntu Project 149
Unesco 64

**V**an der Bijl, HJ 138, 139, 151
Velocipede 116, 117
Velvet worm 240, 245, 246, 259, 288-290
Vertebrates 240, 260-292
Volta, A 22

**W**adley, TL 140, 141, 151-163
    Awards 154, 161, 162
    Education 151-153
    Family life 151-153, 162
    Ionosonde 141, 155
    Personality 152-156
    Radio inventions 141, 154-157
    Tellurometer 141, 157-163
    War experiences 141, 153, 154
Wadley von Hirschberg, M 152, 154, 161
Wallace, AR 28, 30, 294
Warner, B 80, 96, 106
Water conservation 317, 318, 321
Watt, J 82, 83, 86, 87, 91-94
Weaver, sociable 278, 279, 289
Wedgwood, J 87, 91
Wedidi, D el 59
Whales 9, 238, 239, 253-256, 277, 280, 281, 284-289
Wheel in animals 246
Whitehurst, J 83, 91
Wilberforce, S 29, 180
Wilkinson, J 86, 88
Williamson, L 132

Wilson, EO 50
    M 98
Wine, B 61, 62
Wokeness 324
Women inventors 131-133
Women's rights 66-68, 123, 124, 295, 296, 314
Working for Water campaign 317
Wright brothers 93, 101, 121

**Y**ende, P 54
YouNeek Studios 72

**X**-rays 32, 34, 107, 142, 143, 293, 300
Xusa, SL 150, 151

**Z**ebra 241
Zeitz MOCAA 77
Zewail, A 308, 309
Zoanthids 242, 243, 259
*Zoonomia* 82